Polymeric Composites in Road and Bridge Engineering: Characterization, Production and Application

Polymeric Composites in Road and Bridge Engineering: Characterization, Production and Application

Editors

Wensheng Wang
Yongchun Cheng
Heping (Fred) Chen
Guojin Tan

MDPI • Basel • Beijing • Wuhan • Barcelona • Belgrade • Manchester • Tokyo • Cluj • Tianjin

Editors
Wensheng Wang
College of Transportation
Jilin University
Changchun
China

Yongchun Cheng
College of Transportation
Jilin University
Changchun
China

Heping (Fred) Chen
Ingram School of Engineering
Texas State University
San Marcos
USA

Guojin Tan
College of Transportation
Jilin University/Changchun
China

Editorial Office
MDPI
St. Alban-Anlage 66
4052 Basel, Switzerland

This is a reprint of articles from the Special Issue published online in the open access journal *Polymers* (ISSN 2073-4360) (available at: https://www.mdpi.com/journal/polymers/special_issues/Polymeric_Composites_Road_Bridge_Engineering_Characterization_Production_Application).

For citation purposes, cite each article independently as indicated on the article page online and as indicated below:

LastName, A.A.; LastName, B.B.; LastName, C.C. Article Title. *Journal Name* **Year**, *Volume Number*, Page Range.

ISBN 978-3-0365-6778-5 (Hbk)
ISBN 978-3-0365-6779-2 (PDF)

Contents

Wensheng Wang, Yongchun Cheng, Heping Chen and Guojin Tan
Polymeric Composites in Road and Bridge Engineering: Characterization, Production and Application
Reprinted from: *Polymers* **2023**, *15*, 874, doi:10.3390/polym15040874 . . . . **1**

Suhua Chen, Enwei Jin, Gang Xu, Shangzhi Zhuo and Xianhua Chen
Factors Influencing the Low-Temperature Properties of Styrene-Butadiene-Styrene Modified Asphalt Based on Orthogonal Tests
Reprinted from: *Polymers* **2023**, *15*, 52, doi:10.3390/polym15010052 . . . . **5**

Daqian Han, Guosheng Hu and Jingting Zhang
Study on Anti-Aging Performance Enhancement of Polymer Modified Asphalt with High Linear SBS Content
Reprinted from: *Polymers* **2023**, *15*, 256, doi:10.3390/polym15020256 . . . . **19**

Honggang Zhang, Yangpeng Zhang, Jie Chen, Wenchang Liu and Wensheng Wang
Effect of Desulfurization Process Variables on the Properties of Crumb Rubber Modified Asphalt
Reprinted from: *Polymers* **2022**, *14*, 1365, doi:10.3390/polym14071365 . . . . **33**

Gang Xu, Yunhong Yu, Jingyao Yang, Tianling Wang, Peipei Kong and Xianhua Chen
Rheological and Aging Properties of Composite Modified Bitumen by Styrene–Butadiene–Styrene and Desulfurized Crumb Rubber
Reprinted from: *Polymers* **2021**, *13*, 3037, doi:10.3390/polym13183037 . . . . **47**

Zhizhong Zhao, Longlin Wang, Wensheng Wang and Xuanhao Shangguan
Experimental Investigation of the High-Temperature Rheological and Aging Resistance Properties of Activated Crumb Rubber Powder/SBS Composite-Modified Asphalt
Reprinted from: *Polymers* **2022**, *14*, 1905, doi:10.3390/polym14091905 . . . . **65**

Yafeng Gong, Shuzheng Wu, Yuwei Zhang, Yunze Pang and Yulin Ma
Investigation of the High-Temperature and Rheological Properties for Asphalt Sealant Modified by SBS and Rubber Crumb
Reprinted from: *Polymers* **2022**, *14*, 2558, doi:10.3390/polym14132558 . . . . **81**

Chunfeng Zhu, Huijin Luo, Wei Tian, Binbin Teng, Yongmei Qian, Huaxue Ai and Bo Xiao
Investigation on Fatigue Performance of Diatomite/Basalt Fiber Composite Modified Asphalt Mixture
Reprinted from: *Polymers* **2022**, *14*, 414, doi:10.3390/polym14030414 . . . . **99**

Senzhi Ren and Xin Hu
Fatigue Properties and Its Prediction of Polymer Concrete for the Repair of Asphalt Pavements
Reprinted from: *Polymers* **2022**, *14*, 2941, doi:10.3390/polym14142941 . . . . **117**

Yongcheng Ji, Wenhao Ji, Ziyi Zhang and Rui Wang
Road Performance Investigation on Fiber-Reinforced Recycled Cement Base Material
Reprinted from: *Polymers* **2022**, *14*, 4102, doi:10.3390/polym14194102 . . . . **131**

Yafeng Gong, Jianxing Yang, Xin He, Xiang Lyu and Hanbing Liu
Structural Design Calculation of Basalt Fiber Polymer-Modified RPC Beams Subjected to Four-Point Bending
Reprinted from: *Polymers* **2021**, *13*, 3261, doi:10.3390/polym13193261 . . . . **157**

Yafeng Gong, Jiaxiang Song, Yulong He and Guirong Ma
Value Coefficient of Polyethylene Fiber Soil Embankment Slope Based on Response Surface Analysis
Reprinted from: *Polymers* **2022**, *14*, 4295, doi:10.3390/polym14204295 . . . . **173**

Li Luo, Xingang Wang, Chen Xue, Daozheng Wang and Baoqin Lian
Laboratory Experiments and Numerical Simulation Study of Composite-Material-Modified Loess Improving High-Speed Railway Subgrade
Reprinted from: *Polymers* **2022**, *14*, 3215, doi:10.3390/polym14153215 . . . . **187**

Wei Wang, Beifeng Lv, Chen Zhang, Na Li and Shaoyun Pu
Mechanical Characteristics of Lime-Treated Subgrade Soil Improved by Polypropylene Fiber and Class F Fly Ash
Reprinted from: *Polymers* **2022**, *14*, 2921, doi:10.3390/polym14142921 . . . . **205**

Zhichao Wang, Weiqing Zhang, Ping Jiang and Cuihong Li
The Elastic Modulus and Damage Stress–Strain Model of Polypropylene Fiber and Nano Clay Modified Lime Treated Soil under Axial Load
Reprinted from: *Polymers* **2022**, *14*, 2606, doi:10.3390/polym14132606 . . . . **217**

Ping Jiang, Yewen Chen, Xinjiang Song, Na Li, Wei Wang and Erlu Wu
Study on Compressive Properties and Dynamic Characteristics of Polypropylene-Fiber-and-Cement-Modified Iron-Ore Tailing under Traffic Load
Reprinted from: *Polymers* **2022**, *14*, 1995, doi:10.3390/polym14101995 . . . . **235**

Beifeng Lv, Yinuo Zhao, Na Li, Yanfei Yu, Yanting Wu and Miaojie Gu
Triaxial Mechanical Properties and Mechanism of Waterborne Polyurethane-Reinforced Road Demolition Waste as Road Bases
Reprinted from: *Polymers* **2022**, *14*, 2725, doi:10.3390/polym14132725 . . . . **253**

Hassan Ali Alkadhim, Muhammad Nasir Amin, Waqas Ahmad, Kaffayatullah Khan, Mohammed Najeeb Al-Hashem, Sara Houda, et al.
Knowledge Mapping of the Literature on Fiber-Reinforced Geopolymers: A Scientometric Review
Reprinted from: *Polymers* **2022**, *14*, 5008, doi:10.3390/polym14225008 . . . . **271**

polymers

MDPI

Editorial

Polymeric Composites in Road and Bridge Engineering: Characterization, Production and Application

Wensheng Wang [1,*], Yongchun Cheng [1], Heping Chen [2] and Guojin Tan [1]

1 College of Transportation, Jilin University, Changchun 130022, China
2 Ingram School of Engineering, Texas State University, San Marcos, TX 78666, USA
* Correspondence: wangws@jlu.edu.cn

As a result of their rapid development, polymer composites are seeing wider use in transportation infrastructure in China and worldwide. Styrene–butadiene–styrene (SBS), epoxy resin (ER), polyethylene (PE), expanded polyethylene (EP), polyurethane (PU), crumb rubber and other polymer materials can be used to modify asphalt and improve the performance of asphalt mix. Meanwhile, the application of polymer admixtures such as PE fiber, polyvinyl alcohol (PVA) fiber, polypropylene (PP) fiber, etc., in bituminous, cementitious or soil materials also improves their strength and durability. Polymer materials are also widely used in pavement maintenance and treatment; they can quickly restore road function and extend the road's service life.

This Special Issue contains sixteen research papers and one systematic review. The low-temperature properties of SBS-modified asphalt are not satisfactory; thus, Chen et al. analyzed the main factors for improving the low-temperature performance of SBS-modified asphalt base on the orthogonal tests [1]. To solve the insufficient anti-aging performance of modified asphalt with high SBS content, Han et al. added nanomaterials and polyphosphoric acid (PPA) to high SBS content linear-modified asphalt as anti-aging agents [2]. Zhang et al. adopted the mechanochemical method to prepare desulfurized crumb rubber-modified asphalt and analyzed the effects of desulfurization process variables [3]. Furthermore, Xu et al. investigated the modification mechanism, rheological, and aging properties of SBS/desulfurized crumb rubber composite-modified bitumen [4]. Zhao et al. explored the influence of composite modification of activated crumb rubber powder and SBS on asphalt [5]. Gong et al. utilized SBS and rubber crumb to modify the asphalt-based sealants to overcome the poor high-temperature and rheological properties of sealant [6]. SBS in crumb rubber-modified asphalt formed a three-dimensional network structure to improve the performance. Meanwhile, Zhu et al. investigated the fatigue performance of fiber composite modified asphalt mixture [7].

Some research studies are devoted directly to the polymer materials applied in bituminous, cementitious or soil materials. Ren et al. studied the fatigue life and the material design of polymer concrete including ER and PU [8]. Ji et al. analyzed the influence of failure modes of PVA fiber on fiber-reinforced recycled brick powder cementitious composites [9]. Gong et al. investigated the mechanical parameters of basalt fiber polymer-modified active powder concrete (RPC) for the design of structural reinforcement [10]. Due to the high thermal stability, superior mechanical properties, corrosion and chemical resistance of polymer materials, they show great potential for application in road bases, subgrade and embankment slopes, etc. Gong et al. proposed value coefficients to evaluate the cost-effectiveness of PE fiber-reinforced roadbeds and analyzed the effects [11]. Luo et al. used cement, PP fiber and soil curing agent for loess subgrade improvement of a high-speed railway [12]. Wang et al. investigated the improvement effect of PP fiber modified lime-treated soil and fly ash-modified lime-treated soil [13]. Meanwhile, Wang et al. used PP fiber and nano clay to modify lime-treated soil and evaluated the static and dynamic properties of modified lime-treated soil [14]. Jiang et al. used PP fiber and

Citation: Wang, W.; Cheng, Y.; Chen, H.; Tan, G. Polymeric Composites in Road and Bridge Engineering: Characterization, Production and Application. *Polymers* **2023**, *15*, 874. https://doi.org/10.3390/polym15040874

Received: 6 February 2023
Accepted: 9 February 2023
Published: 10 February 2023

cement to modify iron-ore tailing in road engineering as an effective reuse strategy [15]. Lv et al. adopted waterborne polyurethane (WPU) to reinforce the road demolition waste as a promising new polymer reinforcement material [16].

The Special Issue concludes with a review on fiber-reinforced geopolymers and highlights the difficult aspects of current research, in addition to assessing the literature records [17].

Author Contributions: Conceptualization, W.W., Y.C., H.C. and G.T.; formal analysis, W.W., Y.C., H.C. and G.T.; writing—original draft preparation, W.W.; writing—review and editing, H.C.; project administration, W.W.; funding acquisition, W.W. All authors have read and agreed to the published version of the manuscript.

Funding: This research was funded by the Scientific Research Project of Department of Education of Jilin Province (grant number: JJKH20221019KJ), Scientific and Technological Project of Science and Technology Department of Jilin Province (grant number: 20210508028RQ), National Natural Science Foundation of China (grant number: 52208438).

Institutional Review Board Statement: Not applicable.

Data Availability Statement: Not applicable.

Conflicts of Interest: The authors declare no conflict of interest.

References

1. Chen, S.; Jin, E.; Xu, G.; Zhuo, S.; Chen, X. Factors influencing the low-temperature properties of styrene-butadiene-styrene modified asphalt based on orthogonal tests. *Polymers* **2022**, *15*, 52. [CrossRef] [PubMed]
2. Han, D.; Hu, G.; Zhang, J. Study on anti-aging performance enhancement of polymer modified asphalt with high linear sbs content. *Polymers* **2023**, *15*, 256. [CrossRef] [PubMed]
3. Zhang, H.G.; Zhang, Y.P.; Chen, J.; Liu, W.C.; Wang, W.S. Effect of desulfurization process variables on the properties of crumb rubber modified asphalt. *Polymers* **2022**, *14*, 1365. [CrossRef] [PubMed]
4. Xu, G.; Yu, Y.; Yang, J.; Wang, T.; Kong, P.; Chen, X. Rheological and aging properties of composite modified bitumen by styrene-butadiene-styrene and desulfurized crumb rubber. *Polymers* **2021**, *13*, 3037. [CrossRef] [PubMed]
5. Zhao, Z.Z.; Wang, L.L.; Wang, W.S.; Shangguan, X.H. Experimental investigation of the high-temperature rheological and aging resistance properties of activated crumb rubber powder/sbs composite-modified asphalt. *Polymers* **2022**, *14*, 1905. [CrossRef] [PubMed]
6. Gong, Y.; Wu, S.; Zhang, Y.; Pang, Y.; Ma, Y. Investigation of the high-temperature and rheological properties for asphalt sealant modified by sbs and rubber crumb. *Polymers* **2022**, *14*, 2558. [CrossRef] [PubMed]
7. Zhu, C.; Luo, H.; Tian, W.; Teng, B.; Qian, Y.; Ai, H.; Xiao, B. Investigation on fatigue performance of diatomite/basalt fiber composite modified asphalt mixture. *Polymers* **2022**, *14*, 414. [CrossRef] [PubMed]
8. Ren, S.; Hu, X. Fatigue properties and its prediction of polymer concrete for the repair of asphalt pavements. *Polymers* **2022**, *14*, 2941. [CrossRef] [PubMed]
9. Ji, Y.; Ji, W.; Zhang, Z.; Wang, R. Road performance investigation on fiber-reinforced recycled cement base material. *Polymers* **2022**, *14*, 4102. [CrossRef] [PubMed]
10. Gong, Y.; Yang, J.; He, X.; Lyu, X.; Liu, H. Structural design calculation of basalt fiber polymer-modified rpc beams subjected to four-point bending. *Polymers* **2021**, *13*, 3261. [CrossRef] [PubMed]
11. Gong, Y.; Song, J.; He, Y.; Ma, G. Value coefficient of polyethylene fiber soil embankment slope based on response surface analysis. *Polymers* **2022**, *14*, 4295. [CrossRef] [PubMed]
12. Luo, L.; Wang, X.; Xue, C.; Wang, D.; Lian, B. Laboratory experiments and numerical simulation study of composite-material-modified loess improving high-speed railway subgrade. *Polymers* **2022**, *14*, 3215. [CrossRef]
13. Wang, W.; Lv, B.; Zhang, C.; Li, N.; Pu, S. Mechanical characteristics of lime-treated subgrade soil improved by polypropylene fiber and class f fly ash. *Polymers* **2022**, *14*, 2921. [CrossRef] [PubMed]
14. Wang, Z.; Zhang, W.; Jiang, P.; Li, C. The elastic modulus and damage stress-strain model of polypropylene fiber and nano clay modified lime treated soil under axial load. *Polymers* **2022**, *14*, 2606. [CrossRef] [PubMed]
15. Jiang, P.; Chen, Y.; Song, X.; Li, N.; Wang, W.; Wu, E. Study on compressive properties and dynamic characteristics of polypropylene-fiber-and-cement-modified iron-ore tailing under traffic load. *Polymers* **2022**, *14*, 1995. [CrossRef]

16. Lv, B.; Zhao, Y.; Li, N.; Yu, Y.; Wu, Y.; Gu, M. Triaxial mechanical properties and mechanism of waterborne polyurethane-reinforced road demolition waste as road bases. *Polymers* **2022**, *14*, 2725. [CrossRef] [PubMed]
17. Alkadhim, H.A.; Amin, M.N.; Ahmad, W.; Khan, K.; Al-Hashem, M.N.; Houda, S.; Azab, M.; Baki, Z.A. Knowledge mapping of the literature on fiber-reinforced geopolymers: A scientometric review. *Polymers* **2022**, *14*, 5008. [CrossRef]

polymers

MDPI

Article

Factors Influencing the Low-Temperature Properties of Styrene-Butadiene-Styrene Modified Asphalt Based on Orthogonal Tests

Suhua Chen [1], Enwei Jin [2], Gang Xu [1,*], Shangzhi Zhuo [1] and Xianhua Chen [1,*]

1 School of Transportation, Southeast University, 2 Sipailou, Nanjing 210096, China
2 East China Branch of China Railway Bridge Survey and Design Institute Group Co., Ltd., Nanjing 210003, China
* Correspondence: 230198682@seu.edu.cn (G.X.); chenxh@seu.edu.cn (X.C.)

Abstract: Styrene-butadiene-styrene (SBS) is widely used in asphalt modification to obtain superior high-temperature performance. Nevertheless, studies on the low-temperature properties of SBS-modified asphalt are not satisfactory. Orthogonal tests are valid in analysing the results. In this paper, the main factors (SBS content, sulfur content, and the addition of rubber processing oil) for improving the low-temperature performance of SBS-modified asphalt were analyzed base on the orthogonal tests. Firstly, the frequency sweep test, bending beam rheometer (BBR) test, and force-ductility test were conducted to evaluate the low-temperature properties of SBS-modified asphalt. Investigation of low-temperature parameters obtained through these tests was conducted base on the orthogonal analysis method. The G-R parameter was abandoned in the analysis of the orthogonal method for the result that the increase of SBS content was negative to the low-temperature properties by the Glover-Rowe (G-R) parameter, which were contrary to the results of BBR and force-ductility tests. Moreover, the other parameters (ΔT_c and toughness) sorted according to the orthogonal analysis method indicated the effect on low-temperature performance of the SBS-modified asphalt as SBS content > rubber processing oil > sulfur. As shown above that both SBS and rubber processing oil play a critical role in improving the low-temperature properties of SBS-modified asphalt, for SBS could resist the generation and subsequent propagation of cracks while the rubber processing oil could supplement the maltene loss.

Keywords: styrene-butadiene-styrene modified asphalt; low-temperature performance; orthogonal test; Glover-Rowe (G-R) parameter

Citation: Chen, S.; Jin, E.; Xu, G.; Zhuo, S.; Chen, X. Factors Influencing the Low-Temperature Properties of Styrene-Butadiene-Styrene Modified Asphalt Based on Orthogonal Tests. *Polymers* **2023**, *15*, 52. https://doi.org/10.3390/polym15010052

Academic Editors: Wensheng Wang, Yongchun Cheng, Heping (Fred) Chen and Guojin Tan

Received: 3 October 2022
Revised: 2 December 2022
Accepted: 4 December 2022
Published: 23 December 2022

1. Introduction

The rapid development of expressways and the increasing traffic loads have prompted styrene-butadiene-styrene (SBS) modified asphalt to reduce the temperature susceptibility, increase the cohesion, and modify the rheological characteristics, which can alleviate the occurrence of early failure of asphalt pavement [1–4]. Although a considerable amount of research has focused on the SBS-modified asphalt, most of these concentrated on the high temperature properties and performances before and after aging. As the temperature cracks are closely related to the low-temperature cracking resistance of asphalt binders, the low-temperature properties of SBS-modified asphalt have to be comprehensively investigated [5]. Maltenes, which are composed of saturates and aromatics in asphalt binder, can play a significant role in the low-temperature cracking resistance. The addition of sulfur will form stronger molecular bonds with the asphalt molecules to form a three-dimensional lattice structure, thus enhancing the viscosity and storage stability of the modified asphalt [6]. Rubber processing oil is added due to its rich lightweight component, which can promote the swelling of SBS copolymers to Improve the low temperature crack resistance of SBS modified asphalt [5]. Besides, SBS copolymers alter the microstructure and composition of the asphalt, leading to improved low-temperature properties [5–7]. Therefore, it is imperative to study the influence of SBS content, sulfur, and rubber processing oil on the low-temperature properties of SBS-modified asphalt.

Compared with conventional asphalt, the low-temperature properties of SBS-modified asphalt consist of two parts: base asphalt deformation and SBS modifier deformation. It has been shown that polybutadiene (PB) with low glass transition temperatures (T_g) could improve the rheological properties of the base asphalt at low temperatures [8]. However, the stiffness and m-value obtained from bending beam rheometer (BBR) test demonstrate that the low-temperature deformability of modified asphalt would be weakened if the SBS amount exceeds a particular range. Cao et al. [9] defined the temperature corresponding to the phase angle and loss of peak modulus value as the binder glass transition temperature (T_g). Considering T_g as the evaluation parameter, the low-temperature performance of the modified asphalt would be improved with the increasing of SBS amount. Sassan et al. [10] investigated the low-temperature properties of SBS-modified asphalt by the rheological method. It was found that the rheological models were not consistent with different indexes for evaluating the low-temperature properties. Based on these findings, the addition of an SBS modifier, which could improve the low-temperature properties of base asphalt, was considered. However, the influence of SBS content on the low-temperature properties of SBS-modified asphalt is inadequate due to a variety of test methods and evaluation indicators. The polarity, molecular weight, and solubility parameters of SBS are much different from the asphaltenes, resulting in poor compatibility and stability among SBS and base asphalt. Dong et al. (2014) [2] elucidated the SBS with high aromatic content is more compatible with asphalt from the perspective of dispersion. J-F. Masson et al. (2005) [8] elucidated that asphalt with high light maltenes content would have a better miscibility with SBS copolymer. Sun et al. (2006) [11] found that sulfur addition formed a cross-linked network structure with asphalt, thus improved the storage stability of modified asphalt. Liang (2017) [7] discovered that the viscous characteristics decrease with the increase of polymerized sulfur content through the frequency sweep tests. The researches on rubber processing oil and sulfur have primarily focused on promoting the stability between base asphalt and SBS copolymers. Still, there is little research on its low-temperature performance, while the influence of different amounts remain sparse.

From the existing research results, it can be seen that the rules to assess the influence of SBS content, sulfur, and rubber processing oil on the low-temperature performance of base asphalt are inconsistent due to the different evaluation indicators. Therefore, it is necessary to determine reasonable evaluation indicators before exploring the influencing factors of low-temperature properties of SBS-modified asphalt. Traditionally, low-temperature properties of the asphalt binders are tested using low-temperature penetration, low-temperature ductility, and Fraas breaking point test. Recent researches have confirmed that the conventional test methods cannot predict the cracking temperature accurately [12,13]. Meanwhile, contemporary testing procedures including the rheological tests and the correlated rheological data, like complex shear modulus (G^*), phase angle (δ), creep stiffness (S), and creep rate (m), were employed to characterize the low-temperature performances of the asphalt binder [14–18]. This research aims to investigate the low-temperature properties of the SBS-modified asphalt with different amounts of SBS, sulfur, and rubber processing oil. Therefore, the orthogonal tests, which can replace the total tests and analyze the comprehensive influence of multiple factors with fewer tests, were used to design L3*3 tests in three different quantities. After that, the creep stiffness(S), creep stiffness rate (m), G-R constant, maximum force (Fmax), stretched elongation, and other indices obtained from the BBR experiment, DSR frequency sweeping test, and force-ductility test were conducted. In addition, a more reasonable index was determined to fairly evaluate the low-temperature performance of SBS-modified asphalt.

2. Materials and Methodology

2.1. Materials and Preparation

The base asphalt using SK-70 was provided by the SK Petroleum Asphalt Factory, South Korea. The conventional properties of the asphalt matrix, such as penetration, softening point, and ductility, have been given in Table 1. As the literature indicates, the structure type such as linear or star slightly influences the properties of modified asphalt. The SBS used in this research was come from the Yanshan Petrochemical Co., Ltd., Beijing, China, with contents ranging from 2% to 6% [19]. Table 2 presents its fundamental physical performance indices. Sulfur, which was selected as a cross-linking agent to improve the storage stability of SBS-modified asphalt with a content of 0%, 0.15%, and 0.3% respectively [5], came from Beijing Tiansuo Trading Company, and its main components have been given in Table 3. The rubber processing oil produced by Jiangsu Zhonghong Petroleum Asphalt Factory, China, was used to promote swelling of the modifier. Its content ranged from 0% to 4% [5]. Table 4 displays the details of the orthogonal test. It should be noted that different types of SBS-modified asphalt used in this research are termed XL-XS-XR for the convenience of the following charts. For example, the 2L-0.15S-2R contains 2% weight of SBS, 0.15% weight of sulfur, and 2% weight of rubber processing oil. In the Orthogonal experiment, the A, B, C represented the SBS, sulfur and rubber processing oil respectively, the I, II, III represented three degrade of additive content respectively.

Table 1. Main performance indexes of base asphalt.

Physical Properties	Units	SK-70#	Test Method
Penetration (25 °C, 100 g, 5 s)	0.1 mm	67.6	ASTM D5 [20]
Softening point (R&B)	°C	48.6	ASTM D36 [21]
Ductility (15 °C, 5 cm/min)	cm	>100	ASTM D113 [22]
Density	g/cm^3	1.02	ASTM D792 [23]
PG degrade	-	PG64	ASTM M320 [24]

Table 2. SBS basic physical performance indexes.

Structure Type	Styrene-Butadiene Ratio	Breaking Elongation (%)	Tensile Strength (MPa)	Melt Flow Rate (g·10·min^{-1})
Linear	30/70	800	26	0.5

Table 3. Main components and content of sulfur.

Name	Content (%)
Rhombic sulfur	95.71
NH_4SCN	1.33
$(NH)_4S_20_3$	1.13
Others	1.83

Table 4. Orthogonal experimental schemes.

Number	Factor A (SBS/%)	Factor B (Sulfur/%)	Factor C (Rubber Processing Oil/%)	Program	Name
1	2	0.00	0	AIBICI	2L-0S-0R
2	2	0.15	2	AICIIDII	2L-0.15S-2R
3	2	0.30	4	AIBIIICIII	2L-0.3S-4R
4	4	0.15	4	AIIBIICIII	4L-0.15S-4R
5	4	0.30	0	AIIBIIICI	4L-0.3S-0R
6	4	0.00	2	AIIBICII	4L-0S-2R
7	6	0.30	2	AIIIBIIICII	6L-0.3S-2R
8	6	0.00	4	AIIIBICIII	6L-0S-4R
9	6	0.15	0	AIIIBIICI	6L-0.15S-0R

The experiments were conducted in the laboratory of transportation College, Southeast University. A uniform procedure was adopted to prepare the samples listed in Table 1d to minimize the influence of the process on the asphalt [5]. First, SBS was added to the base asphalt and stirred for 15 min under 1000 rpm rotation speed at 175 °C. Then, the corresponding amount of rubber processing oil were added to the mixture and sheared for 90 min under 6000 rpm rotation speed at 175 °C by a mechanical stirrer. Sulfur was then added to the mixture and stirred for another 30 min under 3000 rpm rotation speed. At last, the mixture were stirred for 30 minutes at a speed of 500 rpm to ensure the complete blending of SBS-modified asphalt. Once all samples are prepared, they should be used directly to prevent phase separation. The flow chart is shown in Figure 1.

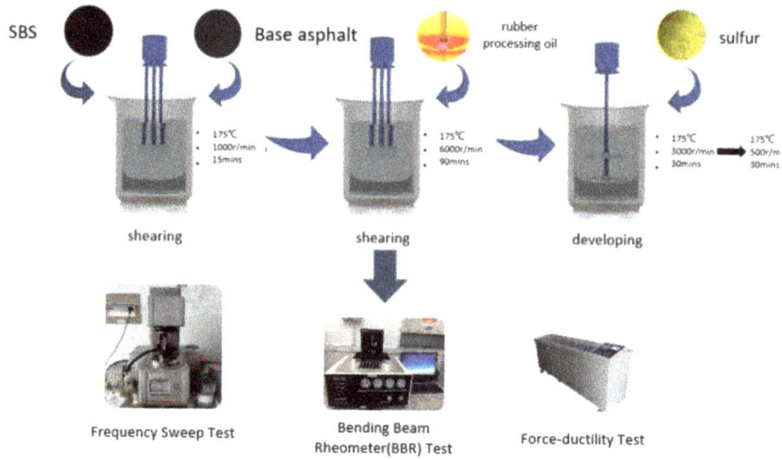

Figure 1. The flow chart of the research.

2.2. *Test Mothodology*

2.2.1. Frequency Sweep Test

An Anton Paar SmartPave 102 dynamic shear rheometer was used for this test at temperatures between 15 °C to 75 °C in interval of 15 °C to obtain the complex shear modulus (G^*) and the phase angle (δ) at 0.1% strain. The test frequency is form 0.1 rad to 100 rads. The tests were performed between 15 °C and 30 °C with 8 mm diameter and 2 mm gap, while the others were undertaken with a 25 mm diameter and 1 mm gap geometry. The time-temperature superposition principle and the 2S2P1D model were also used to produce the master curve at a reference temperature of 15 °C [25,26]. The 2S2P1D model is a generalized model derived from the Huet-Sayegh model, consisting of two springs, two parabolic creep elements and a viscous pot, which can accurately describe the rheological properties of adhesive and asphalt mixtures. According to the established rheological curves, the storage and loss moduli under the condition of 0.005 rad/s can be calculated based on Equations (1) and (2). Finally, the G-R constant is obtained from Equation (3). The 2S2P1D model can be classified into seven parameters, and the expression of $|G^*|$ is shown in Equation (4).

$$\lg\omega_r = \lg\omega + \lg\alpha(T) \quad (1)$$

$$\lg\alpha(T) = \frac{C_1(T - T_r)}{C_2 + (T - T_r)} \quad (2)$$

$$\mathrm{G-R\ Parameter} = G^* \times (\cos\delta)^2/(\sin\delta)\ (15\ ^\circ\mathrm{C},\ 0.005\ \mathrm{rad/s}) \quad (3)$$

$$G^*(\omega) = G_0 + \frac{G_g - G_0}{1 + \mu(i\omega\tau)^{-k} + (i\omega\tau)^{-h} + (i\omega\beta\tau)^{-1}} \quad (4)$$

where, 'T_r' is the reference temperature, 'T' is the test temperature, 'C_1' and 'C_2' are the regression coefficients dependent on the type of asphalt binder, 'ω_r' is the reduced frequency, 'ω' is the actual angular frequency, 'G^*' and 'δ'are complex shear modulus and phase angle, respectively. 'μ' is a correction index, 'β' is related to viscosity, 'k' and 'h' correlate with the material properties ($0 < k < h < 1$). When $\omega \to 0$, 'G_0' is a static modulus. When $\omega \to \infty$, 'G_g' is a glassy modulus.

2.2.2. Bending Beam Rheometer (BBR) Test

According to ASTM D6648 [27], the BBR instrument (Cannon 9732-V31) can be used to determine the creep stiffness (S) and creep rate (m) of SBS modified asphalt. In this research, the stiffness and m-value were obtained under the temperature of −12 °C, −18 °C, and −24 °C, respectively. The critical PG low-temperature of asphalt is determined by the interpolation method. The critical low-temperature grade determined by stiffness (s) is obtained through logarithmic linear interpolation. The critical low-temperature grade determined by creep rate (m) is obtained through linear interpolation. The low-temperature properties of SBS modified asphalt were evaluated with ΔT_c (Equation (5)) [28,29].

$$\begin{aligned} \Delta T_c &= T_{c(\text{stiffness})} - T_{c(\text{m-slope})} \\ T_{c(\text{stiffness})} &= T_1 + \left(\frac{(T_1-T_2)\cdot(log300-logS_1)}{logS_1-logS_2}\right) - 10 \\ T_{c(\text{m-slope})} &= T_1 + \left(\frac{(T_1-T_2)\cdot(0.3-m_1)}{m_1-m_2}\right) - 10 \end{aligned} \tag{5}$$

where 'T_1', 'T_2', are the test temperature (°C) and 'S_1', 'S_2', 'm_1', 'm_2' are the corresponding test results of the BBR test, '$T_{c(\text{stiffness})}$' (°C)is the critical temperature corresponding to the stiffness (s) is 300 MPa, and '$T_{c(\text{m-slope})}$' (°C)is the critical temperature corresponding to the creep rate (m) is 0.3.

2.2.3. Force-Ductility Test

The force-ductility test was conducted in the ductility tester (Hebei Tuofeng Instrument LYY-7F). An 8-shaped standard durability test model was selected with a loading speed of 5 cm/min at 5 °C to investigate the cohesive tensile performance [30]. Before the peak range of tensile deformation was reached, one tensile force every 1 mm was collected; afterward, one tensile force every 5 cm was collected. By calculating the area enclosed by the force-ductility curve, the viscosity and stiffness (which are related to the low-temperature properties of modified asphalt) were estimated.

3. Results and Discussion

3.1. G-R Parameter Analysis

According to the definition of the G-R parameter given above, this parameter corresponds to a very low frequency. In this test, the 2S2P1D model was conducted to deduce the complex modulus and phase angle at 15 °C and 0.005 rad/s.

The model calibration was accomplished by utilizing the Global Optimization in MATLAB software, and the model parameters are shown in Table 5. The reference temperature was 15 °C.

Storage modulus and loss modulus master curves for 6L-0.3S-2R and 2L-0.15S-2R samples are shown in Figure 2. The results implied a good fitting degree between the 2S2P1D model and the test data. In addition, the difference of storage modulus between the two samples at low and high frequencies was more evident than that of loss modulus. This difference may be attributed to the temperature sensitivity of the SBS polymer, which is lower than asphalt, especially at 6% content. Based on the storage and loss modulus master curves, the complex shear modulus and phase angle at 15 °C and 0.005 rad/s were calculated. Following this procedure, the G-R parameter shown in Figure 3a was determined and plotted on the black space diagram. Each point in the black space diagram represents the low-temperature properties of asphalt. The G-R parameter that exceeds 180 kPa indicates a risk of cracking in asphalt pavement, and the pavement will have severe block cracks and reflective cracks if the G-R parameter exceeds 450 kPa [31]. Therefore, with the increase of the G-R parameter, asphalt will become brittle and prone to cracking.

Table 5. The model parameters of all tested binders.

Type	G_e (MPa)	G_g (MPa)	μ	k	h	β	$\tau_0/10^{-3}$	C_1	C_2
2L-0S-0R	0	1420.31	13.26	0.384	0.825	9523.4	0.24	11.846	82.423
2L-0.15S-2R	0	1120.31	10.26	0.414	0.832	8921.9	0.16	11.529	83.295
2L-0.3S-4R	0	866.776	8.632	0.426	0.862	7845.3	0.13	9.824	80.477
4L-0.15S-4R	0	1330.74	9.458	0.389	0.831	10,000.5	0.53	10.992	78.354
4L-0.3S-0R	0	1715.96	9.593	0.351	0.795	11,964.9	0.61	11.924	81.652
4L-0S-2R	0	1545.91	9.135	0.374	0.805	10,974.6	0.69	11.433	80.299
6L-0.3S-2R	0	1499.46	10.94	0.316	0.750	12,286.9	0.56	11.302	81.853
6L-0S-4R	0	1290.49	9.424	0.320	0.790	11,095.2	0.79	12.449	102.402
6L-0.15S-0R	0	1799.87	13.57	0.315	0.775	13,174.2	0.36	12.839	95.286

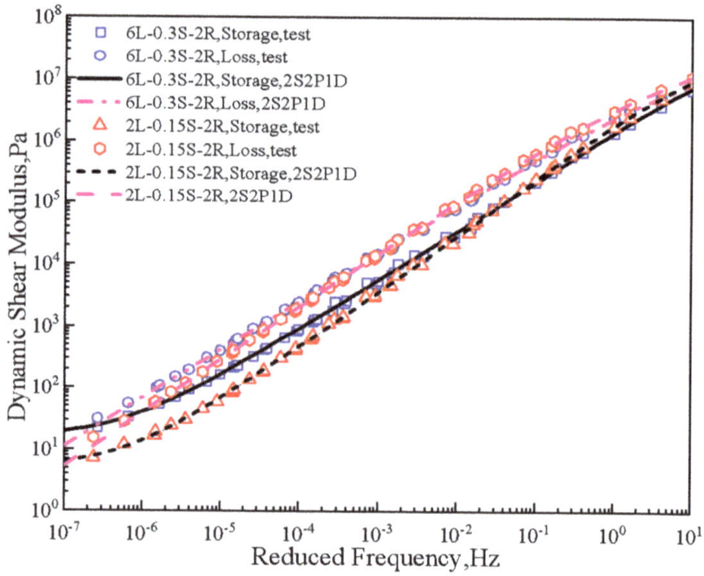

Figure 2. Modulus master curves of 6L-0.3S-2R and 2L-0.15S-2R samples.

It can be seen from Figure 3a that the phase angle decreases sharply with the increase of SBS content, indicating that the elastic part of the asphalt binder increases along with the increase of SBS content. However, the complex modulus remains stable, which may be attributed to the similarity of modulus between SBS and asphalt at this frequency and temperature.

Figure 3b demonstrates the G-R parameter value of SBS modified asphalt under various combinations of additives. The G-R parameter increases, and the increase of SBS content mainly comes from the phase angle(δ). Figure 3b shows that the phase angle(δ) significantly influenced the G-R parameter value. Comparing the influence of rubber processing oil and sulfur with the same amount of SBS, it can be found that adding 4% rubber processing oil can significantly reduce the G-R parameter while adding 0.3% sulfur had no apparent effect. To some extent, rubber processing oil, which was rich in aromatic components that could soften the asphalt, positively affected the base asphalt. Further, the effect of sulfur on the SBS-modified asphalt is similar to rubber vulcanization which could generate a three-dimensional network structure to improve the complex modulus of cementing agent and reduce the phase angle in the process of modifying asphalt. However, this effect was not much pronounced in low frequency and had little influence on the G-R constant.

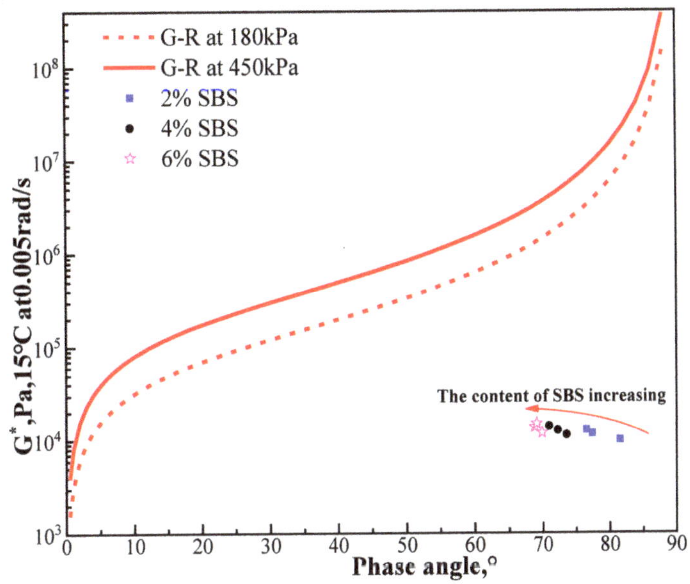

(**a**) Modulus master curves.

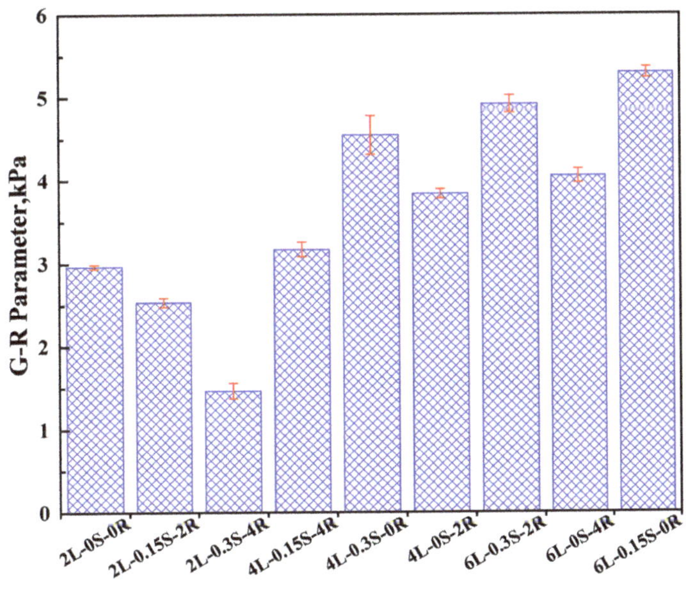

(**b**) Testing results of G-R parameter.

Figure 3. Modulus master curves and testing results of G-R parameter.

The results above illustrated that rubber processing oil could significantly improve the low-temperature properties of the SBS-modified asphalt, while sulfur barely influenced it. Besides, the increase of SBS content was negative to the low-temperature performance.

3.2. *Analysis of BBR Test*

The BBR test results have been assessed in Figure 4 by the stiffness and m-value at different testing temperatures. It could be seen that with an increase of SBS content, the stiffness values decrease relatively gently. When rubber processing oil was added to the base asphalt, the stiffness decreased, and the m-value increased sharply, especially with 4% rubber processing oil content. It was observed that the effect of rubber processing oil on the creep ability of asphalt binder is related to the SBS content. This phenomenon may originate from the absorption of the aromatic components during modification. When the SBS content is low, rubber processing oil will not be absorbed entirely to supply surplus aromatic components to the asphalt, resulting in a better low-temperature performance of SBS-modified asphalt. However, the influence of sulfur under different contents is not obvious and will be further understood through orthogonal test analyses.

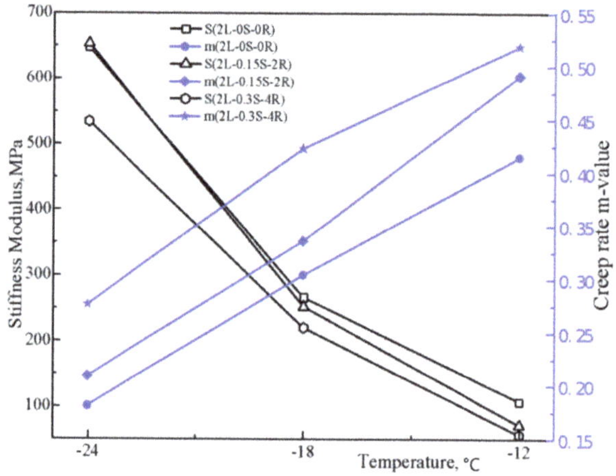

(**a**) SBS content at 2%.

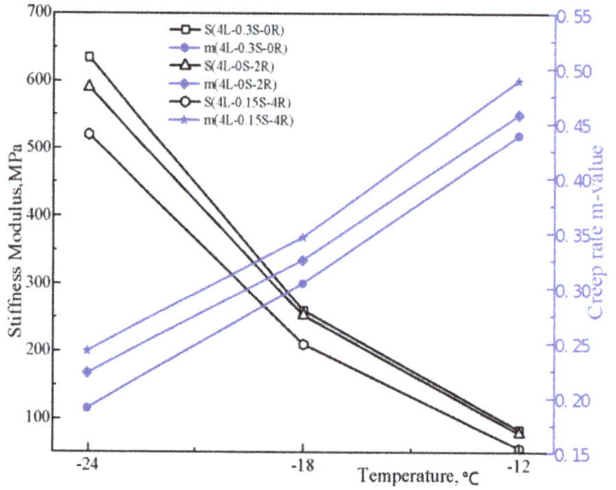

(**b**) SBS content at 4%.

Figure 4. *Cont.*

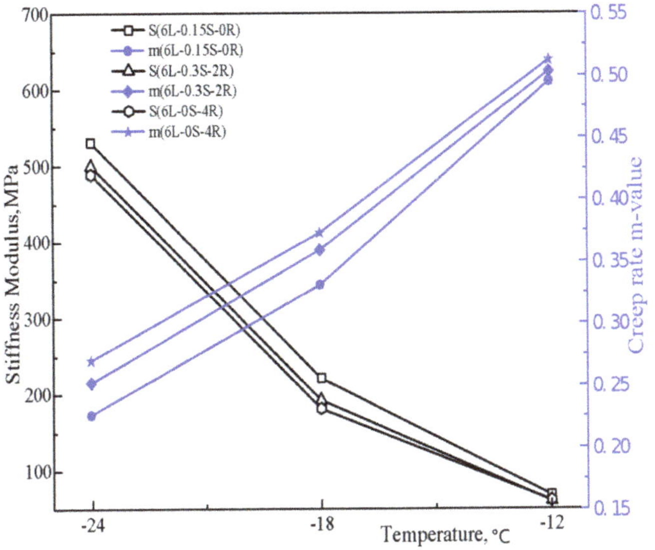

(c) SBS content at 6%.

Figure 4. Test results of BBR.

PG low temperature of BBR test results are shown in Table 6. Since the PG classification rule in SHRP fails to accurately classify the low-temperature degree of SBS-modified asphalt in this test, continuous classification methods have been adopted. It can be found that the low-temperature grade of SBS-modified asphalt decreases with the increase of SBS and rubber processing oil, but this influence was not significant.

Table 6. PG Low temperature of BBR test for all tested binders.

Binder Type	$T_{L,m}$	$T_{L,S}$	T_L	$\Delta T_{L\,(S-m)}$
2L-0S-0R	−28.19	−28.81	−28.19	−0.62
2L-0.15S-2R	−29.70	−29.10	−29.10	0.60
2L-0.3S-4R	−33.02	−30.10	−30.10	2.92
4L-0.15S-4R	−30.65	−30.36	−30.36	0.29
4L-0.3S-0R	−28.21	−28.36	−28.21	−0.15
4L-0S-2R	−29.45	−29.20	−29.20	0.25
6L-0.3S-2R	−31.31	−30.79	−30.86	0.52
6L-0S-4R	−32.25	−31.05	−30.98	1.20
6L-0.15S-0R	−29.77	−30.09	−29.84	−0.32

However, these indexes only consider the stiffness and m-value, whose calculations are one-sided. Therefore, ΔT_c is adopted to better characterize the factors affecting the low-temperature properties of SBS-modified asphalt. According to the research results of Anderson et al. (2011) [28], ΔT_c, the difference between critical PG low temperature determined by 'S' and 'm' values, has a good correlation with cracks. Statistics verified that the critical value of ΔT_c is −2.5 °C; the smaller the ΔT_c value is, the more brittle the asphalt is and the more likely it is to crack. As illustrated in Figure 5, ΔT_c of modified asphalt is all below 0 when SBS content is 2%. It implies that the modified asphalt is controlled by m-value and is easy to crack at this instant. When increasing the rubber processing oil content to 4%, 'ΔT_c' changes from negative to positive, especially at 2% SBS content. It can also be explained that the residual rubber processing oil is rich in aromatic components, which can promote the flowability of asphalt binders at low temperatures.

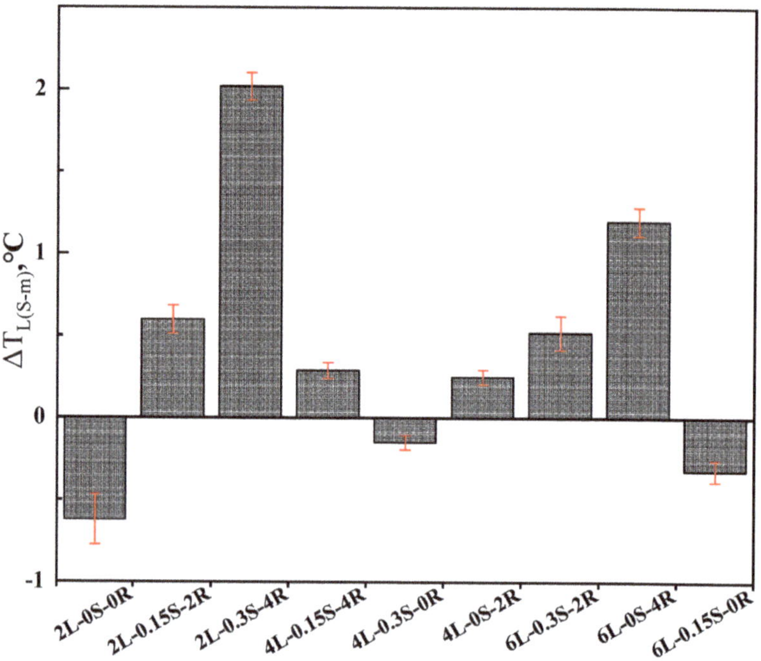

Figure 5. ΔT_c of SBS-modified asphalt.

Comparing the 'ΔT_c'of samples 2L-0S-0R, 4L-0.3S-0R and 6L-0.15S-0R, it can be identified that sulfur addition can make the ΔT_c approach to zero, which implies that sulfur can balance out the stiffness modulus and creep rate. In addition, ΔT_c of 4L-0.3S-0R sample is minimum, contributing to the reasonable amount of SBS and sulfur and resulting in a three-dimensional network structure in the modified asphalt system. The formation of the three-dimensional network system will enhance the intermolecular interaction force, thereby reducing the creep deformation rate.

3.3. *Analysis on Force-Ductility*

The F_{max} and stretched elongation for various types of modified asphalt obtained from the force-ductility test are shown in Figure 6. As seen from the results, no apparent relationship could be ascertained between SBS or sulfur content with F_{max}. However, the maximum force tends to decrease significantly with the increase of rubber processing oil content. It was also observed that the force-ductility test is mainly determined by the composition of the asphalt matrix, while the good compatibility between asphalt and rubber processing oil changes the composition of the asphalt matrix directly. The increase of aromatic components reduces the bonding force of molecules, resulting in the decrease of F_{max}.

Additionally, by evaluating the effect of rubber processing oil with 4% SBS on the reduction of maximum force-ductility, it can be concluded that F_{max} will reduce slightly along with the increase of SBS content. As discussed earlier, it may be attributed to the absorption of aromatic components, which can relieve the effect of F_{max}. Besides, the tensile force and fracture length of modified asphalt increase sharply with increasing SBS content in the late tensile phase, which implies the toughness enhancement of the modified asphalt.

Comparing the addition of rubber processing oil and sulfur at the same SBS content reveals that sulfur impacts the second stage tensile force, whereas the rubber processing oil affects the fracture length. The reasons for such behaviors are that sulfur addition can built a three-dimensional network, and the rubber processing oil addition can provide an aromatic component to asphalt. Overall, the increase of SBS content can improve the low-temperature tensile properties of asphalt and can resist cracking in a more extensive range.

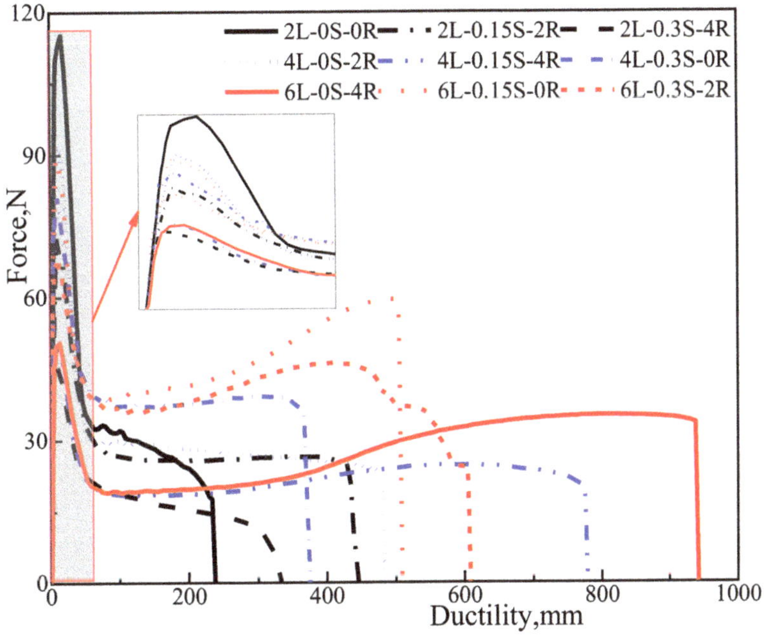

Figure 6. Force-ductility curves for nine different types of polymer modified bitumen.

According to Sun's force-ductility research on modified asphalt [32], Figure 7 illustrates the area surrounded by the ductility curve of samples to quantify the increased ratio of the toughness of SBS-modified asphalt. The results indicate that the SBS increase can strengthen the toughness of modified asphalt, and the sulfur and rubber processing oil have a certain role in it. Excessive rubber processing oil will harm the low-temperature crack resistance, and as such, it is not recommended for low SBS content.

3.4. Sequencing Analysis of Low-Temperature Parameter Based on Orthogonal Test

The experimental data indicate that certain efforts were taken after adding SBS, sulfur, and rubber processing oil, while the results obtained by the G-R parameter are contrary to those of ΔT_c and the toughness. For example, the SBS increase is negative to the low-temperature properties of modified asphalt, whereas the results of other tests are beneficial. It may be attributed to establishing the G-R parameter, which is mainly based on the number of fractures. The G-R parameters of all modified asphalts in this study are far less than the critical cracking value. Therefore, G-R parameter is more suitable to evaluate the low-temperature cracking resistance of aging asphalt. Follow-up orthogonal tests are to be conducted only for ΔT_c and toughness. The orthogonal test results of the multi-factor and single index are to be analyzed by the visual analysis method. Table 7 is the result of the orthogonal tests, in which K_1, K_2, and K_3 are the sum of the indexes below the level I, II, and III, respectively. The range represents the magnitude of the influence of factors on the respective indices. According to Table 5, when ΔT_c is taken as the index, it can be found that the range of SBS content is the largest, followed by rubber processing oil and finally sulfur. Although there is not a very high degree of disparity in these factors, it can still be deduced that the influence of each factor on the low-temperature performance is SBS > rubber processing oil > sulfur. When the evaluation index is toughness, the order is the same, but the range of SBS content is greater than that of sulfur and rubber processing oil. The SBS content has a more significant influence on the low-temperature properties since it can resist the generation and further development of cracks. Meanwhile, the rubber processing oil can delay the generation of the cracks, which can obviously enhance the low-temperature performance of SBS-modified asphalt.

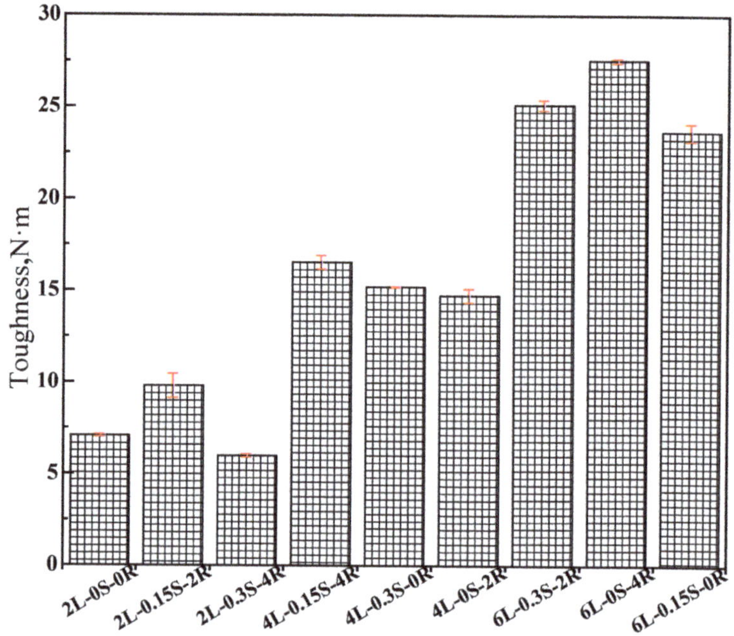

Figure 7. Toughness of all tested binders.

Table 7. Orthogonal test results.

Indexes		**A (SBS)**	**B (Sulfur)**	**C (Rubber Oil)**
$\lvert\Delta T_c\rvert$	K_1	4.140	2.070	1.090
	K_2	0.690	1.210	1.370
	K_3	2.040	3.590	4.410
	Range R	3.450	2.380	3.320
	Primary and secondary factors		A > C > B	
Toughness	K_1	22.843	49.353	45.913
	K_2	46.426	49.976	49.652
	K_3	76.359	46.299	50.063
	Range R	53.516	3.677	4.150
	Primary and secondary factors		A > C > B	

4. Conclusions and Recommendations

In this paper, the effects of SBS content, sulfur, and rubber processing oil on the low-temperature performance of the SBS-modified bitumen were studied based on the orthogonal test. The low-temperature performance was characterized by the frequency sweep test, BBR test, and force-ductility test. The conclusions drawn from the experimental results are as follows.

(1) According to the results of the BBR, force-ductility and G-R test, addition of rubber processing oil can supplement the loss of maltene and improve the low-temperature properties significantly. However, excessive amounts of rubber oil was detrimental to adhesive toughness of SBS modified asphalt.

(2) The BBR test illustrates that low-temperature performance of SBS-modified asphalt is limited by the lack of m-value. The addition of sulfur and rubber processing oil improves its m-value and promotes a balance between stiffness and m-value in SBS-modified asphalt.

(3) According to the direct analysis of the orthogonal test, the influence degree of low-temperature properties of SBS-modified asphalt was SBS > rubber processing oil > sulfur.

Only SBS-modified asphalt binders in long-term aging condition based on one kinds of base asphalt were examined in this study. Additional aging condition, different compositions of base asphalt and asphalt mixtures should be studied in the future to verify the conclusions of this study.

Author Contributions: Conceptualization and methodology, G.X., X.C. and E.J.; Data collection: G.X. and S.C.; Analysis and Interpretation: G.X., E.J. and S.Z.; Draft manuscript preparation: G.X. and S.Z. All authors have read and agreed to the published version of the manuscript.

Funding: This research was funded by the National Natural Science Foundation of China (No. 51778-136), "the Fundamental Research Funds for the Central Universities" and the Scientific Research Foundation of Graduate School of Southeast University (YBPY2160).

Institutional Review Board Statement: Not applicable.

Data Availability Statement: The data will be available on demand.

Conflicts of Interest: The authors declare no conflict of interest.

References

1. Airey, G.D. Rheological properties of styrene butadiene styrene polymer modified road bitumens ☆. *Fuel* **2003**, *82*, 1709–1719. [CrossRef]
2. Dong, F.Q.; Fan, W.Y.; Yang, G.M.; Wei, J.M.; Luo, H.; Wu, M.M.; Zhang, Y.Z. Influence of SBS and asphalt on SBS dispersion and the performance of modified asphalt. *Constr. Build. Mater.* **2014**, *62*, 1–7. [CrossRef]
3. Liu, C.; Lv, S.; Jin, D.; Qu, F. Laboratory investigation for the road performance of asphalt mixtures modified by rock asphalt–styrene butadiene rubber. *J. Mater. Civ. Eng.* **2021**, *33*, 04020504. [CrossRef]
4. Karakas, A.S.; Kuloglu, N.; Kok, B.V.; Yilmaz, M. The evaluation of the field performance of the neat and SBS modified hot mixture asphalt. *Constr. Build. Mater.* **2015**, *98*, 678–684. [CrossRef]
5. Lin, P.; Huang, W.D.; Li, Y.; Tang, N.P.; Xiao, F.P. Investigation of influence factors on low temperature properties of SBS modified asphalt. *Constr. Build. Mater.* **2017**, *154*, 609–622. [CrossRef]
6. Chen, Q.; Wang, C.H.; Song, L. Prediction of low-temperature rheological properties of SBS modified asphalt. *Adv. Civ. Eng.* **2020**, *2020*, 8864766. [CrossRef]
7. Liang, M.; Xin, X.; Fan, W.Y.; Wang, H.; Ren, S.S.; Shi, J. Effects of polymerized sulfur on rheological properties, morphology and stability of SBS modified asphalt. *Constr. Build. Mater.* **2017**, *150*, 860–871. [CrossRef]
8. Masson, J.F.; Polomark, G.; Collins, P. Glass transitions and amorphous phases in SBS–bitumen blends. *Thermochim. Acta* **2005**, *436*, 96–100. [CrossRef]
9. Cao, L.P.; Tan, Y.Q.; Dong, Z.J. Influence of Modifier on the Low Temperature Performance of Styrene-Butadiene-Styrene Modified Asphalt. *J. Tongji Univ.* **2006**, *34*, 607.
10. Aflaki, S.; Hajikarimi, P. Implementing viscoelastic rheological methods to evaluate low temperature performance of modified asphalt binders. *Constr. Build. Mater.* **2012**, *36*, 110–118. [CrossRef]
11. Sun, D.Q.; Ye, F.; Shi, F.; Lu, W.M. Storage stability of SBS-modified road asphalt: Preparation, morphology, and rheological properties. *Pet. Sci. Technol.* **2006**, *24*, 1067–1077. [CrossRef]
12. Lu, X.H.; Isacsson, U.; Ekblad, J. Low-temperature properties of styrene–butadiene–styrene polymer modified bitumens. *Constr. Build. Mater.* **1998**, *12*, 405–414. [CrossRef]
13. Tan, Y.Q.; Zhang, L.; Zhang, X.Y. Investigation of low-temperature properties of diatomite-modified asphalt mixtures. *Constr. Build. Mater.* **2012**, *36*, 787–795.
14. Kimel'blat, V.I.; Muruzina, E.V.; Chebotareva, I.G.; Khakimullin, Y.N.; Murafa, A.V.; Vol'fson, S.I.; Khozin, V.G. Stress relaxation in polymer-modified bitumens. *Colloid J.* **2002**, *64*, 442–446. [CrossRef]
15. Li, X.; Pei, J.; Shen, J.; Li, R. Experimental study on the high-temperature and low-temperature performance of polyphosphoric acid/styrene-butadiene-styrene composite-modified asphalt. *Adv. Mater. Sci. Eng.* **2019**, *2019*, 6384360. [CrossRef]

16. Stangl, K.; Jäger, A.; Lackner, R. The effect of styrene-butadiene-styrene modification on the characteristics and performance of bitumen. *Mon. Für Chem. -Chem. Mon.* **2007**, *138*, 301–307. [CrossRef]
17. Zhan, X.L.; Zhang, X.N.; Tan, Y.Q.; Lu, L. Study on the low temperature performance evaluation indexes of modified asphalt. *J. Highw. Transp. Res. Dev* **2007**, *24*, 42–45.
18. Zhang, C.; Wang, H.N.; You, Z.P.; Gao, J.; Irfan, M. Performance test on Styrene-Butadiene-Styrene (SBS) modified asphalt based on the different evaluation methods. *Appl. Sci.* **2019**, *9*, 467. [CrossRef]
19. Sengoz, B.; Isikyakar, G. Evaluation of the properties and microstructure of SBS and EVA polymer modified bitumen. *Constr. Build. Mater.* **2008**, *22*, 1897–1905. [CrossRef]
20. *ASTM D5M-20*; Standard Test Method for Penetration of Bituminous Materials. ASTM International: West Conshohocken, PA, USA, 2020.
21. *ASTM D36-14*; Standard Test Method for Softening Point of Bitumen (Ring-and-Ball Apparatus). ASTM International: West Conshohocken, PA, USA, 2006.
22. *ASTM D113-17*; Standard Test Method for Ductility of Asphalt Materials. ASTM International: West Conshohocken, PA, USA, 2017.
23. *ASTM D792-20*; Standard Test Method for Density and Specific Gravity. ASTM International: West Conshohocken, PA, USA, 2017.
24. *ASTM M320-05*; Standard Test Method for Performance-Graded Asphalt Binder. ASTM International: West Conshohocken, PA, USA, 2017.
25. Zhou, J.; Chen, X.H.; Xu, G.; Fu, Q.H. Evaluation of low temperature performance for SBS/CR compound modified asphalt binders based on fractional viscoelastic model. *Constr. Build. Mater.* **2019**, *214*, 326–336. [CrossRef]
26. Zhou, J.; Chen, X.H.; Xu, G.; Zhang, H.Y. Evaluation on low temperature characteristics of SBS/CR modified asphalt binder under different aging conditions. In Proceedings of the 4th International Conference on Civil Engineering and Materials Science, Bangkok, Thailand, 17–19 May 2019; Iop Publishing Ltd: Bristol, UK, 2019; Volume 652.
27. *ASTM D6648-08*; Standard Test Method for Determining the Flexural Creep Stiffness of Asphalt Binder Using the Bendin. ASTM International: West Conshohocken, PA, USA, 2017.
28. Anderson, R.M.; King, G.N.; Hanson, D.I.; Blankenship, P.B. Evaluation of the Relationship between Asphalt Binder Properties and Non-Load Related Cracking. In *Asphalt Paving Technology 2011*; Destech Publications, Inc: Lancaster, PA, USA, 2011; Volume 80, pp. 615–663.
29. Ye, Y.S.; Xu, G.; Lou, L.W.; Chen, X.H.; Cai, D.G.; Shi, Y.F. Evolution of rheological behaviors of styrene-butadiene-styrene/crumb rubber composite modified bitumen after different long-term aging processes. *Materials* **2019**, *12*, 2345. [CrossRef] [PubMed]
30. Zhang, J.K.; He, A.Z.; Xie, B.; Wang, J.X.; Yan, C.Q. Characterization of various modified asphalts using force ductility test with "8" shaped mold and straight mold. *Constr. Build. Mater.* **2022**, *356*, 129251. [CrossRef]
31. Glover, C.J.; Davison, R.R.; Domke, C.H.; Ruan, Y.; Juristyarini, P.; Knorr, D.B.; Jung, S.H. Development of a new method for assessing asphalt binder durability with field validation. *Tex. Dept Transp.* **2005**, *1872*, 1–334.
32. Sun, D.Q.; Lv, W.M. Evaluation of low temperature performance of polymer modified asphalt by force ductility test. *J. Build. Mater.* **2007**, *10*, 37–42.

polymers

MDPI

Article

Study on Anti-Aging Performance Enhancement of Polymer Modified Asphalt with High Linear SBS Content

Daqian Han, Guosheng Hu * and Jingting Zhang

School of Materials Science and Engineering, North University of China, Taiyuan 030051, China
* Correspondence: huguosheng@nuc.edu.cn

Abstract: Modified asphalt with high content SBS is widely used in asphalt pavement due to its excellent high and low temperature performance. However, its anti-aging performance is insufficient. In order to improve the anti-aging performance of SBS modified asphalt, nano-ZnO, nano-TiO_2, nano-SiO_2 and polyphosphoric acid (PPA) were added to high content (6.5 wt%) linear SBS modified asphalt as anti-aging agents in this study. Moreover, Dynamic Shear Rheometer (DSR), Fluorescence Microscope, and Fourier Transform Infrared Spectroscopy were employed to reveal the mechanism, through the investigation of the rheological and microscopic properties of modified asphalt before and after aging. The results showed that the influence of nanoparticles on the rutting resistance and fatigue resistance of high content SBS modified asphalt is weak, mainly because there is only weak physical interaction between nanoparticles and the SBS modifier, but no obvious chemical reaction. The significant cross-networking structure of high content SBS modified asphalt even has an adverse effect on the anti-aging performance of nano-modifiers. However, PPA obviously makes the cross-linked network structure of SBS modified asphalt more compact, and significantly improves the performance after short-term aging and long-term aging, mainly due to the chemical reaction between PPA and the active groups in SBS modified asphalt.

Keywords: modified asphalt with high linear SBS content; nanoparticles; PPA; anti-aging properties; fatigue properties; infrared spectroscopy test

Citation: Han, D.; Hu, G.; Zhang, J. Study on Anti-Aging Performance Enhancement of Polymer Modified Asphalt with High Linear SBS Content. *Polymers* **2023**, *15*, 256. https://doi.org/10.3390/polym15020256

Academic Editor: Keon-Soo Jang

Received: 28 November 2022
Revised: 24 December 2022
Accepted: 27 December 2022
Published: 4 January 2023

1. Introduction

Polymer modified asphalt with high content SBS (HCPMA) is a promising asphalt binder in the open-graded friction course (OGFC) pavement due to excellent high temperature shear resistance and adhesion characteristics [1–3]. However, the potential weak aging effect of HCPMA hinders the widespread application [4]. Studies have shown that HCPMA has active chemical properties, which are sensitive to oxygen, and could be degraded easily in high-temperature environments. Moreover, light components volatilize, and the active groups in asphalt react with oxygen during aging, leading to the content increase of carbonyl and sulfoxide functional groups. The average molecular weight of asphalt after aging increases due to intermolecular condensation and polycondensation, the content of heavy components increases, and the elasticity of asphalt decreases. Asphalt becomes brittle and hard, resulting in low-temperature cracking of the pavement [5,6]. Recent studies also showed that after long-term aging, the asphalts were sharply deteriorated and the stripping potential per unit area became overly high, resulting in a remarkable reduction in moisture resistance [7]. As a result, its overall pavement life is only 60~70% of the ordinary dense-graded pavements [8]. Therefore, it is very important to further improve the anti-aging performance of HCPMA.

In recent years, researchers have devoted themselves to improving the anti-aging performance of modified asphalt by adding anti-aging agents [9–11]. Nano materials form stable structures with other atoms due to interfacial effects. Liu selected modified nano-TiO_2 as an anti-aging agent to study the influence of the sunlight and oxygen aging of modified asphalt [12]. The study found that the (colloidal index) of modified nano-TiO_2/modified asphalt decreased after UV aging. In addition, compared with the unmodified nano-TiO_2, the modified nano-TiO_2 had a superior anti-aging effect. Zhang found that nano-ZnO could significantly promote the anti-ultraviolet light aging ability of modified asphalt, but the improvement effect of its thermal oxygen aging resistance was slight [13]. Eillie's research presented how nano-SiO_2 effectively enhanced storage modulus capability and elastic behavior, thereby enhancing rutting resistance [14]. Meanwhile, the variation in rheological properties and the infrared spectrum aging index of aged nano-SiO_2/modified asphalt became slow. Furthermore, many researchers have discussed the anti-aging mechanism of nanoparticles for SBS modified asphalt. Zhang considered how nanoparticles with high surface energy and large active specific surface area have a strong bonding ability to both asphalt and polymer particles. The addition of nanoparticles can form an oriented film between the polymer and asphalt, thus reducing the interfacial tension between the polymer and asphalt and improving the dispersion of polymer particles [15]. However, Guo pointed out that the influence of nano materials on the properties of SBS modified asphalt is not only related to physical changes, but also related to chemical reactions [16]. Besides, in the past 30 years, PPA has attracted much attention from researchers due to its low price and excellent anti-aging performance [17–19]. Ramasamy B.S. has studied the rheological and aging properties of unmodified and modified asphalt [20]. The results showed that PPA enhanced the mechanical and anti-aging properties of modified asphalt. However, PPA was unfriendly to low-temperature and the fatigue properties of modified asphalt. As for the mechanism of the influence of PPA on the performance of SBS modified asphalt, most researchers believed that it is a chemical reaction between PPA and active groups (such as sulfoxide group) in SBS modified asphalt [21,22].

Based on the above analysis, nanoparticles and PPA can enhance the anti-aging performance of low content linear SBS modified asphalt. However, the influence and mechanism of anti-aging agents on the performance of high content linear SBS modified asphalt is almost blank. Therefore, in this study, nano-ZnO, nano-TiO_2, nano-SiO_2 and PPA with 115% orthophosphoric acid were selected as anti-aging modifiers to prepare modified asphalt with a high content of linear SBS. Then, the anti-aging properties were comprehensively evaluated by DSR rheometer, fluorescence microscope and infrared spectrometer. The influencing mechanism of nanoparticles and PPA on the anti-aging properties of high content SBS modified asphalt was revealed.

2. Materials and Methods

2.1. Experiment Materials

Zhonghai 70# (PG64-16) acted as the base asphalt. Linear SBS 791-H was produced by Sinopec. The solubilizer was selected from rubber oil with 3.0 wt% of base binder produced in Iran. The sulfur was employed as a stabilizing agent with 0.1 wt% of Zhonghai 70#. The content of SBS was 6.5 wt% of Zhonghai 70#. Nano-ZnO, nano-TiO_2 and nano-SiO_2 were chemically surface-modified to reduce the specific surface area and promote dispersion. PPA was used with a mass fraction of H_3PO_4 of 115%.

There were main three steps involved in the preparation of composite modified asphalt with a high content of linear SBS (6.5 wt%). Firstly, the SBS and rubber oil were blended to Zhonghai 70# and sheared for 35 min at the rate of 5000 r·min^{-1} and 185 °C with a high-speed disperser. Secondly, weighed inorganic nanoparticles, such as nano-ZnO or PPA, were added to the asphalt and sheared continuously for 30 min. The content of all anti-aging agents in this paper was 2.0% of the total amount of base asphalt. Thirdly, the sulfur was added to the mixture and stirred for 120 min at the rate of 2000 r·min^{-1}. The final samples were represented by SBS, ZnO/SBS, TiO_2/SBS, SiO_2/SBS or PPA/SBS, respectively.

2.2. Binder Test Descriptions

According to the AASHTO standard, the short-term aging (rotating film oven aging (RTOT)) and long-term aging (pressure aging (PAV)) were carried out for the original asphalt in this paper, respectively. Then, the rheological properties of the asphalt were analyzed by DSR. The anti-aging mechanism of the high content SBS polymer composite modified asphalt was expounded from a microscopic perspective through fluorescence microscopic images and Fourier transform infrared spectroscopy.

2.2.1. Dynamic Shear Rheological Test

The rheological properties of the modified asphalt were investigated by the HR10 DSR rheometer (TA Instruments, New Castle DE, United States). The dynamic shear modulus G* and phase angle δ are taken as the basic parameters to characterize the rheological properties of asphalt. The G*/sinδ and G*·sinδ characterize the rutting resistance and fatigue resistance of asphalt, respectively. At the same time, the high temperature PG of the original asphalt and short-term aging asphalt were based on the rutting factor.

The high-temperature rutting resistance of modified asphalt was characterized by the multiple stress creep recovery (MSCR) test. The method was performed continuously at 0.1 kPa and 3.2 kPa, respectively. Every pressure condition was carried out for ten cycles, and each process was 10 s. Deformation recovery rate (R) and non-recoverable compliance (J_{nr}) are the main parameters. The test temperature was 82 °C, 88 °C and 94 °C.

According to the AASHTO standard, the linear amplitude sweep (LAS) test was conducted by DSR. The test temperature was 25 °C.

2.2.2. Fluorescence Microscopy Image Acquisition

Fluorescence microscopy can obtain reproducible microstructure images without destroying the internal structure of modified asphalt. The different preparation methods can significantly affect its microstructure under fluorescence microscopy. The molding method and steps of the observation sample in this paper are as follows. Firstly, place a metal container with a diameter of 5~10 mm and a height of 15~20 mm on a horizontal table top. Then, pour different SBS composite modified asphalt samples evenly into metal containers and store them at room temperature for 0.5 h. Finally, absorb the water on the surface of the asphalt sample with absorbent paper for observation; this prevents the distortion of the observed image of the sample caused by the reflection of water. This method can avoid the interference of the surface and internal morphology of the sample in conventional ways.

2.2.3. Infrared Spectroscopy Test

The infrared spectroscopy test was conducted by Bruker Tensor 27 (Thermo Fisher Scientific, Waltham, United States) infrared spectrometer equipped with the full scattering ATR accessory. The test was scanned 32 times in the wavenumber range of 4000 cm^{-1}~400 cm^{-1}. Each test was repeated 3 times, and the average value was the result. All raw spectra were normalized and baseline corrected.

The semi-quantitative analysis was used to explore the variation in functional groups of the modified asphalt during aging. The carbonyl index (I_{CA}) and a new characteristic index were selected to analyze the effect of long-term aging (PAV) on the chemical composition of modified asphalt by calculating. The new characteristic index was the SBS degradation index ($I_{B/S}$), which uses the polybutadiene peak as the characteristic peak and the polystyrene peak as the reference peak [23]. The study showed that the $I_{B/S}$ is reasonable for presenting the degradation degree of SBS under an ATR scanning environment [24–26]. The different characteristic indices were calculated as follows [23]:

$$I_{CA} = A_{1700} / \sum AR \times 100\% \quad (1)$$

where A_{1700} is the infrared absorption peak belonging to the 1700 cm^{-1}; the ΣAR are the peak areas of the asphalt functional groups in the range of 680~3600 cm^{-1}.

$$I_{B/S} = A_{965} / A_{700} \times 100\% \quad (2)$$

where the A_{965} is the absorption peak belonging to the 965 cm^{-1}; and where the A_{700} is the absorption peak belonging to the 700 cm^{-1}.

3. Results

3.1. Effects of Aging on Rheological Properties of Different High Content SBS Polymer Composite Modified Asphalts

3.1.1. Rheological Analysis

The mechanical and deformation properties of the high content SBS polymer modified asphalt with various anti-aging agents after RTFO aging at 82 °C are shown in Table 1. As we can see from Table 1, the addition of nanoparticles has a slight effect on the failure temperature and rutting factor of modified asphalt, and the PG of each nanocomposite modified asphalt is 82 °C. However, it is worth mentioning that the failure temperature of the PPA/SBS composite modified asphalt is as high as 125.7 °C after adding polyphosphoric acid, which is 46.0% higher than that of SBS modified asphalt. This shows that it possesses extremely outstanding high temperature and aging resistance.

Table 1. Rheological properties of all asphalt samples in RFTO aging state.

Samples	δ/°	G*/Kpa	G*/Sin δ/Kpa	Failure Temperature/°C
SBS	63.10	2.74	3.07	86.1
ZnO/SBS	64.31	2.39	2.65	84.4
TiO_2/SBS	56.80	2.32	2.78	85.6
SiO_2/SBS	57.20	2.41	2.87	85.9
PPA/SBS	40.20	16.05	24.86	125.7

Within the two indicators, phase angle and complex modulus, the complex shear modulus G* characterizes the deformation resistance of asphalt under continuous shear load. The lower the G*, the weaker the deformation resistance. The phase angle δ characterizes the viscoelastic properties of modified asphalt. The δ is 0° for pure elastomers and 90° for pure viscous materials. Table 1 shows that various nanomaterial anti-aging agents, such as nano-ZnO, have no noticeable effect on the G* and δ of SBS modified asphalt. However, adding PPA can significantly reduce the δ and improve G*. Therefore, the elasticity and shear deformation resistance of modified asphalt are enhanced considerably, which means that it will improve the permanent deformation and aging resistance remarkably. The reason is that, PPA can react with small molecules in asphalt and generate a macromolecular structure, which makes the asphalt more elastic [18].

3.1.2. Analysis of MSCR Test Results

(1) R Analysis of deformation recovery

To further evaluate the asphalt binder actual rutting resistance level and rheological properties at high-temperature, this study used the R and J_{nr} of the MSCR test. Modified asphalt samples after RTFO aging were selected to test. Furthermore, three different temperature gradients of 82 °C, 88 °C and 94 °C were selected in the test to study the dependence of modified asphalt on temperature. The results are presented in Figure 1.

The higher the R is, the stronger the proportion of the recoverable strain of the asphalt will be, and the smaller the ratio of the residual strain will be. When no anti-aging agent is added at the 0.1% stress level, the R of SBS modified asphalt is relatively low, and the value reduces remarkably with the increase in temperature, as depicted in Figure 1. After adding nanomaterials, such as nano-ZnO or PPA, the R has been significantly increased, indicating that anti-aging agents can promote the elastic deformation ability of the modified asphalt, thereby improving the resistance to rutting deformation, which is identical to the result of PG. It is noteworthy that the R of composite modified asphalt with the addition of PPA is highest, and the rutting deformation resistance is strongest. The R variation trend of modified asphalt at 3.2 kPa is consistent with 0.1 kPa. When anti-aging agents were added into the modified asphalt, its deformation recovery ability was improved to varying degrees. Moreover, with the increase in temperature, the effect of different anti-aging effects is more obvious. In addition, the R of PPA/SBS modified asphalt at three temperatures is much higher than that of nanoparticles, which will provide it with the strongest elastic deformation ability and show excellent anti-rutting performance.

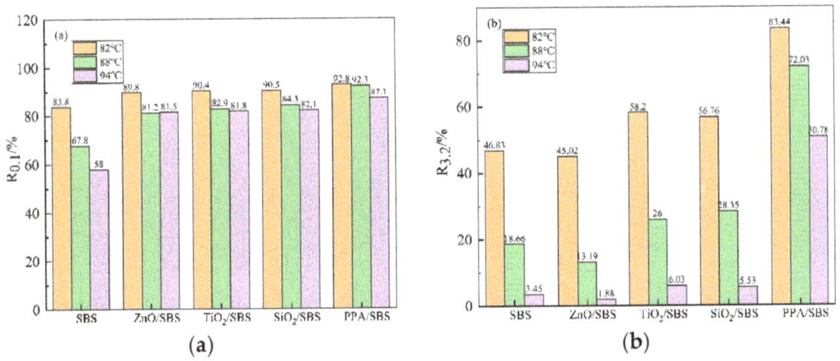

Figure 1. The deformation recovery rate of modified asphalt under different conditions after RTFO aging: (a) $R_{0.1}$; (b) $R_{3.2}$.

The above analysis means that under the same temperature and stress conditions, adding PPA can provide SBS modified asphalt with improved elastic deformation ability and more resistance to rutting deformation. The reason why PPA can increase the recovery rate R of modified asphalt is that PPA can promote the transformation of resin in asphalt to asphaltene, which is conducive to enhancing the content of asphaltene [21]. Therefore, the high-temperature deformation resistance of modified asphalt increases.

To explore the temperature sensitivity of R to stress, the recovery rate difference index R_{diff} was introduced. The lower the R_{diff} is, the more weakly sensitive the recovery rate is to stress. The calculation results of R_{diff} are shown in Table 2. Compared with the SBS modified asphalt, the R_{diff} difference slightly increases after adding nano-ZnO, indicating that the stress sensitivity rises, as shown in Table 2. However, R_{diff} decreases slightly with the addition of nano-TiO_2 or nano-SiO_2. This phenomenon is because nano-ZnO mainly focuses on improving the anti-ultraviolet aging properties, while it is weaker than other nanomaterials in reducing stress sensitivity. The lowest value is PPA/SBS composite modified asphalt. R_{diff} difference at three temperatures is much lower than that of other modified asphalts, which indicates that adding PPA can effectively reduce the asphalt binder sensitivity of the recovery rate to stress, and will help to delay the rapid permanent deformation trend of asphalt caused by increased stress.

Table 2. Recovery rate difference R_{diff} of various modified asphalts at different temperatures.

Samples	R_{diff} (%)		
	82 °C	88 °C	94 °C
SBS	44.1	72.49	94.04
ZnO/SBS	49.87	83.76	97.70
TiO_2/SBS	35.70	68.60	92.62
SiO_2/SBS	37.31	66.37	90.30
PPA/SBS	10.10	21.93	41.73

(2) J_{nr} Analysis of Non-recoverable Compliance

Figure 2 shows the J_{nr} of the MSCR test under different stress and temperature conditions after RTFO aging of different modified asphalts. The lower the J_{nr} of the asphalt is, the less strain will be generated under stress, and the stronger the deformation resistance the asphalt will have. Compared with SBS modified asphalt, adding various anti-aging materials can reduce the irrecoverable creep compliance J_{nr} of asphalt to varying degrees under the same stress or temperature conditions, as shown in Figure 2. This means that adding different anti-aging materials will improve the ability of modified asphalt to resist rutting deformation. It is worth noting that under the same conditions, the J_{nr} of the PPA/SBS modified asphalt is two orders of magnitude lower than that of other modified asphalt. Its resistance to rutting deformation is highly excellent, which is consistent with the above analysis of the recovery rate R.

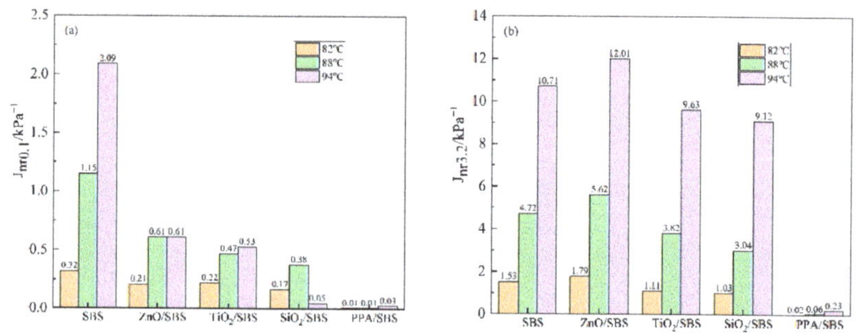

(a) **(b)**

Figure 2. Irreversible creep compliance of asphalt at different stress levels and temperatures after RTFO aging: (**a**) $J_{nr0.1}$; (**b**) $J_{nr3.2}$.

Studies have shown that the MSCR test results at higher stress correlate more with the resistance rutting test of asphalt mixtures than the standard stress required by ASTM D7405. Therefore, in order to analyze the asphalt binder's stress dependence, the MSCR test was supplemented; it was conducted at 10 kPa and 70 °C. The results are shown in Figure 3. After increasing the stress level, there is no clear difference in the R of various asphalts added to anti-aging agents. Furthermore, compared with the SBS modified asphalt, it has a certain degree of improvement, indicating that the resistance deformation ability of the asphalt has been promoted. In addition, the non-recoverable compliance J_{nr10} index of different asphalts is quite different, and the anti-aging properties of asphalts are significantly decreased, which will greatly improve their rutting resistance. PPA/SBS composite modified asphalt has the strongest rutting resistance among different modified asphalts.

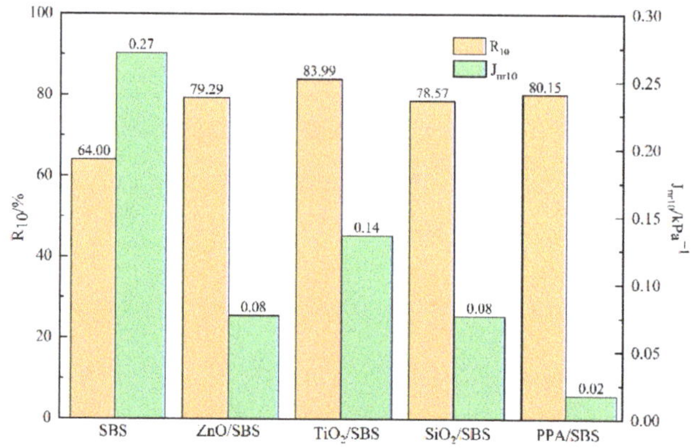

Figure 3. R_{10} and J_{nr10} of asphalt at 10 KPa and 70 °C after RTFO aging.

3.2. Effects of PAV Aging on Medium Temperature Fatigue Properties of Different High Content SBS Polymer Composite Modified Asphalts

G*·sinδ is the loss shear modulus, representing the energy dissipated by the internal friction of asphalt during deformation. The higher the G*·sinδ, the greater the energy loss rate. Therefore, the lower G*·sinδ indicates better fatigue resistance. In the DSR test, the modified asphalt has fatigue failure when the fatigue factor value exceeds 5000 kPa. Figure 4 shows the fatigue factor and failure temperature of modified asphalt after PAV aging. It can be seen, from Figure 4, that the fatigue factor and the failure temperature of various anti-aging composite modified asphalts at the same temperature are significantly reduced due to the addition of anti-aging agents. Among them, PPA has the strongest fatigue resistance. Its failure temperature is as low as 10.9 °C, which is 33.9% lower than that of asphalt without any anti-aging agent, and fatigue resistance is greatly improved.

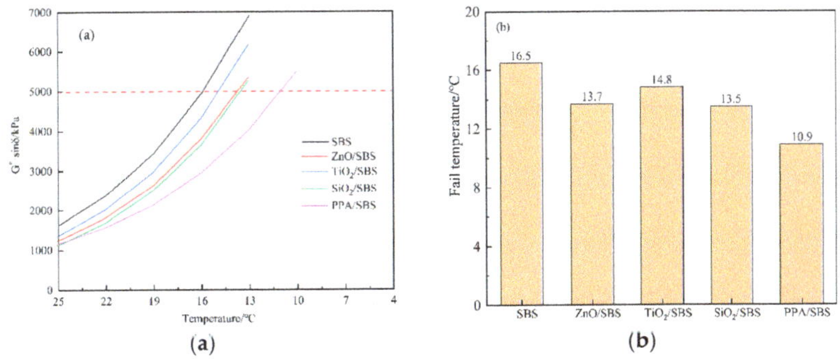

Figure 4. Fatigue factor (**a**) and failure temperature (**b**) of different SBS composite modified asphalts after PAV aging.

Since the fatigue factor is a linear viscoelastic index, its essence belongs to the linear viscoelastic dissipation energy. This index only uses the dissipated energy information of the specimen obtained in 10 loadings of the DSR test; however, it is hard to fully show the material state of the asphalt after tens of thousands of loading processes. So, this study introduced a DSR-based LAS. This method characterizes the fatigue life of pavement as a function of strain and uses progressively increasing pressure for periodic loading to accelerate the fatigue failure of asphalts. The test results of various modified asphalts after PAV aging are depicted in Table 3.

Table 3. LAS test results of different asphalts after long-term aging.

Samples	A	B	α	2.5%N_f	5%N_f	10%N_f
SBS	8.341×10^6	3.648	1.824	294,850	23,522	1877
ZnO/SBS	2.524×10^7	3.775	1.888	793,882	57,976	4234
TiO_2/SBS	9.978×10^6	3.588	1.794	372,679	30,997	2578
SiO_2/SBS	1.385×10^7	3.483	1.741	569,319	50,919	4554
PPA/SBS	1.252×10^{10}	5.423	2.712	87,004,865	2,027,422	47,244

As shown in Table 3, the fatigue performance parameter N_f of composite modified asphalt at three strain levels of 2.5%, 5% and 10% increases due to the addition of anti-aging agents, meaning that their resistance fatigue performance has been improved to some extent. In addition, the N_f of PPA/SBS modified asphalt is clearly higher than that of other modified asphalts by two orders of magnitude, and the fatigue resistance is excellent. The reason is that, adding PPA to asphalt increases the content of asphaltenes, which is conducive to enhance its mechanical properties, and thus improves the fatigue resistance. The fatigue resistance of different modified asphalts is ranked as follows: PPA/SBS > ZnO/SBS > SiO_2/SBS > TiO_2/SBS > SBS modified asphalt.

3.3. Micromorphological Analysis of Different Polymer Modified Asphalts with a High Content of Linear SBS

The micro-morphology of different modified asphalts was observed by fluorescence microscope to evaluate the influence of different anti-aging agents on the dispersion state of SBS. The results are shown in Figure 5. The fluorescent region represents the SBS phase. Different modified asphalts have apparent differences in particle size and distribution as shown in Figure 5. Figure 5a presents how the SBS dispersion in modified asphalt with a high content of linear SBS is relatively poor. In addition, there are large particle sizes and significant particle size differences, meaning that the stability of the asphalt is low. After adding nanoparticles, such as nano-ZnO, the large size particles of SBS in the composite modified asphalt decreased, indicating that these nanomaterials play a certain role in promoting dispersion after special surface modification. Among all of the modified asphalts, the PPA/SBS composite modified asphalt has the most uniform SBS distribution and the tiniest particles, forming an apparent cross-network structure, meaning that PPA reacts with SBS modified asphalt. This structure encapsulates the asphalt in the network to create a single-phase microstructure, within which SBS and asphalt are used as the continuous and dispersed phase, respectively.

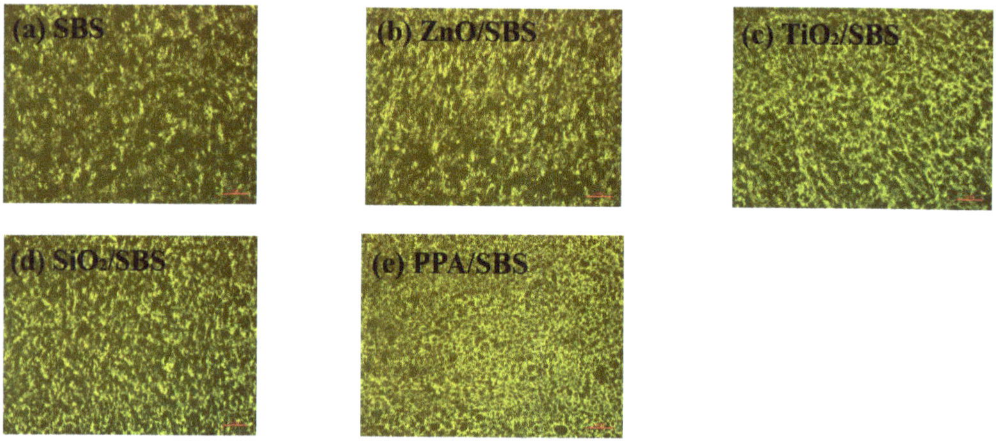

Figure 5. Fluorescence microscopy results of different polymer modified asphalts with a high content of linear SBS.

In summary, SBS dispersed into finer particles and was more evenly distributed in asphalt due to the addition of anti-aging agents. Among the antioxidants, PPA has the most potent dispersing effect on SBS particles, which can significantly enhance the swelling effect of SBS. Furthermore, PPA can also improve the spatial network structure, thereby improving the storage and anti-aging performance of SBS modified asphalt.

3.4. Analysis of Infrared Spectroscopy of Different High Content SBS Polymer Composite Modified Asphalts

The infrared spectra of the samples were qualitatively compared at first, to research the influence of the different anti-aging agents on the chemical components of modified asphalt with a high content of linear SBS and the anti-aging effect. Considering that there is no significant difference in spectra before and after short-term aging, the samples before aging and after long-term aging are drawn here. Figures 6 and 7 present the results of the infrared spectrum, and the characteristic peaks commonly used in the asphalt have also been marked in the figures.

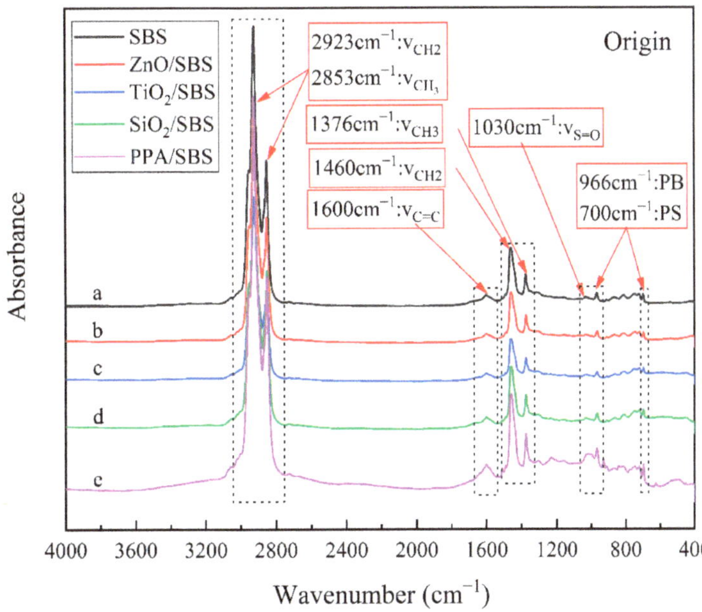

Figure 6. Infrared spectra of original high content SBS composite modified asphalt.

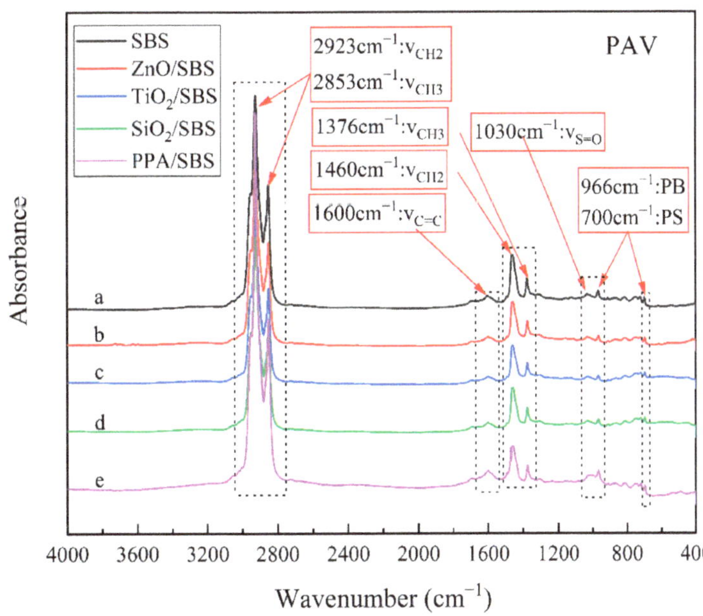

Figure 7. Infrared spectra of different high content SBS composite modified asphalts under a PAV aging state.

Figure 6a shows the infrared spectrum corresponding to the original high content SBS polymer modified asphalt. In the figure, the peaks of 2923 cm^{-1} and 2853 cm^{-1} correspond to the asymmetric stretching characteristic peak of -CH_3 and the symmetrical stretching characteristic peak of -CH_2-, respectively. The peak at 1700 cm^{-1} belongs to the characteristic peak of the carbonyl C=O stretching vibration, and the peak at 1600 cm^{-1} belongs to the C=C stretching vibration, while two characteristic peaks near 1460 cm^{-1} and 1376 cm^{-1} originate from the asymmetric vibration of -CH_3- and the symmetrical vibration of -CH_2-, respectively. At the same time, there is an evident vibration characteristic peak at 1030 cm^{-1} belonging to the sulfoxide group S=O. The characteristic peaks at 966 cm^{-1} and 700 cm^{-1} originate from the bending vibration of polybutadiene PB and polystyrene PS, respectively.

Figure 6b–d correspond to the infrared spectra of nanoparticles/SBS modified asphalt. Compared with Figure 6a, it is easy to notice that the infrared spectrum basically overlaps between high content SBS modified asphalt with and without nano materials. There are no new characteristic peaks generated in their spectra, and the intensities of the existing characteristic peaks do not change significantly, which indicates that the addition of nanoparticles does not generate new substances in modified asphalt, and they are simply physically blended with the asphalt. Therefore, the enhancement of the anti-aging performance of modified asphalt after adding nanoparticles may be due to the synergistic effect of surface effect, volume effect, quantum size effect, and macroscopic quantum tunneling effect of nanoparticles.

Figure 6e depicts that the mid-infrared absorption peak of the PPA/SBS composite modified asphalt has an -OH vibration at 3900~3400 cm^{-1}, which is the strong vibration region of PPA. In addition, most of the absorption peaks are in agreement with SBS modified asphalt, indicating a physical modification phenomenon. The intensity of the peak at 1100~1000 cm^{-1} corresponded to an -OH change, and the relative intensity of the two peaks changed with the addition of PPA, which indicated the existence of physical modification and partial chemical modification. It is mainly the out-of-plane bending vibration absorption peak of unsaturated C-H (=C-H) at less than 1000 cm^{-1} and the S=O vibration in the sulfoxide group near 1030 cm^{-1}. There are relatively great and weak changes in the intensity of these regions, and their peak types are unchanged. However, the contents are different, indicating that the addition of PPA forms chemical cross-links. Therefore, after adding PPA to SBS modified asphalt, physical and chemical modifications are caused. The IR spectrum of asphalt show slight absorption at the strong absorption of SBS and PPA due to physical modification; it is a pity that the change is insignificant. The chemical modification is reflected in that the relative absorption intensity of the peak changes after adding PPA.

Figure 7 shows the infrared spectra of different composite modified asphalts under a PAV aging state. Compared with the original modified asphalt with a high content linear SBS, the absorption peak intensity of C=O (1700 cm^{-1}) and S=O (1030 cm^{261}) increases, which corresponds to the hardening of the asphalt phase. It is also reflected in the reduction of the polystyrene (700 cm^{-1}) and polybutadiene peaks (966 cm^{-1}) due to the degradation of the polymer phase. By comparison, it is found that when the modified asphalt was aged by PAV, the SBS is seriously degraded, leading to insufficient anti-aging ability.

At the same time, comparing different nanoparticle composite modified asphalts, it can be found that although the area of the polybutadiene peak (966 cm^{-1}) decreased after aging, it was slightly higher than that of modified asphalt without nanoparticles, indicating that the SBS degradation degree becomes slow due to the addition of nanoparticles. In addition, the peak area of different asphalts decreased to a similar extent, indicating that nanoparticles, such as nano-ZnO, have no obvious difference in anti-aging property. The degree of reduction in the peak area of different pitches is relatively close, indicating that the anti-aging ability is not much different. All of them can slightly enhance the anti-aging ability, but the effect is insignificant. It should be pointed out that their polystyrene characteristic peak (700 cm^{-1}) does not change significantly. This is mainly because the first hydrogen atom (alpha hydrogen) connected to the carbon–carbon double bond functional group on the polybutadiene segment is chemically active and easily attacked by oxygen to generate oxides, which results in the bond breaking. In contrast, polystyrene is relatively stable in thermal-oxidative aging.

The previous research proved that the anti-aging performance of low content SBS (<5.0 wt%) modified asphalt is effectively improved by nanoparticles. This is because nanoparticles with a high surface energy and large active specific surface area have a strong bonding ability to both asphalt and polymer particles. The addition of nanoparticles can form an oriented film between the polymer and asphalt, thus reducing the interfacial tension between the polymer and asphalt and improving the dispersion of polymer particles, which is consistent with the results of the fluorescence microscopy test in this paper [15]. However, for high content SBS modified asphalt with a strong network cross-link structure, SBS as a continuous phase, and asphalt as a dispersed phase, the movement of SBS molecules is blocked, and the adsorption effect of the nano materials is weakened. Therefore, nanoparticles slightly improve the performance of high content SBS modified asphalt. Interestingly, compared with the original modified asphalt, after PAV aging, the infrared spectrum difference of SBS modified asphalt with nano materials is reduced, which indicates that nano materials have no significant impact on aging resistance. This result is consistent with the fatigue resistance of modified asphalt with or without nano materials after PAV aging, as shown in Table 3.

By comparing the spectra of PPA/SBS composite modified asphalt before and after aging, it can be seen that the infrared spectrum of the aged asphalt almost coincides with the original one. Its carbonyl characteristic peaks, polystyrene peaks and polybutadiene peaks hardly change, and its macroscopic performance is far superior to other modified asphalts. Combined with the fluorescence microscope results and infrared results, it is not difficult to find that the mechanism of PPA within high content SBS modified asphalt is that PPA reacts with the S=O in asphaltene, which enhances the cross-linking effect of the SBS in the asphalt.

To further evaluate the influence of anti-aging agents on the anti-aging performance of modified asphalt, the aging behavior of samples were semi-quantitatively analyzed by calculating the carbonyl index and the $I_{B/S}$ index. The results are depicted in Figure 8. During the aging process, asphalt undergoes an oxygen absorption reaction, producing oxygen-containing components such as aldehydes and ketones, which manifest the -OH (at 1700 cm^{-1}) content increases. It is also accompanied by the degradation of SBS, reflected in the reduction of the polystyrene and polybutadiene peaks However, the decline of the peak at 700 cm^{-1} is much lower than that of 966 cm^{-1} due to the lower sensitivity of polystyrene to aging. Therefore, the smaller the carbonyl index and the larger the $I_{B/S}$ index, the lower the oxidation degree of the asphalt and the higher the integrity of the SBS. According to the test results in Figure 8, the addition of nanomaterials, such as nano-ZnO, can appropriately reduce the carbonyl index and increase the $I_{B/S}$ index, and the anti-aging performance of the asphalt is slightly improved. The carbonyl index of the PPA/SBS modified asphalt is the lowest, and the $I_{B/S}$ index is clearly higher than that of others. SBS content is the highest after aging, and its anti-aging ability is exceptionally outstanding. This is because adding PPA will produce a chemical cross-linking reaction, which can effectively prevent the oxidation of the asphalt and the decomposition of the SBS.

The above qualitative analysis of the infrared spectra of modified asphalts suggests that there is only a simple physical blending of nanoparticles and SBS modified asphalt, while PPA can play a role in physical and chemical modification at the same time. The addition of PPA can effectively reduce the influence of thermal-oxidative aging on asphalt properties and significantly improve the anti-aging properties.

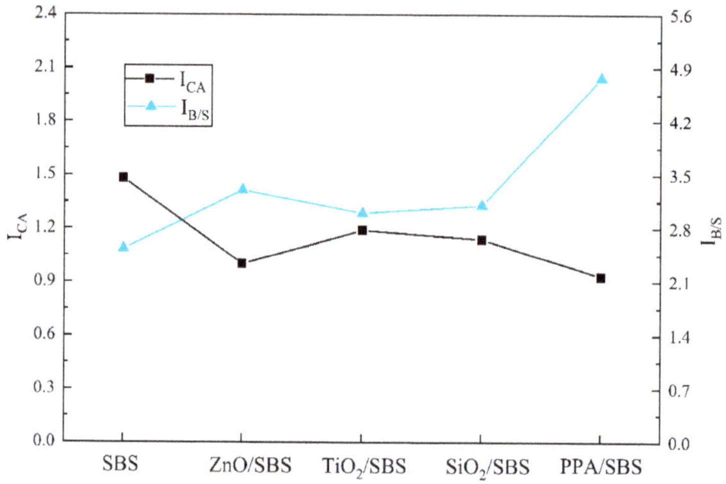

Figure 8. Quantitative analysis results of infrared spectra of different high content SBS composite modified asphalts under PAV aging state.

4. Conclusions

In this paper, from the perspective of rheology and microchemical analysis, the influence of nano-ZnO, nano-TiO_2 and nano-SiO_2 and PPA on the anti-aging properties of modified asphalt with a high content of linear SBS were evaluated, and the modification mechanism was explained. The major conclusions were drawn as follows:

(1) No matter whether nanoparticles are added or not, the PG temperature of high content SBS modified asphalt after short-term aging is 82 °C, which implies that the nanoparticles have no significant effect on high-temperature rutting resistance. After PAV aging, the anti-fatigue factor is in the same order of magnitude. However, PPA significantly improved PG temperature (124 °C) and high-temperature rutting resistance. Compared with other SBS modified asphalts, the fatigue property parameter N_f was improved by two orders of magnitude. In other words, nano materials cannot significantly improve the anti-aging performance of high content SBS modified asphalt. On the contrary, PPA has excellent anti-aging performance.

(2) The infrared spectrum of high content SBS modified asphalt with and without nanoparticles basically overlaps after aging, indicating that nanoparticles have hardly any anti-aging performance. The reason is that the nanoparticles and high content SBS modified asphalt are simply mixing physically; the movement of the SBS molecules is blocked by the strong network cross-linking structure, which leads to a reduction of the adsorption effect of the nanoparticles.

(3) By comparing the spectra of the PPA/SBS composite modified asphalt before and after aging, it can be found that the carbonyl characteristic peaks, polystyrene peaks and polybutadiene peaks have little change, and its macroscopic mechanical performance is far superior to other modified asphalts, meaning that PPA has outstanding anti-aging performance. This is because PPA reacts with the S=O in asphaltene, which enhances the cross-linking effect of the SBS in asphalt.

Author Contributions: Methodology, investigation and writing—original draft preparation, D.H.; validation, resources, data curation and supervision, G.H.; writing—review and editing, J.Z. All authors have read and agreed to the published version of the manuscript.

Funding: This research was funded by the Natural Science Foundation of Shanxi Province (No.20210302123041).

Institutional Review Board Statement: Not applicable.

Informed Consent Statement: Not applicable.

Data Availability Statement: The data that supports the findings of this study are available within the article.

Acknowledgments: This work was supported by the Natural Science Foundation of Shanxi Province (No.20210302123041) and Open Fund Project of Sichuan Province Key Laboratory of Material Corrosion and Protection (2021CL22).

Conflicts of Interest: The authors declare no conflict of interest.

References

1. Li, L.; Wu, C.; Cheng, Y.; Ai, Y.; Li, H.; Tan, X. Comparative analysis of viscoelastic properties of open graded friction course under dynamic and static loads. *Polymers* **2021**, *13*, 1250. [CrossRef] [PubMed]
2. Yuan, D.; Xing, C.; Jiang, W.; Xiao, J.; Wu, W.; Li, P.; Li, Y. Viscoelastic behavior and phase structure of high-content SBS-modified asphalt. *Polymers* **2022**, *14*, 2476. [CrossRef]
3. Cuciniello, G.; Leandri, P.; Filippi, S.; Lo Presti, D.; Losa, M.; Airey, G. Effect of ageing on the morphology and creep and recovery of polymer-modified bitumens. *Mater. Struct.* **2018**, *51*, 136. [CrossRef]
4. Cortizo, M.S.; Larsen, D.O.; Bianchetto, H.; Alessandrini, J.L. Effect of the thermal degradation of SBS copolymers during the ageing of modified asphalts. *Polym. Degrad. Stabil.* **2004**, *86*, 275–282. [CrossRef]
5. Alsolieman, H.A.; Babalghaith, A.M.; Memon, Z.A.; Al-Suhaibani, A.S.; Milad, A. Evaluation and comparison of mechanical properties of polymer-modified asphalt mixtures. *Polymers* **2021**, *13*, 2282. [CrossRef]
6. Abdalrhman, M.; Ahmed, S.B.A.; Nur, I.M.Y. A review of the utilisation of recycled waste material as an alternative modifier in asphalt mixtures. *Civ. Eng. J.* **2020**, *6*, 42–60. [CrossRef]
7. Xiao, R.; Huang, B.S. Moisture damage mechanism and thermodynamic properties of hot-mix asphalt under aging conditions. *ACS Sustain. Chem. Eng.* **2022**, *10*, 14865–14887. [CrossRef]
8. Zhang, Y. Extending the Lifespan of Porous Asphalt Concrete. Ph.D. Thesis, Delft University of Technology, Delft, The Netherlands, 2015. [CrossRef]
9. Apeagyei, A.K. Laboratory evaluation of antioxidants for asphalt binders. *Constr. Build. Mater.* **2011**, *25*, 47–53. [CrossRef]
10. Sun, Z.Y.; Jin, L.; He, S.; Zhao, Y.F.; Evans, G.D.; Duan, X. A structured catalyst based on cobalt phthalocyanine/calcined Mg-Al hydrotalcite film for the oxidation of mercaptan. *Green Chem.* **2012**, *14*, 1909–1916. [CrossRef]
11. Yamaguchi, K.; Sasaki, I.; Meiarashi, S. Mechanism of asphalt binder aging by ultraviolet irradiation and aging resistance by adding carbon black. *J. Jpn. Pet. Inst.* **2004**, *47*, 266–273. [CrossRef]
12. Liu, C.H.; Zhang, H.L.; Li, S. The effect of surface modified nano-titania on the ultraviolet aging properties of bitumen. *Pet. Sci. Technol.* **2014**, *32*, 2995–3001. [CrossRef]
13. Zhang, H.B.; Zhang, H.L.; Ke, N.X. The effect of different nanomaterials on the long-term aging properties of bitumen. *Pet. Sci. Technol.* **2015**, *33*, 388–396. [CrossRef]
14. Ellie, H.F.; Pouria, H.; Mohammad, R. Physiochemical, rheological, and oxidative aging characteristics of asphalt binder in the presence of mesoporous silica nanoparticles. *J. Mater. Civil. Eng.* **2017**, *28*, 04015133.
15. Zhang, H.L.; Su, M.M.; Zhao, S.F.; Zhang, Y.P.; Zhang, Z.P. High and low temperature properties of nano-particles/polymer modified asphalt. *Constr. Build. Mater.* **2016**, *114*, 323–332. [CrossRef]
16. Guo, Q.; Xue, Y.B.; Ma, J.S.; Liu, Z.M. Research progress of nano-material/SBS composite modified asphalt. *Contemp. Chem. Ind.* **2022**, *51*, 2709–2712.
17. Ramayya, V.V.; Ram, V.V.; Krishnaiah, S.; Sandra, A.K. Performance of VG30 paving grade bitumen modified with polyphosphoric acid at medium and high temperature regimes. *Constr. Build. Mater.* **2016**, *105*, 157–164. [CrossRef]
18. Feng, Z.; Yu, J. A study on the aging kinetics of PPA modified asphalt. *Liq. Fuels Technol.* **2010**, *28*, 1338–1344. [CrossRef]
19. Xiao, F.; Serji, A.; Hainian, W.; Hao, P. Rheological property investigations for polymer and polyphosphoric acid modified asphalt binders at high temperatures. *Constr. Build. Mater.* **2014**, *64*, 316–323. [CrossRef]
20. Naresh, B.; Ramasamy, B.S. Effect of Polyphosphoric Acid on Aging Characteristics of PG 64-22 Asphalt Binder. Master's Thesis, University of North Texas, Denton, TX, USA, 2010.
21. Masson, J.F.; Collins, P.; Woods, J.R.; Bundalo, S.; Margeson, J. Chemistry and effects of polyphosphoric acid on the microstructure, molecular mass, glass transition temperatures and performance grades of asphalts. *J. Assoc. Asph. Pav. Technol.* **2009**, *78*, 403–430.
22. Liu, S.; Zhou, S.; Peng, A. Evaluation of polyphosphoric acid on the performance of polymer modified asphalt binders. *J. Appl. Polym. Sci.* **2020**, *137*, 34. [CrossRef]
23. Yan, C.Q.; Huang, W.D.; Xiao, F.P.; Wang, L.F.; Li, Y.W. Proposing a new infrared index quantifying the aging extent of SBS-modified asphalt. *Road Mater. Pavement* **2018**, *19*, 1406–1421. [CrossRef]
24. Yan, C.Q.; Huang, W.D.; Tang, N.P. Evaluation of the temperature effect on rolling thin film oven aging for polymer modified asphalt. *Constr. Build. Mater.* **2017**, *137*, 485–493. [CrossRef]

25. Xing, C.; Liu, L.; Cui, Y.; Ding, D. Analysis of base bitumen chemical composition and aging behaviors via atomic force microscopy-based infrared spectroscopy. *Fuel* **2020**, *264*, 116845. [CrossRef]
26. Xing, C.W.; Jiang, W.; Li, M.C.; Wang, M.; Xiao, J.J.; Xu, Z.C. Application of atomic force microscopy in bitumen materials at the nanoscale: A review. *Constr. Build. Mater.* **2022**, *342*, 128059. [CrossRef]

polymers

MDPI

Article

Effect of Desulfurization Process Variables on the Properties of Crumb Rubber Modified Asphalt

Honggang Zhang [1,2,3,*], Yangpeng Zhang [2,3,*], Jie Chen [2,3], Wenchang Liu [4] and Wensheng Wang [5]

1 School of Traffic and Transportation Engineering, Changsha University of Science and Technology, Changsha 410114, China
2 Guangxi Transportation Science and Technology Group Co., Ltd., Nanning 530007, China; cj_engineering@163.com
3 Guangxi Key Lab of Road Structure and Materials, Nanning 530007, China
4 School of Materials Science and Engineering, Chang'an University, Xi'an 710064, China; lwc_engineering@163.com
5 College of Transportation, Jilin University, Changchun 130025, China; wangws@jlu.edu.cn
* Correspondence: zhg_engineering@163.com (H.Z.); zyp_engineering@outlook.com (Y.Z.)

Abstract: A large number of waste tires are in urgent need of effective treatment, and breaking waste tires into crumb rubber powder for modifying asphalt has been proved as a good idea to solve waste tires. Crumb rubber modified asphalt not only has good high and low temperature performance, durability, and aging resistance but can also reduce pavement noise and diseases, which has wide application prospects. In this study, crumb rubber powder was desulfurized by mechanochemical method to prepare desulfurized crumb rubber modified asphalt. During the desulfurization process of crumb rubber, the effects of desulfurization process variables including desulfurizer type, desulfurizer content, and desulfurization mixing temperature and time were considered, and then the physical properties of modified asphalt were tested. The test results showed that after mixing crumb rubber powder with desulfurizer, the viscosity of crumb rubber powder modified asphalt can be reduced. Moreover, the storage stability of crumb rubber powder modified asphalt could also be improved by mixing crumb rubber with desulfurizer. Based on the physical properties of crumb rubber powder modified asphalt, the desulfurization process of selected organic disulfide (OD) desulfurizer was optimized as follows: the OD desulfurizer content was 3%, the desulfurization mixing temperature was 160 °C, and the mixing time was 30 min. In addition, Fourier infrared spectroscopy analysis was carried out to explore the modification mechanism of desulfurized crumb rubber powder modified asphalt. There is no fracture and formation of chemical bonds, and the modification of asphalt by crumb rubber powder is mainly physical modification.

Keywords: asphalt; waste crumb rubber; desulfurization process; mechanochemical method

Citation: Zhang, H.; Zhang, Y.; Chen, J.; Liu, W.; Wang, W. Effect of Desulfurization Process Variables on the Properties of Crumb Rubber Modified Asphalt. *Polymers* **2022**, *14*, 1365. https://doi.org/10.3390/polym14071365

Academic Editor: Changwoon Nah

Received: 28 January 2022
Accepted: 25 March 2022
Published: 28 March 2022

1. Introduction

With the rapid development of the rubber industry and automobile industry, a large number of waste rubber tire products have piled up [1–4]. In China, the amount of waste tires has reached 13.07 million tons by 2016, and it is still growing rapidly every year. It is expected that the accumulation of waste tires in China will rank first in the world [5–7]. The accumulation of waste tires is not only a waste of rubber resources but also a serious harm to the environment [8–10]. Therefore, the recycling of waste rubber has important social significance and economic value.

It has been found that adding crumb rubber powder as a modifier into asphalt for modification can effectively improve the technical properties of bitumen, thus that the road performance and service quality of the modified asphalt mixture can meet the requirements of road traffic operations [11–13]. Crumb rubber modified asphalt mixture has excellent high and low temperature performance, weatherability, elasticity, and anti-aging properties, can reduce pavement thickness, reduce traffic noise, and effectively prevent freezing in cold areas, and prolong the service life of the pavement [14–16]. On the other hand, there is excellent field performance exhibited by the dry process rubber modification of asphalt mixtures, which is equivalent to polymer-modified mixtures [17]. Picado-Santos et al. studied the functional characteristics, including roughness, skid resistance, texture, and a quality index for a crumb rubber asphalt pavement produced by the dry process, that had been open to traffic for eight years after rehabilitation. The comparison results showed that crumb rubber asphalt pavement produced by the dry process behaved in a very robust way [18]. Rodriguez-Fernandez et al. analyzed the dry process for the incorporation of crumb rubber from waste tires and found that mixtures with crumb rubber have adequate performance, being less susceptible to aging than a conventional polymer-modified mixture [19]. Rath et al. investigated the effects of a chemically engineered dry-process ground tire rubber modification in bitumen and mixtures, and the cracking and rutting performance of all the field sections was good-to-excellent [20,21]. However, for wet-process modified rubber asphalt, because the crumb rubber powder used in ordinary crumb rubber modified asphalt was treated by a vulcanization process, it had a relatively solid three-dimensional network structure and low activity, which makes the crumb rubber powder difficult to be compatible in asphalt and often accompanied by a large number of crumb rubber particles [22–24]. The ordinary crumb rubber modified asphalt prepared from vulcanized crumb rubber powder has a series of shortcomings, such as poor high temperature storage stability, easy segregation, high viscosity, and difficult construction, which limits its development and application [25–27]. Relevant research shows that rubber powder with high surface activity can be obtained by using desulfurization technology, leading to the reduction of viscosity of vulcanized crumb rubber powder modified asphalt [23]. The desulfurized crumb rubber modified asphalt has been proved to have good storage stability, low-temperature performance, and workability. In addition, the harmful gas emission of desulfurized crumb rubber modified asphalt is lower, thus it is more environmentally friendly [24,28]. Therefore, it is necessary to desulfurize and activate the vulcanized crumb rubber powder to improve the stability of crumb rubber modified asphalt.

In view of the shortcomings of ordinary crumb rubber modified asphalt, researchers carried out desulfurization activation treatment on crumb rubber powder to a certain extent, and then prepared desulfurized crumb rubber modified asphalt [28,29]. The earliest desulfurized crumb rubber modified asphalt can be traced back to the 1970s, which was tested and prepared in America. Later, a large number of studies were involved in this aspect; Ye et al. observed the microstructure of ordinary crumb rubber powder, dynamic desulfurized crumb rubber powder, and high-speed shear desulfurized crumb rubber powder in asphalt through fluorescence microscope, and detected the properties of three modified asphalt [30]. The results showed that dynamic desulfurization could destroy the internal vulcanization structure of crumb rubber and improve the compatibility between crumb rubber powder and asphalt, but its performance is lower than that of ordinary crumb rubber modified asphalt. High-speed shear desulfurization can not only improve the swelling capacity of asphalt crumb rubber powder but also avoid agglomeration. Therefore, the performance of asphalt can be improved by high-speed shear desulfurized crumb rubber powder. Ibrahim et al. used the waste rubber treated by gamma rays to produce crumb rubber asphalt [31]. Through comparison, it was found that the properties of crumb rubber modified asphalt made by gamma radiation treated waste rubber was increased significantly at a higher temperature, lower temperature, and aging resistance than that of untreated crumb rubber. Lin et al. studied the effect of the dissolution of crumb rubber on the physical properties of asphalt under the reaction conditions of high temperature and high shear rate [32]. The results showed that under the condition of high temperature reaction, crumb rubber molecules dissolved rapidly, degraded, and depolymerized, thus as to reduce its average molecular weight. It is speculated that ordinary crumb rubber powder may produce desulfurization reaction under high temperature, but when the temperature is too high, it will lead to the aging of crumb rubber modified asphalt. Ma et al. studied the modification mechanism of desulfurized crumb rubber modified asphalt by using scanning electron microscope, component analysis, and infrared spectroscopy, and tested the road performance of its modified asphalt and mixture [33]. The results showed that the modification mechanism of desulfurized crumb rubber modified asphalt was different from that of ordinary crumb rubber modified asphalt, and the desulfurized crumb rubber powder can have an obvious chemical reaction with asphalt, which makes the performance of the two have significant differences. Compared with ordinary crumb rubber modified asphalt, desulfurized crumb rubber modified asphalt has a lower viscosity and better storage stability and is suitable for dense mixtures. Juganaru et al. modified asphalt with partially desulfurized crumb rubber [34]. Through fluorescence microscope observation, the distribution of desulfurized crumb rubber in asphalt was more uniform. Compared with vulcanized crumb rubber modified asphalt, the adhesion of desulfurized crumb rubber powder modified asphalt was significantly increased, which can improve the workability of rubber asphalt.

Many scholars have studied the desulfurization methods and desulfurization conditions of waste tire rubber powder, covering the desulfurization process and common desulfurization methods. The research indicated that high temperature, high-speed shear, gamma radiation, and other conditions could make the desulfurization of ordinary crumb rubber powder. After desulfurization, the spatial structure of crumb rubber powder was changed, and the molecular weight was reduced, which can increase the reaction degree with asphalt, significantly improve the compatibility between crumb rubber powder and asphalt, and effectively improve the defects of crumb rubber modified asphalt, such as easy segregation and poor high temperature storage stability. However, there are few studies on the production process of desulfurized crumb rubber powder modified asphalt and the road performance of modified asphalt mixture. At present, the commonly used rubber desulfurization activation methods include heat treatment based on the thermal environment, a chemical method using chemical adjuvants, and a mechanochemical method using chemical adjuvants combined with mechanical force to accelerate the chemical reaction [27,35,36]. The mechanochemical method can effectively play the role of desulfurization activation of additives through rubber broken to increase the surface roughness and activity of rubber powder. Moreover, the mechanochemical method has the advantages of simple equipment, high production efficiency, low cost, and small smell, and is very suitable for large factory production. Therefore, in this study, the crumb rubber powder was desulfurized by the mechanochemical method, and the desulfurized crumb rubber powder modified asphalt was prepared. The effects of the desulfurization process on the physical properties and micro characterization of crumb rubber powder modified asphalt were systematically studied.

2. Materials and Methods

2.1. Raw Materials

In this study, Maoming 70# road petroleum asphalt (Guangdong, China) was adopted, with a penetration of 72 (0.1 mm), softening point of 46.0 °C, ductility at 15 °C more than 100 cm, and density of 1.08 g/cm^3. The crumb rubber powder was 30 mesh made of the waste tire of wheels, in which rubber hydrocarbon content was 48.9%, total organic content was 54.5%, sulfur content was 1.896%, and its density was 1.33 g/cm^3. The original appearance and scanning electron microscope (SEM) of crumb rubber used in this study are shown in Figure 1 to better describe the macro and micromorphology of crumb rubber. The desulfurizer used in this study (Hebei Richway Technology Co. Ltd., Shijiazhuang, China) includes adjuvant (organometallic complexes) labeled as OMC, adjuvant (organic disulfide) labeled as OD, and self-made desulfurizer (deep eutectic solvents) labeled as DES, which have the advantages of better desulfurization effect, stable desulfurizer source, and simple desulfurization process [37,38].

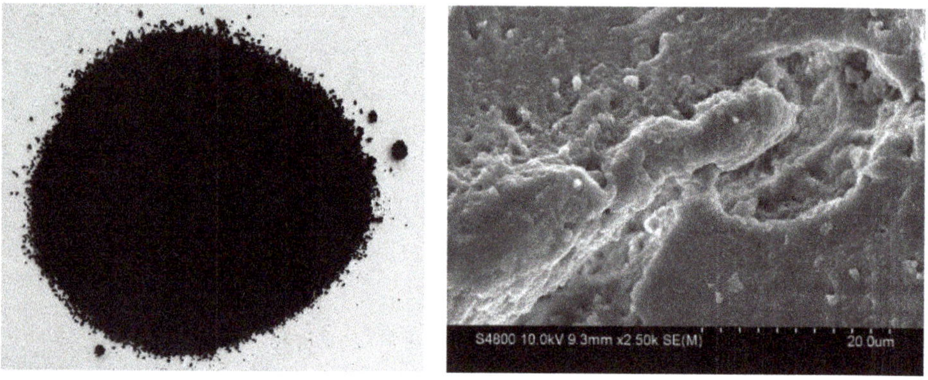

(**a**) (**b**)

Figure 1. Crumb rubber used in this study: (**a**) original appearance; (**b**) SEM.

2.2. Sample Preparation

The modified crumb rubber powder by mechanochemical method was prepared using an internal mixer according to previous studies [37,38]. After adding crumb rubber powder into the internal mixing chamber, appropriately increasing the temperature and maintaining low-speed rotation, the desulfurizer was also added into the internal mixing chamber until it was mixed evenly. Then, the mixing temperature and mixing time was set, and the desulfurized crumb rubber powder could be prepared at a higher rotation speed after the internal mixer stopped.

In order to prepare desulfurized crumb rubber modified asphalt, the following steps were adopted:

1. Place the base asphalt in an oven of 140 °C and heat it to the flowing state, and slowly add the weighed desulfurized crumb rubber powder (25% of asphalt by mass) with a mixing speed of 300 r/min. During the addition process, the temperature was raised to 170 °C and maintained for 10 min;
2. Put the mixed asphalt an oven of 170 °C and swell for 40 min;
3. The swelled rubber asphalt was placed to a high-speed shear equipment, the shear rate was gradually increased to about 4500 r/min, and the temperature was controlled at 185 °C, and then taken it out after shearing for 1 h;
4. Put the prepared crumb rubber modified asphalt into an oven of 175 °C for 30 min, and then the crumb rubber modified asphalt samples could be used for test according to the specification requirements.

2.3. Experimental Methods

In this study, according to the Chinese standard "Standard Test Methods of Bitumen and Bituminous Mixtures for Highway Engineering" (JTG E20-2011), the penetration, softening point, ductility at 5 °C, and rotational viscosity at 180 °C was tested for crumb rubber powder modified asphalt [39]. In addition, the Fourier infrared spectrum of these materials was also tested [40].

3. Results and Discussion

3.1. Comparison and Selection of Desulfurizer Type Based on Physical Properties of Crumb Rubber Modified Asphalt

Referring to the recommended process in the desulfurizer product manual, the mixing time of crumb rubber modified asphalt was 60 min, the mixing temperature was 120 °C, and the desulfurizer content was 3%. The penetration, ductility, softening point, rotational viscosity at 180 °C, elastic recovery, and softening point difference (Δ) of the crumb rubber modified asphalt were tested according to the "Standard Test Methods of Bitumen and

Bituminous Mixtures for Highway Engineering" (JTG E20-2011), and the effect of desulfurizer types on the physical properties of crumb rubber modified asphalt could be studied to compare and select a better adjuvant for in-depth and detailed process variable analysis. The results are shown in Table 1.

Table 1. The physical properties of asphalt modified by crumb rubber treated with different desulfurizer types.

Types	Penetration (0.1 mm)	Softening Point (°C)	Ductility at 5 °C (cm)	Viscosity at 180 °C (Pa·s)	Elastic Recovery (%)	Softening Point Δ (°C)
Non-desulfurized crumb rubber modified asphalt	42.0	70.0	9.0	2.75	84.5	1.6
Crumb rubber modified asphalt (Desulfurized by OMC)	43.5	71.2	7.3	2.11	84.5	0.6
Crumb rubber modified asphalt (Desulfurized by OD)	44.5	66.3	11.5	1.83	82.0	1.0
Crumb rubber modified asphalt (Desulfurized by OMC + OD)	38.0	68.4	10.0	2.24	86.0	2.0
Crumb rubber modified asphalt (Desulfurized by DES)	37.0	69.0	6.5	2.27	80.0	1.6

It can be seen from Table 1 that the viscosity results at 180 °C of modified asphalt prepared by desulfurized crumb rubber powder treated with OMC, OD, and DES were reduced. The penetration values of modified asphalt with desulfurized crumb rubber powder treated with OMC and OD were higher than that of ordinary non-desulfurized crumb rubber modified asphalt, while the desulfurizer (OMC + OD or DES) could reduce the penetration of modified asphalt. Except that the softening point value of modified asphalt with desulfurized crumb rubber powder treated with OMC increased, the other softening point results of modified asphalt were all reduced. Besides, crumb rubber powder treated with desulfurizers OMC and DES could reduce the ductility at 5 °C of modified asphalt. The segregation softening point difference of desulfurized crumb rubber powder modified asphalt using OMC and OD alone were decreased after 48 h of storage. Thus, the purpose of improving the compatibility between crumb rubber powder and asphalt and reducing the viscosity of crumb rubber modified asphalt through crumb rubber powder desulfurization technology can be achieved. According to the comparison results in the radar chart (Figure 2), OMC increased the penetration and softening point of desulfurized crumb rubber modified asphalt, while the ductility, viscosity at 180 °C, and softening point Δ were greatly reduced. For desulfurized crumb rubber modified asphalt by OD, except for the penetration and ductility, other physical properties decreased. However, for desulfurized crumb rubber modified asphalt by OMC + OD, the improvement effect of modified asphalt properties was not a simple superposition of modified asphalt using OMC and OD alone. DES reduced the overall physical properties of crumb rubber modified asphalt to varying degrees. Considering the two indexes of ductility at 5 °C and viscosity at 180 °C, the desulfurizer OD was selected as the desulfurizer used for the follow-up study.

3.2. *Effect of Desulfurizer Content on the Properties of Crumb Rubber Modified Asphalt*

The desulfurized crumb rubber powder modified asphalt prepared by crumb rubber treated with different OD desulfurizer content (i.e., 1.5%, 3%, 4.5%, 6%) were prepared with the mixing time of 60 min and mixing temperature of 120 °C. The changes of physical properties of desulfurized crumb rubber powder modified asphalt were compared and analyzed. The comparison results are shown in Figure 3.

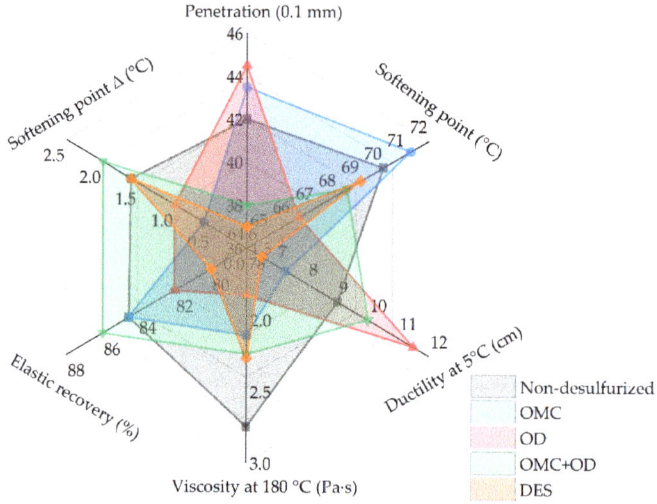

Figure 2. Radar chart for of desulfurizer type comparison based on physical properties of crumb rubber modified asphalt.

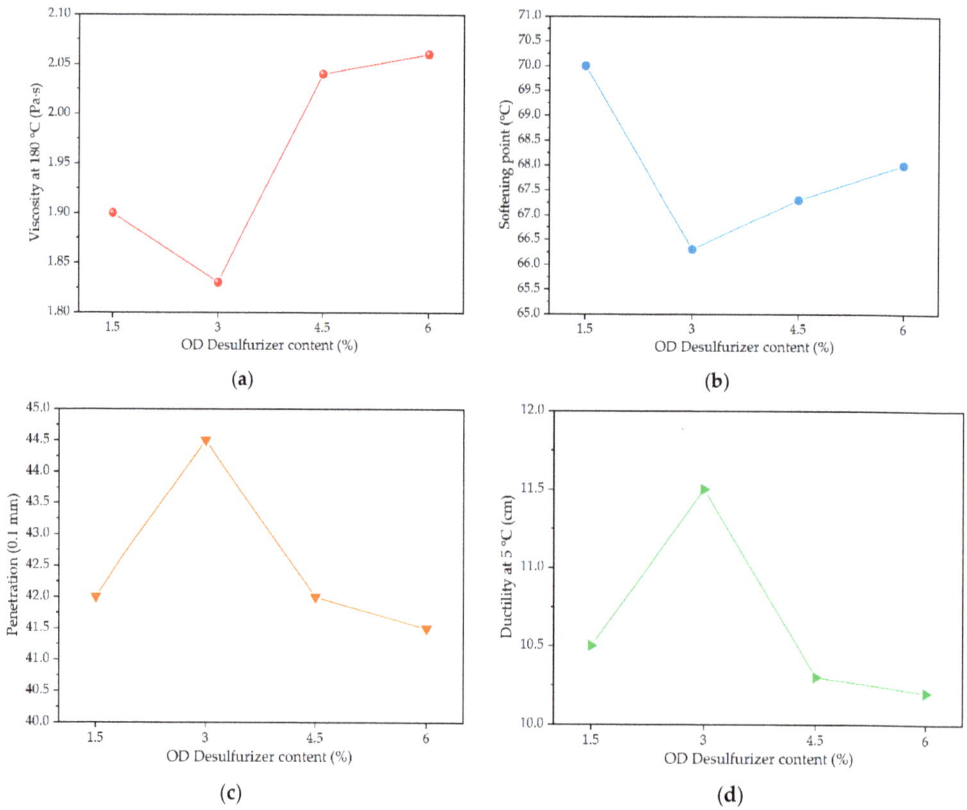

Figure 3. The physical properties of asphalt modified by crumb rubber treated with different OD desulfurizer contents: (**a**) viscosity at 180 °C; (**b**) softening point; (**c**) penetration; (**d**) ductility at 5 °C.

From Figure 3a,b, it can be seen that with the increase of OD desulfurizer content, the viscosity at 180 °C, and softening point of desulfurized crumb rubber powder modified asphalt had similar change laws; these two physical properties first decreased and then increased. When the OD desulfurizer content was 3%, the viscosity at 180 °C and softening point of desulfurized crumb rubber powder modified asphalt reached the minimum. On the other hand, with the increase of OD desulfurizer content, the penetration and ductility at 5 °C of desulfurized crumb rubber powder modified asphalt had similar change laws. In Figure 3c,d, the values of penetration and ductility increased first and then decreased. It was also found that the penetration and ductility at 5 °C of desulfurized crumb rubber powder modified asphalt reached the maximum at the OD desulfurizer content of 3%.

The reason for the above phenomena is that the OD desulfurizer can promote the desulfurization and activation of crumb rubber powder by opening the S-S and C-S keys in the cross-linked network, and partially restore the properties of raw rubber to improve the compatibility with asphalt [41]. The greater the amount of OD desulfurizer content, the more obvious the desulfurization effect of crumb rubber [42]. When the content of OD desulfurizer is too high, some OD desulfurizer will not participate in the desulfurization reaction with crumb rubber and have a surplus of OD desulfurizer. This surplus part of OD desulfurizer would interact with asphalt during the modification process of asphalt, resulting in a poor modification effect of crumb rubber modified asphalt, i.e., the too low penetration and ductility as well as too high viscosity as shown in Figure 3. Therefore, the content of OD desulfurizer needs to be controlled within a reasonable range. According to the above analysis results of the physical properties of crumb rubber modified asphalt, the recommended content of OD desulfurizer was set as 3%.

3.3. Effect of Mixing Temperature on the Properties of Crumb Rubber Modified Asphalt

Based on the selected desulfurizer type of OD desulfurizer and its content of 3% of crumb rubber powder by mass, the effects of mixing process on the performances of crumb rubber modified asphalt were also studied. In order to ensure the reaction between crumb rubber powder and the OD desulfurizer at a lower temperature, the mixing time of crumb rubber powder was fixed for 90 min, the asphalt modified by crumb rubber powder treated with OD desulfurizer were prepared at different desulfurization temperatures (i.e., 60 °C, 80 °C, 100 °C, 120 °C, 140 °C, and 160 °C). The changes of properties of desulfurized crumb rubber powder modified asphalt were compared and analyzed, and the comparison results are shown in Figure 4.

In Figure 4a,b, it can be seen that with the increase of mixing temperature, the viscosity at 180 °C and softening point of desulfurized crumb rubber powder modified asphalt have similar change trends, and the viscosity at 180 °C and softening point of modified asphalt gradually decrease. For the mixing temperature between 60 °C and 120 °C, the decline of viscosity at 180 °C and softening point were moderate. While the mixing temperature reached 120 °C, the viscosity at 180 °C and softening point of desulfurized rubber powder modified asphalt were significantly reduced. This is because under the action of OD desulfurizer and mechanical force, the crumb rubber powder was desulfurized and activated, resulting in some properties of raw crumb rubber being restored to improve the compatibility with asphalt, and the particle core of crumb rubber powder became smaller [43]. Therefore, the high temperature performance of desulfurized crumb rubber powder modified asphalt became worse, and the corresponding viscosity decreased to a certain extent.

As shown in Figure 4c,d, there were also some similar variation trends of the penetration and ductility at 5 °C for desulfurized crumb rubber powder modified asphalt. When the mixing temperature was lower than 120 °C, the penetration of desulfurized crumb rubber powder modified asphalt gradually increased with the increase of mixing time. While the mixing temperature was higher than 120 °C, the penetration of desulfurized crumb rubber powder modified asphalt sharply increased. With the increase of mixing time, the ductility at 5 °C of desulfurized crumb rubber powder modified asphalt increased gradually.

Through analysis, it can be known that under the action of OD desulfurizer and mechanical force, crumb rubber powder was desulfurized and activated with restored properties of raw crumb rubber to improve the compatibility with asphalt. After the desulfurized crumb rubber powder was used for modifying base asphalt, the consistency of desulfurized crumb rubber powder modified asphalt would be reduced and the corresponding ductility could be enhanced [44].

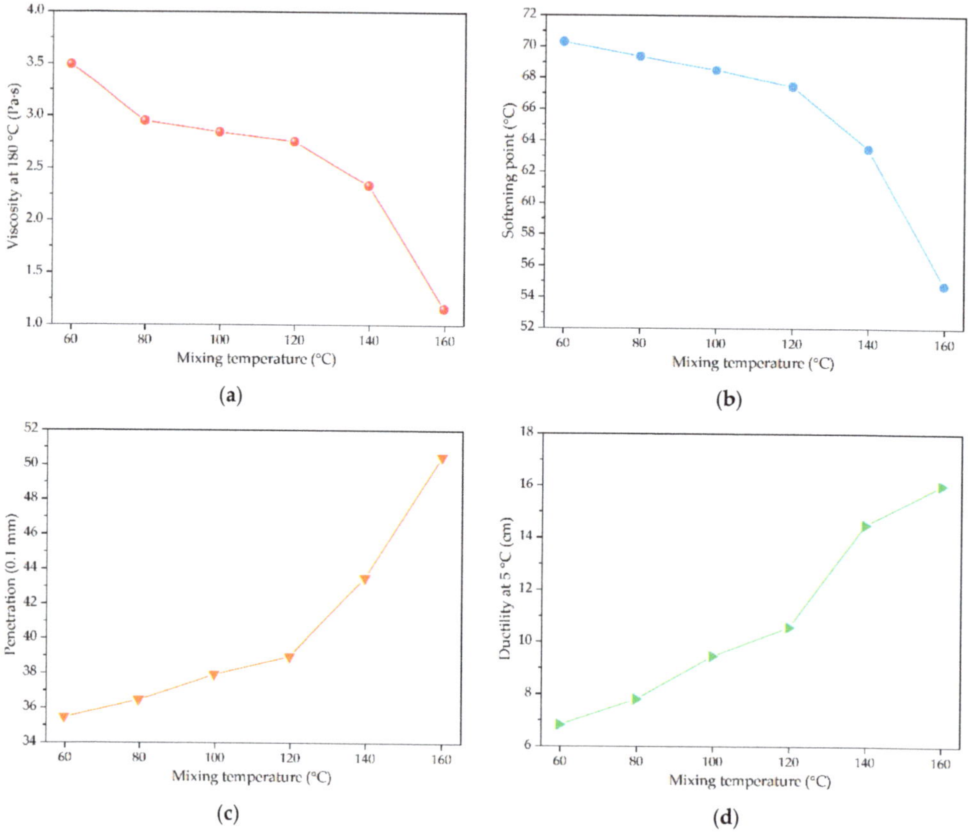

Figure 4. The physical properties of asphalt modified by crumb rubber treated with different mixing temperatures (i.e., 60, 80, 100, 120, 140, 160 °C): (**a**) viscosity at 180 °C; (**b**) softening point; (**c**) penetration; (**d**) ductility at 5 °C.

To sum up, when the mixing temperature was low, the desulfurization effect of crumb rubber powder was not obvious, and the viscosity reduction of desulfurized crumb rubber powder modified asphalt was too small. At the same time, the mixing temperature needs to be higher than 120 °C. Therefore, the mixing temperatures of 140 °C and 160 °C were selected as the mixing temperature for subsequent tests. Subsequently, desulfurized crumb rubber powder was prepared and modified base asphalt under the test conditions of mixing temperature of 140 °C and mixing time of 120 min. It was found that the viscosity of the modified asphalt decreased significantly, but the softening point also decreased significantly. At the same time, considering that the mixing time was too long, it was not suitable for large-scale production [45]. Therefore, the mixing temperature of 160 °C was selected as the preferred mixing temperature.

3.4. Effect of Mixing Time on the Properties of Crumb Rubber Modified Asphalt

According to the above analysis results, the desulfurizer type was selected as OD desulfurizer, and its content was determined as 3% of crumb rubber powder by mass, the effects of mixing process on the performances of crumb rubber modified asphalt were further studied. The mixing temperature for desulfurized crumb rubber powder was preferably 160 °C, and the desulfurized crumb rubber powder modified asphalt under different desulfurization mixing times (i.e., 10 min, 20 min, 30 min, 60 min, and 90 min) were prepared. The changes of properties of desulfurized crumb rubber powder modified asphalt were compared and analyzed, and the corresponding comparison results are shown in Figure 5.

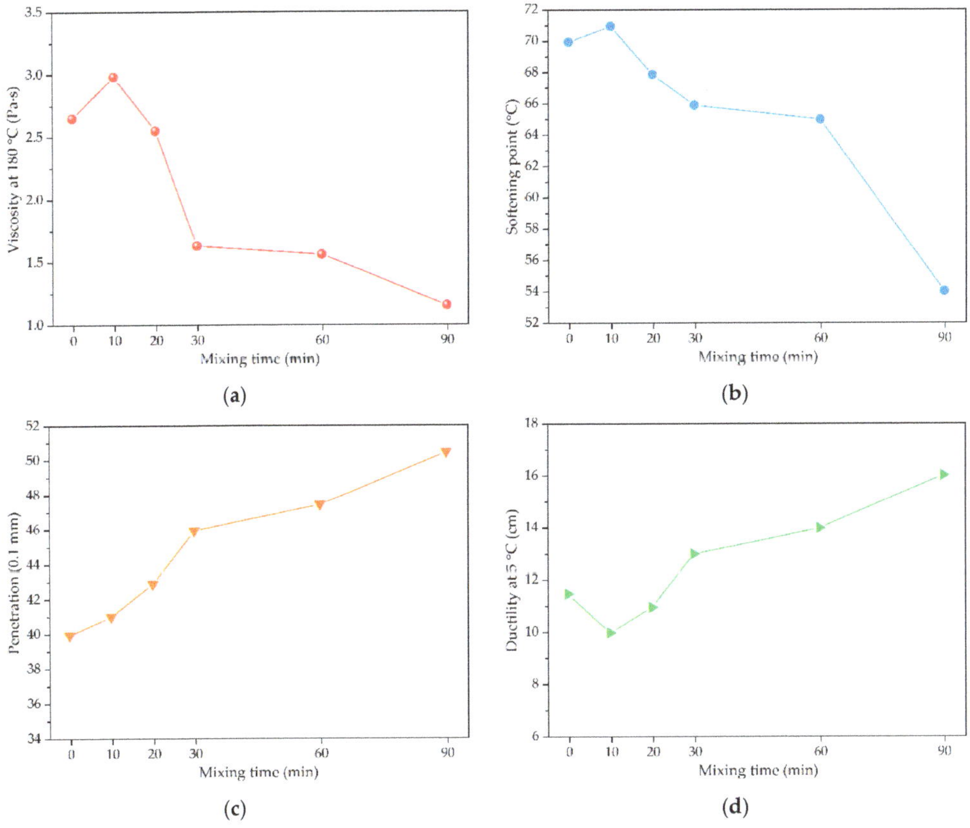

Figure 5. The physical properties of asphalt modified by crumb rubber treated with different mixing time (i.e., 0, 10, 20, 30, 60, 90 min): (**a**) viscosity at 180 °C; (**b**) softening point; (**c**) penetration; (**d**) ductility at 5 °C.

It can be seen from Figure 5a,b that with the increase of mixing time, the viscosity at 180 °C and softening point of desulfurized crumb rubber powder modified asphalt first increased and then decreased. However, when the mixing time was 30 min, the viscosity at 180 °C of desulfurized crumb rubber powder modified asphalt decreased greatly, while the softening point decreased only slightly. Therefore, the desulfurization mixing time of crumb rubber powder was determined as 30 min. The corresponding analysis shows that the existence of mechanical force can promote the chemical reaction between crumb rubber powder and the selected OD desulfurizer, and the crumb rubber powder was activated by OD desulfurization, partially restoring the properties of raw crumb rubber to improve the compatibility with asphalt [45]. When the mixing time was too short, the reaction between the OD desulfurizer and crumb rubber powder was insufficient, and some OD desulfurizer remained in desulfurized crumb rubber powder, which will react with asphalt, resulting in a poor modification effect of asphalt. However, when the mixing time is too long, the crumb rubber powder will be excessively reduced under the action of mechanical force, and the core of crumb rubber powder particles will lose strength, which will also lead to the deterioration of high temperature performance of modified asphalt [46].

As illustrated in Figure 5c,d, the penetration of desulfurized crumb rubber powder modified asphalt gradually increased with the increase of mixing time. The ductility at 5 °C of desulfurized crumb rubber powder modified asphalt first decreased and then increased. Through analysis, under the action of the OD desulfurizer and mechanical force, crumb rubber powder was desulfurized and activated, and some properties of raw rubber were restored to improve the compatibility with asphalt. After being used for the modification of asphalt, the consistency of desulfurized crumb rubber powder modified asphalt decreased and the ductility was enhanced. With the increase of mixing time, the degree of desulfurized activation of crumb rubber powder increased. After the modification of asphalt, the changes of the properties of desulfurized crumb rubber powder modified asphalt became more and more obvious with mixing time [47].

Combined with the analysis results of deterioration mixing temperature and mixing time, when the mixing temperature was 160 °C and the mixing time was 30 min, the overall properties of desulfurized crumb rubber powder modified asphalt could be regarded as the best, its viscosity decreased significantly, and the reduction of softening point was limited. Therefore, the preferred mixing process is a mixing temperature of 160 °C and mixing time of 30 min.

3.5. Micro Characterization of Desulfurized Crumb Rubber Modified Asphalt

In order to explore the modification mechanism of desulfurized crumb rubber powder modified asphalt, Fourier infrared spectroscopy analysis was carried out on desulfurized crumb rubber powder modified asphalt under different mixing temperatures and mixing times. The Fourier infrared spectroscopy results are shown in Figure 6.

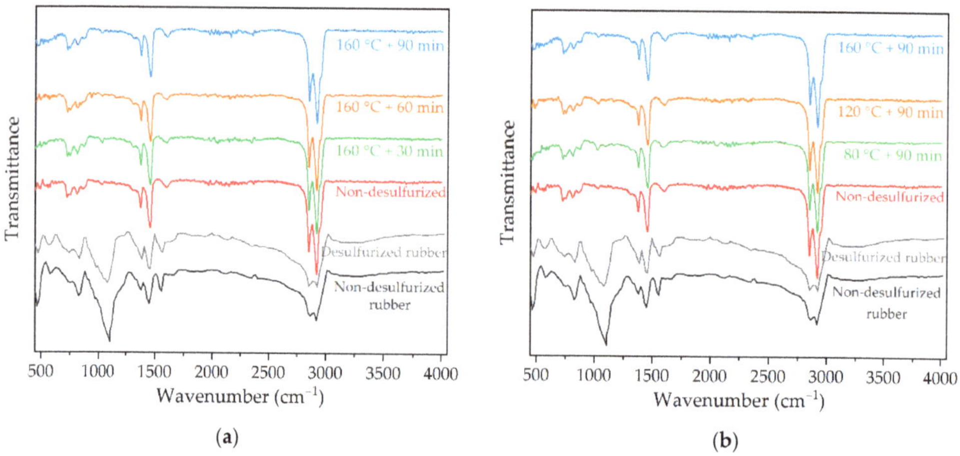

Figure 6. Fourier infrared spectroscopy of desulfurized crumb rubber powder modified asphalt: (**a**) with different mixing time; (**b**) with different mixing temperature.

From Figure 6, it can be seen that the Fourier infrared spectrum of crumb rubber powder modified asphalt mainly shows three concentrated strong absorption peaks [29], of which the absorption peaks at 1925 cm^{-1} and 2843 cm^{-1} were mainly caused by the stretching vibration of "-CH_2-" in cycloalkanes and alkanes of asphalt. The absorption peak at 1350~1470 cm^{-1} was caused by the bending vibration in the "C-H" bond plane, which was the shear vibration absorption peak of the "-CH_3" alkane group in asphalt. The absorption peak at 1611 cm^{-1} was caused by the "C=C" bond and "C=O" bond, which is a typical characteristic peak of aromatic components in asphalt. The C-S bond was destroyed in the desulfurization process, and the "S=O" bond at 1082 cm^{-1} and the "C=O" bond at 1710 cm^{-1} were changed in the spectrum in varying degrees shown in Figure 6. However, the "S=O" bond and "C=O" bond do not exist in the spectrum of modified asphalt, which shows that some functional groups in the rubber reacted with asphalt, making their combination more stable. In addition, compared with different mixing time and mixing temperature, the position and peak of the characteristic peaks of the Fourier infrared spectrum of the desulfurized crumb rubber powder modified asphalt did not change significantly, which indicates that the fracture and formation of chemical bonds were not found in the modification process of desulfurized crumb rubber powder modified asphalt, and the modification process of asphalt by desulfurized crumb rubber powder was mainly physical blending, which is also consist with previous studies [4,48,49].

4. Conclusions

In this study, crumb rubber powder was desulfurized by mechanochemical method to prepare desulfurized crumb rubber modified asphalt. During the desulfurization process of crumb rubber, the effects of the desulfurization process variables including desulfurizer type, desulfurizer content, and desulfurization mixing temperature and time were considered, and then the physical properties of modified asphalt were tested.

(1) Two commercial regenerant labeled as OMC, activators labeled as OD, as well as self-made desulfurizer labeled as DES and crumb rubber powder were selected for mixing the desulfurization treatment. The three desulfurizers can reduce the viscosity of desulfurized crumb rubber powder modified asphalt, among which OD and OMC can reduce the softening point at the same time;

(2) The desulfurization process of selected OD desulfurizer is optimized as follows: the OD desulfurizer content is 3%, the desulfurization mixing temperature is 160 °C and the mixing time is 30 min;

(3) Based on the Fourier infrared spectrum analysis, the desulfurization treatment did not change the functional groups of crumb rubber powder modified asphalt, and the modification of asphalt by crumb rubber powder was mainly physical modification.

Future research on the performing and defining grade of each modified asphalt binder as well as the effects of desulfurization agents on rubber is highly warranted.

Author Contributions: Conceptualization, H.Z. and Y.Z.; methodology, H.Z., J.C. and W.L.; validation, Y.Z. and W.W.; formal analysis, H.Z., Y.Z., J.C. and W.W.; investigation, H.Z., Y.Z., J.C. and W.L.; writing—original draft preparation, H.Z. and W.W.; writing—review and editing, Y.Z., J.C. and W.L.; project administration, H.Z., Y.Z. and W.W.; funding acquisition, H.Z., Y.Z. and W.W. All authors have read and agreed to the published version of the manuscript.

Funding: This research was funded by the Central Guidance on Local Science and Technology Development Fund of Guangxi Autonomous Region (grant number: ZY21195043), Guangxi Science and Technology Major Project (grant number: AA18242032), Nanning innovation and entrepreneurship leading talents "Yongjiang plan" funded project (grant number: 2019009). This research was also supported by the Scientific and Technological Project of Science and Technology Department of Jilin Province (grant number: 20210508028RQ), Scientific Research Project of Department of Education of Jilin Province (grant number: JJKH20221019KJ), China Postdoctoral Science Foundation (grant number: 2021T140262).

Institutional Review Board Statement: Not applicable.

Informed Consent Statement: Not applicable.

Data Availability Statement: Not applicable.

Conflicts of Interest: The authors declare no conflict of interest.

References

1. Wang, W.S.; Cheng, Y.C.; Chen, H.P.; Tan, G.J.; Lv, Z.H.; Bai, Y.S. Study on the performances of waste crumb rubber modified asphalt mixture with eco-friendly diatomite and basalt fiber. *Sustainability* **2019**, *11*, 5282. [CrossRef]
2. Saberi, K.F.; Fakhri, M.; Azami, A. Evaluation of warm mix asphalt mixtures containing reclaimed asphalt pavement and crumb rubber. *J. Clean. Prod.* **2017**, *165*, 1125–1132. [CrossRef]
3. Ge, D.D.; Zhou, X.D.; Chen, S.Y.; Jin, D.Z.; You, Z.P. Laboratory evaluation of the residue of rubber-modified emulsified asphalt. *Sustainability* **2020**, *12*, 8383. [CrossRef]
4. Chen, T.; Ma, T.; Huang, X.M.; Guan, Y.S.; Zhang, Z.X.; Tang, F.L. The performance of hot-recycling asphalt binder containing crumb rubber modified asphalt based on physiochemical and rheological measurements. *Constr. Build. Mater.* **2019**, *226*, 83–93. [CrossRef]
5. Chen, M.Z.; Zheng, J.; Li, F.Z.; Wu, S.P.; Lin, J.T.; Wan, L. Thermal performances of asphalt mixtures using recycled tyre rubber as mineral filler. *Road Mater. Pavement Des.* **2015**, *16*, 379–391. [CrossRef]
6. Ding, X.H.; Ma, T.; Zhang, W.G.; Zhang, D.Y. Experimental study of stable crumb rubber asphalt and asphalt mixture. *Constr. Build. Mater.* **2017**, *157*, 975–981. [CrossRef]
7. Yu, H.Y.; Leng, Z.; Zhou, Z.Y.; Shih, K.M.; Xiao, F.P.; Gao, Z.M. Optimization of preparation procedure of liquid warm mix additive modified asphalt rubber. *J. Clean. Prod.* **2017**, *141*, 336–345. [CrossRef]
8. Marini, S.; Lanotte, M. Waste rubber from end-of-life tires in 'lean' asphalt mixtures-a laboratory and field investigation in the arid climate region. *Polymers* **2021**, *13*, 3802. [CrossRef]
9. Chen, S.Y.; Ge, D.D.; Jin, D.Z.; Zhou, X.D.; Liu, C.C.; Lv, S.T.; You, Z.P. Investigation of hot mixture asphalt with high ground tire rubber content. *J. Clean. Prod.* **2020**, *277*, 124037. [CrossRef]
10. de Almeida, A.F.; Battistelle, R.A.; Bezerra, B.S.; de Castro, R. Use of scrap tire rubber in place of sbs in modified asphalt as an environmentally correct alternative for brazil. *J. Clean. Prod.* **2012**, *33*, 236–238. [CrossRef]
11. Kukielka, J.; Bankowski, W.; Mirski, K. Asphalt-cement concretes with reclaimed asphalt pavement and rubber powder from recycled tire. *Materials* **2021**, *14*, 2412. [CrossRef] [PubMed]
12. Wang, S.Q.; Gao, Y.; Yan, K.Z.; You, L.Y.; Jia, Y.S.; Dai, X.W.; Chen, M.; Diab, A. Effect of long-term aging on waste tire rubber and amorphous poly alpha olefin compound modified asphalt binder and its mixtures. *Constr. Build. Mater.* **2021**, *272*, 121667. [CrossRef]
13. Li, D.N.; Leng, Z.; Zou, F.L.; Yu, H.Y. Effects of rubber absorption on the aging resistance of hot and warm asphalt rubber binders prepared with waste tire rubber. *J. Clean. Prod.* **2021**, *303*, 127082. [CrossRef]
14. Wang, H.N.; You, Z.P.; Mills-Beale, J.; Hao, P.W. Laboratory evaluation on high temperature viscosity and low temperature stiffness of asphalt binder with high percent scrap tire rubber. *Constr. Build. Mater.* **2012**, *26*, 583–590. [CrossRef]
15. Kocak, S.; Kutay, M.E. Fatigue performance assessment of recycled tire rubber modified asphalt mixtures using viscoelastic continuum damage analysis and aashtoware pavement me design. *Constr. Build. Mater.* **2020**, *248*, 118658. [CrossRef]
16. Yan, C.; Lv, Q.; Zhang, A.A.; Ai, C.; Huang, W.; Ren, D. Modeling the modulus of bitumen/sbs composite at different temperatures based on kinetic models. *Compos. Sci. Technol.* **2022**, *218*, 109146. [CrossRef]
17. Buttlar. *State of Knowledge Report on Rubber Modified Asphalt*; US Tires Manufacturers' Association (USTMA): Columbia City, SC, USA, 2021.
18. Picado-Santos, L.G.; Capitao, S.D.; Dias, J.L.F. Crumb rubber asphalt mixtures by dry process: Assessment after eight years of use on a low/medium trafficked pavement. *Constr. Build. Mater.* **2019**, *215*, 9–21. [CrossRef]
19. Rodriguez-Fernandez, I.; Cavalli, M.C.; Poulikakos, L.; Bueno, M. Recyclability of asphalt mixtures with crumb rubber incorporated by dry process: A laboratory investigation. *Materials* **2020**, *13*, 2870. [CrossRef]
20. Rath, P.; Majidifard, H.; Jahangiri, B.; Chen, S.; Buttlar, W.G. Laboratory and field evaluation of pre-treated dry-process rubber-modified asphalt binders and dense-graded mixtures. *Transp. Res. Rec.* **2021**, *2675*, 381–394. [CrossRef]
21. Rath, P.; Meister, J.; Jahangiri, B.; Buttlar, W. Evaluation of the effects of engineered crumb rubber (ecr) on asphalt mixture characteristic. *J. Test. Eval.* **2022**, *50*, 20210077. [CrossRef]
22. Ghabchi, R.; Arshadi, A.; Zaman, M.; March, F. Technical challenges of utilizing ground tire rubber in asphalt pavements in the united states. *Materials* **2021**, *14*, 4482. [CrossRef] [PubMed]
23. Liu, W.H.; Xu, Y.S.; Wang, H.J.; Shu, B.N.; Barbieri, D.M.; Norambuena-Contreras, J. Enhanced storage stability and rheological properties of asphalt modified by activated waste rubber powder. *Materials* **2021**, *14*, 2693. [CrossRef] [PubMed]
24. Xu, P.; Gao, J.P.; Pei, J.Z.; Chen, Z.; Zhang, J.P.; Li, R. Research on highly dissolved rubber asphalt prepared using a composite waste engine oil addition and microwave desulfurization method. *Constr. Build. Mater.* **2021**, *282*, 122641. [CrossRef]
25. Liu, B.Q.; Li, J.; Han, M.Z.; Zhang, Z.Q.; Jiang, X.Y. Properties of polystyrene grafted activated waste rubber powder (ps-arp) composite sbs modified asphalt. *Constr. Build. Mater.* **2020**, *238*, 117737. [CrossRef]
26. Chen, Z.X.; Pei, J.Z.; Wang, T.; Amirkhanian, S. High temperature rheological characteristics of activated crumb rubber modified asphalts. *Constr. Build. Mater.* **2019**, *194*, 122–131. [CrossRef]
27. Yang, X.L.; Shen, A.Q.; Li, B.; Wu, H.S.; Lyu, Z.H.; Wang, H.; Lyu, Z.F. Effect of microwave-activated crumb rubber on reaction mechanism, rheological properties, thermal stability, and released volatiles of asphalt binder. *J. Clean. Prod.* **2020**, *248*, 119230. [CrossRef]

28. Wang, J.R.; Zhang, Z.Q.; Li, Z.L. Performance evaluation of desulfurized rubber asphalt based on rheological and environmental effects. *J. Mater. Civil. Eng.* **2020**, *32*, 04019330. [CrossRef]
29. Ma, T.; Zhao, Y.L.; Huang, X.M.; Zhang, Y. Characteristics of desulfurized rubber asphalt and mixture. *Ksce J. Civ. Eng.* **2016**, *20*, 1347–1355. [CrossRef]
30. Ye, Z.G.; Kong, X.M.; Yu, J.Y.; Wei, L.Q. Microstructure and properties of desulfurized crumb rubber modified bitumen. *J. Wuhan Univ. Technol.* **2003**, *18*, 83–85. [CrossRef]
31. Ibrahim, I.M.; Fathy, E.S.; El-Shafie, M.; Elnaggar, M.Y. Impact of incorporated gamma irradiated crumb rubber on the short-term aging resistance and rheological properties of asphalt binder. *Constr. Build. Mater.* **2015**, *81*, 42–46. [CrossRef]
32. Lin, P.; Huang, W.D.; Tang, N.P.; Xiao, F.P. Performance characteristics of terminal blend rubberized asphalt with sbs and polyphosphoric acid. *Constr. Build. Mater.* **2017**, *141*, 171–182. [CrossRef]
33. Ma, T.; Wang, H.; He, L.; Zhao, Y.L.; Huang, X.M.; Chen, J. Property characterization of asphalt binders and mixtures modified by different crumb rubbers. *J. Mater. Civil. Eng.* **2017**, *29*, 04017036. [CrossRef]
34. Juganaru, T.; Bombos, M.; Vasilievici, G.; Bombos, D. Devulcanized rubber for bitumen modification. *Mater. Plast.* **2015**, *52*, 336–339.
35. Guo, L.; Lv, D.J.; Ren, D.H.; Qu, L.N.; Wang, W.C.; Hao, K.F.; Guo, X.R.; Chen, T.C.; Sun, J.Y.; Wang, C.S.; et al. Effectiveness of original additives in waste rubbers for revulcanization after reclamation with a low-temperature mechanochemical devulcanization method. *J. Clean. Prod.* **2021**, *297*, 126620. [CrossRef]
36. Valdes, C.; Hernandez, C.; Morales-Vera, R.; Andler, R. Desulfurization of vulcanized rubber particles using biological and couple microwave-chemical methods. *Front Environ. Sci.-Switz.* **2021**, *9*, 633165. [CrossRef]
37. Zhang, B.; Chen, H.X.; Zhang, H.G.; Kuang, D.L.; Wu, J.Y.; Zhang, X.L. A study on physical and rheological properties of rubberized bitumen modified by different methods. *Materials* **2019**, *12*, 3538. [CrossRef]
38. Zhang, B.; Chen, H.X.; Zhang, H.G.; Wu, Y.C.; Kuang, D.L.; Guo, F.J. Laboratory investigation of aging resistance for rubberized bitumen modified by using microwave activation crumb rubber and different modifiers. *Materials* **2020**, *13*, 4230. [CrossRef]
39. Wang, W.S.; Cheng, Y.C.; Tan, G.J.; Liu, Z.Y.; Shi, C.L. Laboratory investigation on high- and low-temperature performances of asphalt mastics modified by waste oil shale ash. *J. Mater. Cycles Waste* **2018**, *20*, 1710–1723. [CrossRef]
40. Zhou, P.L.; Wang, W.S.; Zhu, L.L.; Wang, H.Y.; Ai, Y.M. Study on performance damage and mechanism analysis of asphalt under action of chloride salt erosion. *Materials* **2021**, *14*, 3089. [CrossRef]
41. Zheng, W.H.; Wang, H.N.; Chen, Y.; Ji, J.; You, Z.P.; Zhang, Y.Q. A review on compatibility between crumb rubber and asphalt binder. *Constr. Build. Mater.* **2021**, *297*, 123820. [CrossRef]
42. Geng, J.; Chen, M.; Xia, C.; Liao, X.; Chen, Z.; Chen, H.; Niu, Y. Quantitative determination for effective rubber content in aged modified asphalt binder. *J. Clean. Prod.* **2022**, *331*, 129978. [CrossRef]
43. Thives, L.P.; Pais, J.C.; Pereira, P.A.A.; Triches, G.; Amorim, S.R. Assessment of the digestion time of asphalt rubber binder based on microscopy analysis. *Constr. Build. Mater.* **2013**, *47*, 431–440. [CrossRef]
44. Liu, S.J.; Zhou, S.B.; Peng, A.H.; Xuan, W.A.; Li, W. Analysis of the performance and mechanism of desulfurized rubber and low-density polyethylene compound-modified asphalt. *J. Appl. Polym. Sci.* **2019**, *136*, 48194. [CrossRef]
45. Zhao, M.Z.; Dong, R.K.; Chi, Z.H.; Aljarmouzi, A.; Li, J.R. Effect of process variables on the chemical characteristics of crumb rubber desulfurized by waste cooking oil and its desulfurization mechanism. *Constr. Build. Mater.* **2021**, *311*, 125361. [CrossRef]
46. Liu, X.H.; Bi, Y.F.; An, Y.N.; Zhou, S.Q.; Li, F. Secondary-reaction method to improve performance of a compound modified rubber asphalt after short-term aging. *J. Transp. Eng. Part B Pavements* **2021**, *147*, 04021038. [CrossRef]
47. Yi, X.Y.; Dong, R.K.; Tang, N.P. Development of a novel binder rejuvenator composed by waste cooking oil and crumb tire rubber. *Constr. Build. Mater.* **2020**, *236*, 117621. [CrossRef]
48. Lv, S.T.; Ma, W.B.; Zhao, Z.G.; Guo, S.C. Improvement on the high-temperature stability and anti-aging performance of the rubberized asphalt binder with the lucobit additive. *Constr. Build. Mater.* **2021**, *299*, 124304. [CrossRef]
49. Zhang, J.W.; Chen, M.Z.; Wu, S.P.; Zhou, X.X.; Zhao, G.Y.; Zhao, Y.C.; Cheng, M. Evaluation of vocs inhibited effects and rheological properties of asphalt with high-content waste rubber powder. *Constr. Build. Mater.* **2021**, *300*, 124320. [CrossRef]

polymers

MDPI

Article

Rheological and Aging Properties of Composite Modified Bitumen by Styrene–Butadiene–Styrene and Desulfurized Crumb Rubber

Gang Xu, Yunhong Yu, Jingyao Yang, Tianling Wang, Peipei Kong and Xianhua Chen *

National Demonstration Center for Experimental Road and Traffic Engineering Education, School of Transportation, Southeast University, Southeast University Road, Nanjing 210096, China; xugang619@hotmail.com (G.X.); iiyyhong@163.com (Y.Y.); yangjingyao19@foxmail.com (J.Y.); 220183061@seu.edu.cn (T.W.); xc_kong@seu.edu.cn (P.K.)

* Correspondence: chenxh@seu.edu.cn

Abstract: Taking advantage of crumb rubber from waste tires to modify bitumen is widely for the environmentally friendly and sustainable development of pavement. This study investigated the modification mechanism, rheological, and aging properties of styrene–butadiene–styrene (SBS)/desulfurized crumb rubber (DCR) composite modified bitumen (SBS/DCRMB). Morphological features and chemical characteristics were assessed by fluorescence intensity measurement and gel permeation chromatography (GPC), respectively, and results demonstrated that the DCR and SBS modifier in SBS/DCRMB had been vulcanized and formed a three-dimensional network structure. Moreover, a comparison of the GPC elution curve showed the residual bitumen hardly changed due to carbon black released from DCR of SBS/DCRMB during the aging process of SBS/DCRMB, and the polymer molecules condensed to larger units. However, the remaining bitumen in SBSMB had changed evidently and the polymer degraded to smaller molecules. Meanwhile the rheological testing results, including multiple stress creep recovery, linear amplitude sweep and bending beam rheometer, declared that the SBS/DCRMB is superior to SBSMB before and after aging.

Keywords: desulfurized crumb rubber; styrene–butadiene–styrene; composite modified bitumen; rheological properties; aging process

Citation: Xu, G.; Yu, Y.; Yang, J.; Wang, T.; Kong, P.; Chen, X. Rheological and Aging Properties of Composite Modified Bitumen by Styrene–Butadiene–Styrene and Desulfurized Crumb Rubber. *Polymers* **2021**, *13*, 3037. https://doi.org/10.3390/polym13183037

Academic Editor: Heping Chen

Received: 10 August 2021
Accepted: 6 September 2021
Published: 8 September 2021

1. Introduction

Nowadays, the noticeable increase in traffic volume and the dramatic increase in heavy vehicle axle load has led to the necessity to attention focus on pavement performance. At the same time, base bitumen cannot meet the requirements of the structure and function of bitumen concrete pavement [1,2]. To address this issue, more and more modified bitumens, such as styrene–butadiene–styrene (SBS) modified bitumen (SBSMB), have been applied to high-grade pavement, which significantly enhances the performance of bitumen pavement [3,4]. However, the high cost of SBSMB indicates its incapability to achieve resource conservation and environmental friendliness. Therefore, bitumen modified with crumb rubber (CR) prepared from waste tires has been considered by many researchers due to its advantages of energy thriftiness, environmental protection, and economic significance [5,6]. Currently, CR modified bitumen (CRMB) has been utilized as a binder of asphalt mixtures on many roads, both in China and internationally, which achieves promising performance and effectively reduced pollution of the tire. However, due to the three-dimensional structure of rubber molecules and vulcanization, it is difficult to swell during processing. Furthermore, crumb rubber rarely reacts with bitumen, resulting in rubber and bitumen prone to natural segregation, poor storage stability under high temperature, and other defects [7–9]. These issues restrict the further development of CRMB in road-building technology. Hence, researchers have been pre-treating the crumb rubber to improve the performance of CRMB.

It has been investigated that the desulfurization of crumb rubber could significantly improve the performance of CRMB. Desulphurization promotes the activity of crumb rubber, which can substantially improve the performance of rubber bitumen and its mixture [10–12]. As indicated by the literature, desulfurized rubber particles have a low crosslinking degree and more active chemical bonds, enabling more apparent chemical effects between the rubber particles and bitumen. Furthermore, the high-temperature viscosity is noticeably decreased, and the preparation temperature also decreased slightly. In addition, it is more environmentally friendly, as less odor exhausts during the preparation process [13]. Moreover, the solvents and other reactive agents are added in the preparation process of CRMB, which can ameliorate the phase dissolution rate of rubber and bitumen, as well as improve the stability of CRMB. Desulfurized crumb rubber modified bitumen (DCRMB) and reactive rubber modified bitumen exhibits high elastic response, which can reduce non-recoverable deformation, thus yielding a better low-temperature crack resistance and fatigue resistance [14–17].

Additionally, CRMB has increased flexibility and better resistance to heat and oxygen aging, and features many other beneficial properties compared with base bitumen. However, it has been proven that the rutting resistance and storage stability of bitumen modified by desulfurized rubber still remain poor. In this regard, it can be concluded that bitumen single modifier is hard to satisfy the multiple performance requirements of bitumen pavement [12,18–20]. Consequently, researchers began to combine two or more kinds of modifiers and methods for modifying bitumen, to combine the advantages different modifiers offer. For example, tire crumb rubber combined with recycled polyethylene extruded melt blend as a composite additive was conducted to enhance the performance of binder and results found recycled polyethylene played a key role to improving the high-temperature performance of composite modified bitumen [21].

SBS copolymers are widely used for bitumen modification. However, the main drawbacks of SBSMB are high cost and poor aging resistance. Hence, researchers began to examine the application of SBS/DCRMB, as not only can it unite the performance advantages of DCR and SBS, but it can also reduce the SBS modifier dosage to reduce construction costs as well as the environmental pollution caused by the utilization of crumb tires in bitumen [22–24]. According to the results of morphology observation, the addition of SBS to CRMB leads to the formation of a staggered polymer network. At the same time, vulcanization or aging would improve the compatibility between polymer and bitumen, enhances their interaction and increases the storage stability of SBS/DCRMB [25,26]. The results also indicate that carbon black will be released during the process, which can improve the flexibility of network structures and boosts their anti-aging property. Furthermore, black carbon increases the adhesion of bitumen, so as to improve the general performance of bitumen mixture [27–29]. In summary, the performance of the SBS/DCRMB has been improved over a wide temperature range, as the SBS in the bitumen enhances the high-temperature properties, DCR improves the fatigue and low-temperature cracking resistance due to its excellent elasticity. However, current studies on SBS/DCRMB were merely the simple addition of two single modifiers by different orders, where there is no further pretreatment of the two types of modifiers. In addition, detailed descriptions about the morphology and the systematic discussion of the combined modification mechanism of the two modifiers are still unclear.

In this sense, this research proposes a type of SBS/DCRMB by SBS/DCR composite additive, which was prepared by force-chemical reactor. The modification mechanism of SBS/DCRMB was characterized by fluorescence intensity measurement and gel permeation chromatography (GPC). The rheological properties, including rutting, fatigue, and low-temperature performance of SBS/DCR and SBS modified bitumen were analyzed and compared before and after thin film oven test (TFOT) and pressurized aging vessel (PAV) aging. Based on the results of GPC, the relationship between the molecular distribution and macroscopic properties was explored. Then, their ability to resist aging was compared between SBS/DCRMB and SBSMB.

2. Methods

2.1. Materials and Preparation

Pen70 was selected as the neat binder for modified bitumen, and the physical properties are shown in Table 1. Two conventional polymers were selected, including SBS and desulfurized crumb rubber (DCR). The average molecule weight of linear SBS was 80,000 g/mol, and the weight percentage of styrene was 30%. The DCR modifier was devulcanized by twin-strew extruder and the principle of desulphurization is shown in Figure 1. Meanwhile, rubber processing oil, which is rich in aromatics and saturates, was utilized to achieve sufficient DCR and SBS swelling in bitumen. The content of rubber processing oil was 4% by mass of base bitumen. Sulfur powder acted as crosslinking agent, and an amount of 5% of the modifiers weight was applied.

Table 1. Basic properties of bitumen.

Properties	Unit	Test Results			Test Method
		Neat	SBS	SBS/DCR	
Penetration (25 °C, 100g, 5 s)	(0.1 mm)	68.9	58.9	72.4	ASTM D5 [30]
Softening point (ring and ball method)	°C	47.2	66.5	85.8	ASTM D36 [31]
Ductility (15 °C, 5 cm/s)	Cm	>100	–	–	ASTM D113 [32]
Ductility (5 °C, 5 cm/s)	Cm	–	20.4	44.5	ASTM D113 [32]
Change in mass TFOT	%	−0.2	−0.11	−0.08	ASTM D2872 [33]
Flashpoint, Cleveland open cup	°C	289	291	310	ASTM D92 [34]

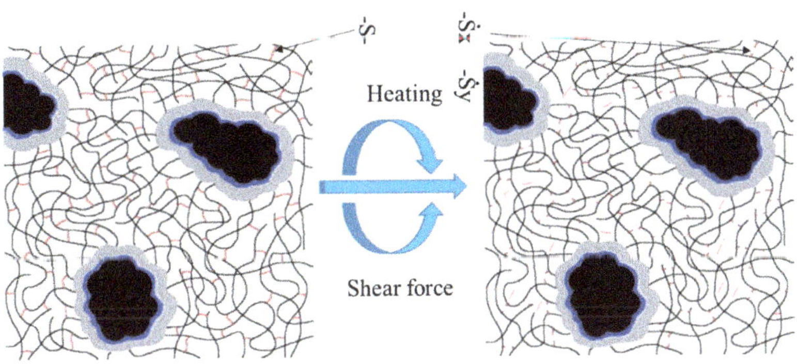

Figure 1. The principle of desulphurization by twin-strew extruder.

DCR/SBSMB was prepared in a laboratory at 180 °C. Firstly, the DCR/SBS composite modifier was added to the base asphalt and the blend was sheared at a speed of 6000 r/min for 1.5 h. The speed was then reduced to 1500 r/min for 1h until fully swelled. The preparation process of the DCR/SBSMB is illustrated in Figure 2. Conventional CR/SBS composite modified asphalt with 20% CR and 4% SBS was prepared using the same method.

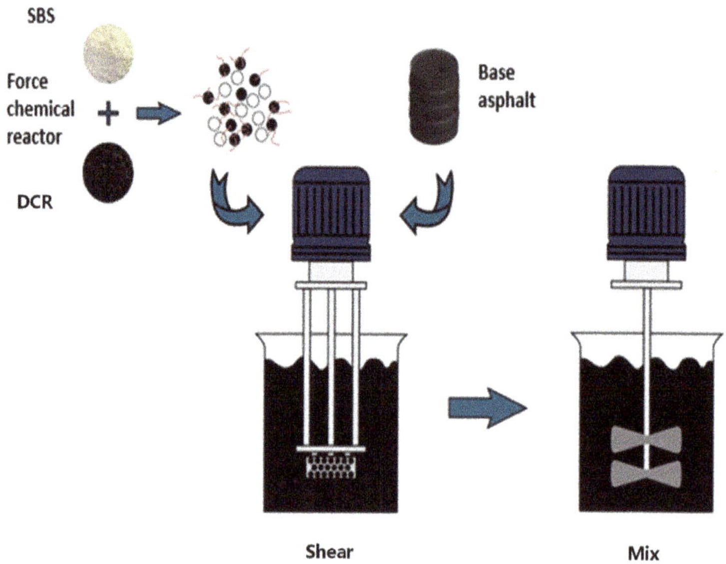

Figure 2. Flowchart of material preparation.

As mentioned above, there was a type of bitumen considered as references for the comparison of SBS/DCRMB, namely SBSMB with 4% SBS (of the weight of base bitumen). It should be noted that the afore-mentioned preparation conditions were also used for preparation of SBSMB.

2.2. Aging Method

The bitumen samples were assessed using TFOT aging at 163 °C for 5 h according to ASTM D1754 to simulate the short-term aging of bitumen during mixing, transportation, and paving [35]. Furthermore, long-term aged bitumen was obtained by conducting TFOT (for 5 h at 163 °C), followed by a PAV test. ASTM D6531 recommends imitating the age of the road's service life [36]. The aging temperature in the PAV test was 100 °C. The pressure was set to 2.1 MPa and the aging time was 20 h.

2.3. Fluorescence Intensity Measurement

Fluorescence microscopy is a standard method to observe the state of asphalt binders during their modification with polymers [37,38]. The polymer-rich phase that is swelled by the oily aromatic part of the asphalt appears to be yellow-colored, while the asphaltene phase is dark [39,40]. A Nikon Y-IDP Japan fluorescence microscope (Nikon, Tokyo, Japan) was employed to measure the fluorescence intensity of the modified bitumen, which supports up to 400× *g*. Under fluorescence, when excited by short-wave light, the polymer phase, formed by the swelling of the polymer, will emit a longer wavelength of light, while the asphalt phase emits no light. Therefore, it is easy to distinguish the yellow polymer phase from the black asphalt phase under a fluorescence microscope, and the distribution of modifiers in the asphalt can be clearly seen.

2.4. Molecular Distribution Analysis Using Gel Permeation Chromatography

GPC (HLC-8320, Tosoh, Tokyo, Japan) was employed to characterize changes of the molecular dimension. A total of five columns (TSK gel Super HM-M W0054, TSK gel Super HM-M W0055, TSK gel Super HM-M W0056 TSK gel Super H-RC W0109, and TSK Guard Column Super HH W0114, Tosoh, Tokyo, Japan) are required for GPC to separate asphalt binders based on their molecular size. Before the test, tetrahydrofuran was used to dissolve a small amount of asphalt to prepare a 0.5–5 mg/mL solution. The THF flow rate was controlled at 0.2 mL/min. After the test, software configured by a GPC device was used to calculate the number average molecular weight (Mn) and weight average molecular weight (Mw). Then, their ratio was used to represent the molecular distribution length and investigate the molecular distribution of modified asphalt before and after aging [29,41,42].

2.5. Multiple Stress Creep Recovery Test

The multiple stress creep recovery (MSCR) test was conducted by a dynamic shear rheometer (DSR) device (Anton-paar, Graz, Austria) based on AASHTO T350 [43]. The diameter of samples used for testing was 25 mm, the thickness was 1 mm and the testing temperature was 64 °C, which can simulate the anti-rutting performance. The results of the MSCR test were obtained based on two replicates. The MSCR test consisted of 20 cycles, a 1 s creep period, and a 9 s recovery period at a stress level of 0.1 kPa. This was followed by another 10 cycles of creep and recovery at 3.2 kPa according to AASHTO T350. The three main parameters obtained in the experiment mainly used the following Equations (1)–(3).

$$\text{Recovery\% at } 100Pa \text{ or } 3200Pa = \frac{1}{10}\left[\sum_{i=1}^{10}\frac{\gamma_{(r)i}}{\gamma_{(t)i}}\right] \times 100 \quad (1)$$

$$J_{nr\ at\ 100Pa}(1/KPa) = \frac{1}{10}\left[\sum_{i=1}^{10}\frac{\gamma_{(nr)i}}{0.1}\right] \quad (2)$$

$$J_{nr\ at\ 3200Pa}(1/KPa) = \frac{1}{10}\left[\sum_{i=1}^{10}\frac{\gamma_{(nr)i}}{3.2}\right] \quad (3)$$

where $\gamma_{(r)i}$ is recoverable strain, $\gamma_{(nr)i}$ is non-recoverable strain at the end of 9 s rest, and $\gamma_{(t)i}$ is total creep strain at the end of the loading time of 1 s in each cycle.

2.6. Linear Amplitude Scanning Test

Current asphalt binder specifications lack the ability to characterize the damage resistance of asphalt binder to fatigue loading. Multiple accelerated testing procedures are currently being investigated in an attempt to describe the contribution of asphalt binders to mixture fatigue efficiently and accurately [17,40]. One of these tests, which has received acceptance by experts and has been submitted as a draft AASHTO standard, is the linear amplitude sweep (LAS) test. The LAS test was conducted at 25 °C. According to AASHTO TP 101-14, the asphalt samples (8 mm in diameter and with a 2-mm thickness) were tested in two stages [44]. In the first stage, a frequency sweep test with a strain level of 0.1% was performed at different frequencies (0.2 to 30 Hz). This test was used to obtain the undamaged material parameter (α). At the second stage, the oscillating shear was used at a frequency of 10 Hz in strain control mode at the selected temperature. The scanning time was 300 s, and the loading amplitude increased linearly from 0.1% to 30%.

2.7. Bending Beam Rheometer Test

The bending beam rheometer (BBR) test was employed to characterize the low-temperature performance of SBSMB and SBS/DCRMB before and after aging. The test temperatures were −12, −18, and −24 °C, and the average results of three replicates were used as the testing results. The two key parameters of bitumen at low temperature, as obtained by Equations (4) and (5), respectively, are creep stiffness modulus (S) and creep rate (m).

$$\mathrm{S(t)} = \frac{PL^3}{4bh^3\delta(t)} \quad (4)$$

$$m = a + 2blg(t) \quad (5)$$

where S(t)(MPa) is flexural creep stiffness and P(N) is constant load. h, L, and b (mm) are thickness, length, and width of a thin beam sample, respectively. $\delta(t)(mm)$ is deformation at the mid-span of a thin beam sample.

Using the interpolation method according to ASTM D7643-16 can determine the low service temperature (T_L) of asphalt binders from BBR test in multiple testing temperatures [45]. The critical temperature $T_{L,s}$ and $T_{L,m}$ corresponding to stiffness = 300 MPa and m value = 0.3 were obtained by the regression Equations (6) and (7), respectively. The low service temperature (T_L) was defined as Equation (8).

$$\log_{10}(s) = a_1 + b_1 T \tag{6}$$

$$\mathrm{m} = a_2 + b_2 T \tag{7}$$

$$T_L = \max(T_{L,s}, T_{L,m}) - 10 \tag{8}$$

where a_1, a_2, b_1, and b_2 are the regression coefficients; T is the test temperature (°C); $T_{L,s}$ is the critical temperature when S = 300 MPa (°C); $T_{L,m}$ is the critical temperature when m = 0.3 (°C); and T_L is the low service temperature (°C).

3. Results and Discussion

3.1. Fluorescence Microscopy (FM) Analysis of SBS and DCR in Bitumen

The fluorescence micrographs of SBSMB and SBS/DCRMB are shown in Figure 3. Figure 3a shows that under the excitation of violet light, the yellow spots are SBS components, and the orange parts are asphaltenes. The dot interface of SBS is clear and has a poor affinity with asphalt, resulting in poor compatibility. However, Figure 3c presents more black carbon components, and an irregular distribution shape of SBS. The interface is fuzzy and the interaction with asphalt is enhanced, thereby increasing the stability of the interface.

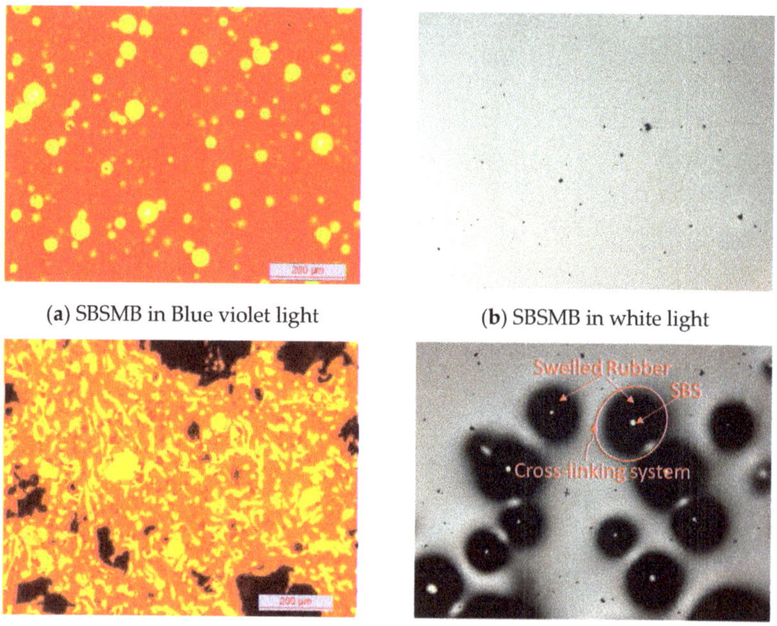

(**a**) SBSMB in Blue violet light (**b**) SBSMB in white light

(**c**) SBS/DCRMB in Blue violet light (**d**) SBS/DCRMB in white light

Figure 3. Micrography of SBSMB (**a**,**c**)and SBS/DCRMB (**b**,**d**) with different magnifications and different light sources in Fluorescence.

The fluorescence microscopic images of SBSMB and SBS/DCRMB are different under white light. As illustrated in Figure 3b, a few black spots (which might be impurities in the bitumen) were scattered in the white setting (base bitumen), while in Figure 3d, clear white dots are either surrounded by or connected to black patches. Comparing Figure 3b,d indicates that the aromatic component of bitumen or rubber processing oil is absorbed by DCR and the border of black patches are blurred and thick. Due to the addition of crosslinking agents during the preparation, SBS and DCR are crosslinked by non-directional vulcanization. In the process of DCR swelling, SBS is wrapped up between the DCR, and the polymer network structure of SBS/DCR forms [46].

These results indicate that during the preparation of SBS/DCRMB, aromatic components and rubber processing oil are absorbed during swelling, depolymerization, and dispersion. Under the interaction with sulfur, SBS, DCR, and the SBS–DCR polymer network, i.e., the cohesive forces between polymer, as shown in Figure 4, which briefly present the mechanism of composite modified bitumen.

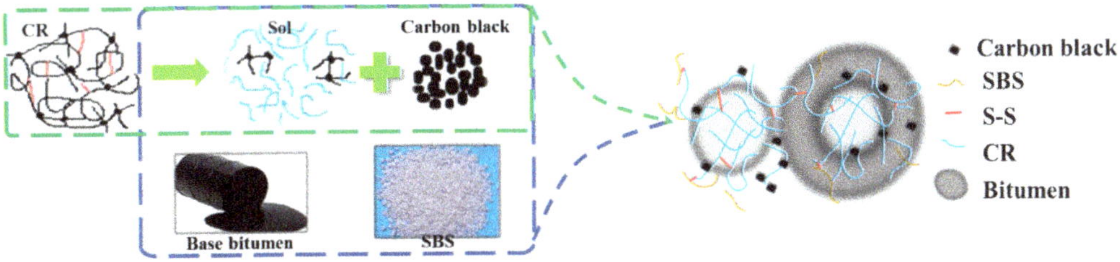

Figure 4. Modification mechanism of SBS/DCRMB.

3.2. Molecular Distribution Analysis with GPC

GPC analysis can effectively characterize the molecular structure and the decomposition of polymers during the aging process [42]. According to Figure 4, two regions can be observed from the GPC chromatogram of SBS/DCRMB and SBSMB. These are the residual matrix bitumen region on the left (with smaller molecular weight) and the polymer region on the right (with larger molecular weight). The appearance of the two molecular weight regions in SBS/DCRMB indicates that a mixed crosslinking system formed between SBS and DCR. The existence of a crosslinking system is verified via fluorescence microscopic images under white light.

Focusing on the left region, Figure 5a shows that the aging effect leads to significant differences among the three peaks of the molecular weight distribution of residual asphalt. In contrast, the SBS/DCRMB residual bitumen molecular weight distribution curve shown in Figure 5b approximates an overlap and with good consistency. This may be ascribed to the following two reasons: On the one hand, the aromatics component is absorbed by the swelling of polymer components during the preparation of SBS/DCRMB. On the other hand, the residual bitumen is protected by the black carbon released during the preparation of DCR from thermo-oxidative aging.

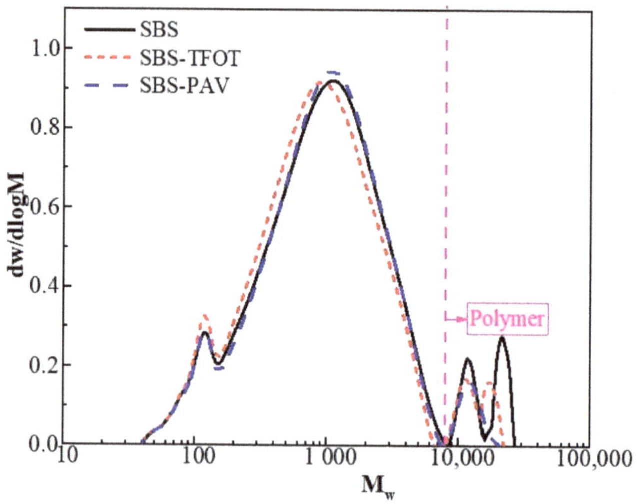

(**a**) Molecular weight distribution of SBS/MB before and after aging.

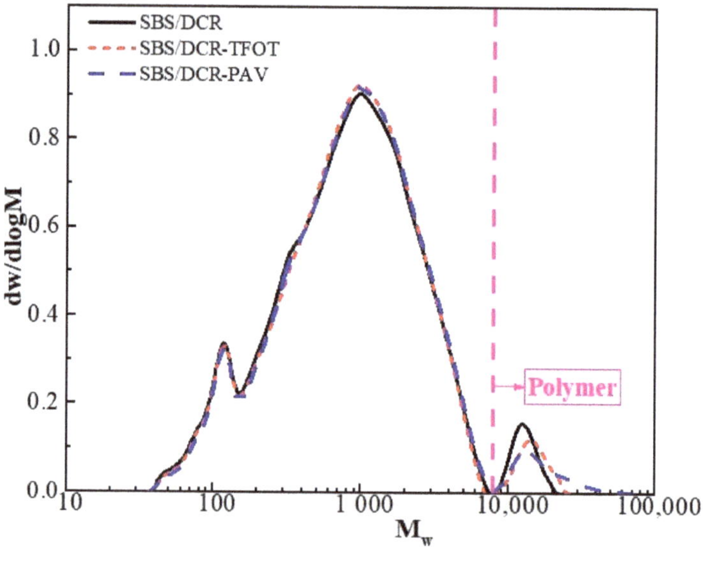

(**b**) Molecular weight distribution of SBS/DCRMB before and after aging.

Figure 5. Molecular distribution of SBSMB and SBS/DCRMB in different aging conditions.

Furthermore, focusing on the right region, the results of GPC also show the significant variation trend of the polymer region during the aging process. In SBSMB, the polybutadiene chains in SBS were destroyed, and the polymer molecular weight decreased, which leads to a left shift of the molecular weight distribution curve and damages the crosslinking structure. Dissimilarly in the polymer area of SBS/DCRMB, in response to aging, causes the curve to move to the right, indicating that some chemical reactions (such as condensation) are caused between the two polymers (SBS and DCR). Condensation is another variation of the reaction of bitumen aging that will generate larger molecules.

Ultimately, Mn can be used to characterize the variation of small molecules, while Mw reflects the macromolecule. Table 2 presents the detailed values of Mn, Mw, and Mn/Mw obtained by GPC test. Given the amount of macromolecular component, it is conflicting that the value of Mw would decrease in response to the dissociation of polymer modifier. The value of Mw would increase by the condensation of polymer and the conversion of aromatic components. As the indicator of Mn is free from such conflict, Mn can adequately reflect the aging resistance of modified bitumen. According to the variation of Mn, the Mn value of SBS/DCRMB is higher than that of SBSMB before aging, while SBSMB is higher than SBS/DCRMB after PAV aging. This implies that SBS/DCRMB has a better anti-aging ability.

Table 2. Molecular weight calculation of SBSMB and SBS/DCRMB before and after aging.

Binder Type	Aging Conditions	M_n	M_w	M_n/M_w
SBSMB	Virgin	850	2270	2.67
	TFOT	885	2303	2.60
	PAV-20 h	928	2354	2.54
SBS/DCRMB	Virgin	860	2577	3.00
	TFOT	884	2664	3.01
	PAV-20 h	910	2790	3.07

In addition, the ratio of Mn to Mw in Table 2 shows that the ratio of SBS becomes smaller. This corresponds to the polymer distribution in Figure 5a, where the dispersion of the molecular distribution becomes smaller due to degradation. However, the ratio of SBS/DCRMB follows an increasing trend, and there is a significant increase between short-term aging and PAV. This corresponds to the molecular weight distribution of the polymer part, which increases due to shown polycondensation in Figure 5b.

3.3. Multiple Stress Creep Recovery Results

Figure 6a shows the time–strain curve of un-aged SBSMB and SBS/DCRMB in a MSCR test. With increasing test time, the original bitumen strain also increased, and the strain of SBSMB became significantly larger than that of SBS/DCRMB. Additionally, the unrecoverable creep compliance (J_{nr}) of SBS/DCRMB is much smaller than that of SBSMB. This illustrates that SBS/DCRMB possesses much higher resilience and resistance to the high-temperature deformation than SBSMB under the same loading conditions. The remarkable anti-rutting performance of SBS/DCRMB benefits from the high elastic property of DCR and the strong three-dimensional network structure developed by DCR and SBS.

Non-recoverable compliance (J_{nr}), stress sensitivity ($J_{nr-diff}$), and the creep recovery rate are the main parameters for evaluating the high-temperature properties of bitumen binders in MSCR testing [19,47]. Figure 6b–d present the results of the MSCR test for different aging conditions. All obtained results show differences between the two modified types of bitumen. These results show that the rate of creep recovery of SBSMB decreased significantly with increasing loading stress and aging, but the rate of creep recovery of SBS/DCRMB varies slightly. This might be related to the three-dimensional network structure formed by SBS and DCR during the modification process (as shown in the fluorescence micrographs). The complete three-dimensional network structure can be gradually recovered with the superior elastic properties provided from DCR when the stress diminishes. In contrast, the polymer network structure in SBSMB is difficult to rebuild due to a lack of elastic recovery and sensitivity to stress (as shown in Figure 3a).

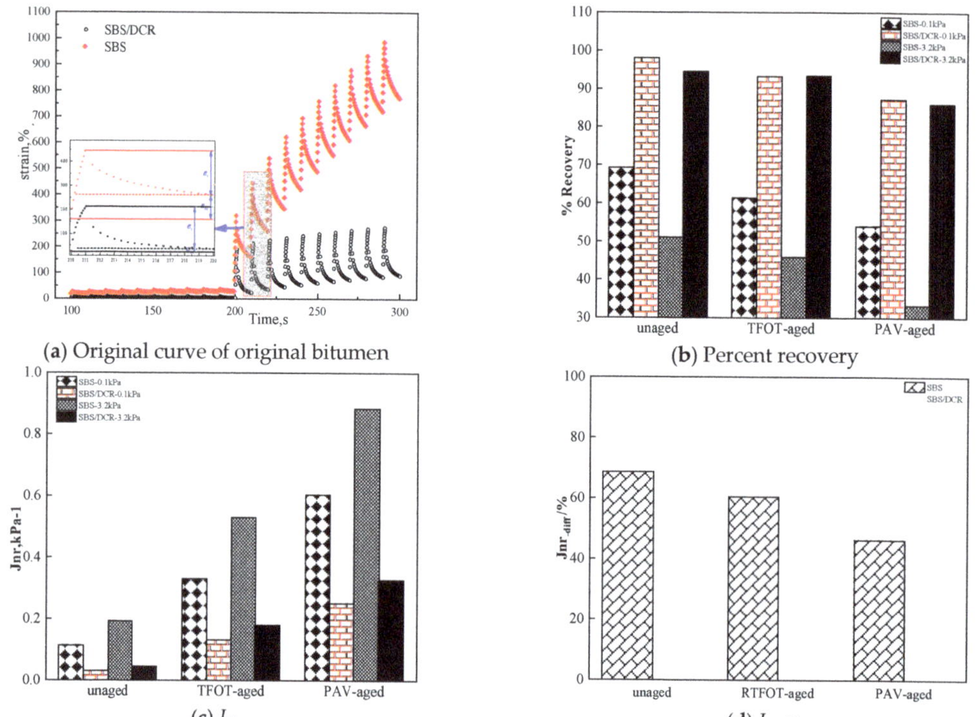

(a) Original curve of original bitumen (b) Percent recovery (c) J_{nr} (d) $J_{nr\text{-}diff}$

Figure 6. MSCR test result in different aging conditions of two modified bitumen.

However, the J_{nr} of the two modified bitumens contrasts with the rate of creep recovery before and after aging. As shown in Figure 6c, J_{nr} increases with increasing unrecoverable deformation at a higher stress level and with the degree of aging. This contrasts with the conclusion by Zhou et al., who claimed that the stiffness of base bitumen is enhanced after adding RAP, which resulted in a reduction in J_{nr} [48]. After short-term and long-term aging, the modifiers decomposed as indicated by the GPC test. The decomposition of modifiers leads to the weakening of the polymer network structure and the interaction between the modifiers, thus reducing the resistance to deformation. Figure 6d shows the stress sensitivity of two types of bitumen under different aging conditions through $J_{nr\text{-}diff}$. To a certain extent, the stress sensitivity of the modified bitumen is reduced by aging and all values were less than 75%. The GPC testing results of SBSMB indicate that in the process of thermo-oxidative aging, the aromatic component of SBSMB is reduced, and the stiffness is improved, thus, bitumen became insensitive to stress.

Based on the non-recoverable compliance, stress sensitivity, and the rate of creep recovery obtained by the MSCR test, the elastic properties of bitumen improved effectively by adding DCR. Additionally, as the polymer network formation by the DCR and SBS and the black carbon release by the DCR during the preparation process, the high-temperature performance of the bitumen was significantly improved, resulting in excellent road performance of the SBSMB.

3.4. Linear Amplitude Scanning Test Results

Table 3 displays the fatigue damage values obtained by LAS test based on the VECD model. Figure 7a,b present the rapid fatigue damage curve and the relationship between damage intensity (D) and integrity parameters (C), respectively. Table 3 shows that the C_1 and C_2 of SBSMB bitumen increase with aging, while the values of SBS/DCRMB show a completely different variation. Here, C_1 decreases while C_2 increases. The lower C1 and C2 values, the better the fatigue performance of bitumen will become [47]. During the aging process, the magnitude of the bitumen modulus, the increase in the polymer content, and the strength of the network will increase C_1 and decrease C_2. Combined with the results of GPC testing, the decomposition of the polymer and network structure during aging increases C_2. For C_1 parameters, in SBSMB, the positive effect from the increasing modulus is not counteracted by the weak or negative impact caused by the decomposition of the polymer. However, with the addition of DCR, the polymer proportion in SBS/DCRBM increases and the condition is improved. The impact of the damage intensity and the integrity parameter combined with C_1 and C_2 should be comprehensively considered.

Table 3. Linear amplitude sweep test results of all tested binders based on viscoelastic continuum damage analysis.

Binder Type	Aging Conditions	C_1	C_2	A	B	α	τ_{max}
SBS	Virgin	0.050	0.473	3,142,334	2.894	1.447	0.236
	TFOT	0.059	0.496	1,856,421	2.888	1.444	0.258
	PAV-20 h	0.066	0.521	1,533,497	2.922	1.461	0.318
SBS/DCR	Virgin	0.058	0.490	8,010,102	2.978	1.489	0.199
	TFOT	0.054	0.504	5,165,232	3.026	1.513	0.210
	PAV-20 h	0.050	0.513	3,456,893	3.042	1.521	0.249

As shown in Figure 7a, during the aging process, SBSMB is more sensitive to aging compared with SBS/DCRMB. During aging, the D–C curves of SBSMB indicated a large difference between before and after the same short-term aging, which is similar to SBS/DCRMB. Additionally, the loss rate of the integrity parameter increased sharply after PAV aging. The reason for this change may also be consistent with the result of the GPC test. In the aging process of SBSMB, a number of aromatic compounds are converted into asphaltenes, which play a dominant role in fatigue performance. Moreover, the degradation of modifier and the damage of the polymer network also accelerated the loss of fatigue performance. Therefore, aging severely impacts the fatigue performance of asphalt. SBS/DCRMB offers an advantage in resisting short-term aging. After PAV aging, the fatigue performance of asphalt cementation material decreases sharply. Therefore, the fatigue performance of asphalt is severely affected by aging and SBS/DCRMB offers an advantage in resisting short-term aging.

Figure 7b demonstrates that the loss rate of asphalt sensitivity decreases gradually with increase in strain level and aging degree. Comparing the loss of the fatigue life before and after aging shows that the fatigue life of SBS/DCR modified bitumen is longer than that of SBSMB bitumen regardless of the underlying aging conditions. This is caused by the addition of DCR, which can improve the fatigue resistance. It can also be found that aging conditions affect the bitumen fatigue life, and its effect is the same than that of the loss rate of the integrity parameter.

In conclusion, the effect of aging on the fatigue life of SBSMB is mainly the result of the hardening of bitumen. In contrast, the fatigue performance loss of SBS/DCRMB is primarily caused by polymer condensation and the disintegration of the structure of the polymer network.

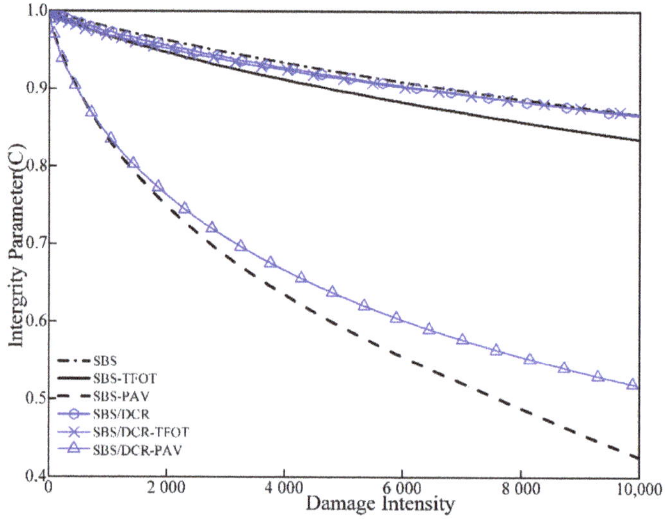

(**a**) Integrity Parameter VS. Damage Intensity.

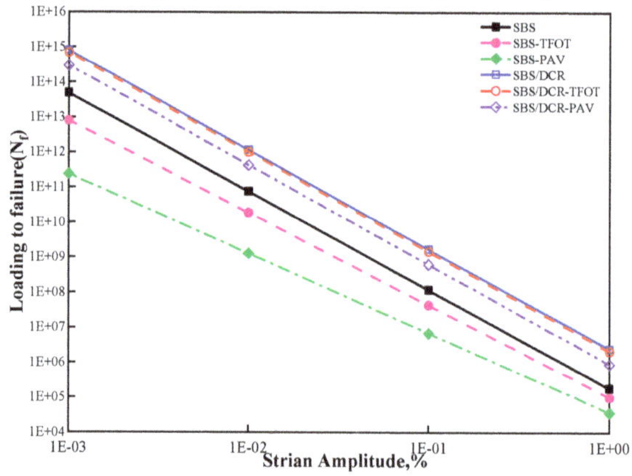

(**b**) Number of cycles for loading to failure.

Figure 7. Results Linear Amplitude Scanning Test in Different aging conditions. (**a**): Integrity Parameter VS. Damage Intensity; (**b**): Number of cycles for loading to failure.

3.5. Low-Temperature Performance

The change in stiffness modulus and the creep rate of SBSMB and SBS/DCRMB are presented in Figure 8a,b and the low service temperature (T_L) of SBSMB, and SBS/DCRMB were compared in Figure 8c. The depicted results indicate that the stiffness modulus increased, and the creep rate decreased, after long-term aging at low temperature. Existing research indicates that low stiffness modulus and high creep rate are conducive to crack resistance, implying that the binder has better low-temperature performance. A comparison of the stiffness modulus and creep rate with temperature changes for different aging conditions shows that the degree of aging will increase the sensitivity of stiffness modulus and creep rate. Furthermore, regardless of the aging state, the S value of SBS/DCRMB is smaller than that of SBSMB at the same temperature, while the m value is larger. According to the AASHTO T313 standard, the minimum S value and the maximum m-value for

PAV-aged asphalt binders should be 300 MPa and 0.3, respectively. From Figure 6, the stiffness values of SBSMB-PAV at 18 °C is larger than 300 MPa and m-value is smaller than 0.3, but SBSMB-PAV can meet the S value and m-value requirement at 12 °C. Therefore, the low-temperature grade of SBSMB is PG-22 and the low-temperature continuous grade is −23.64 °C. Following the same analysis, the low-temperature grade of SBS/DCRMB is PG-28 and the low-temperature continuous grade is −28.59 °C. The main reason for this is that the SBS and DCR formed a three-dimensional network; when the strain reached the limit, the fracture stress can be rapidly concentrated on the surface of DCR. Then, DCR absorbs and consumes the energy, which prevents the formation and expansion of cracks, increasing low-temperature creep deformation. After thermal-oxygen aging and pressure aging, the original bitumen developed brittle and rigid characteristics and the polymer network broke down with the degradation and condensation. These effects further increase the S value and decrease the m-value. Comparing the decay laws of low-temperature properties for different aging conditions shows that the growth of the stiffness modulus and the creep rate of SBSMB before and after pressure aging both exceed that of SBS/DCRMB.

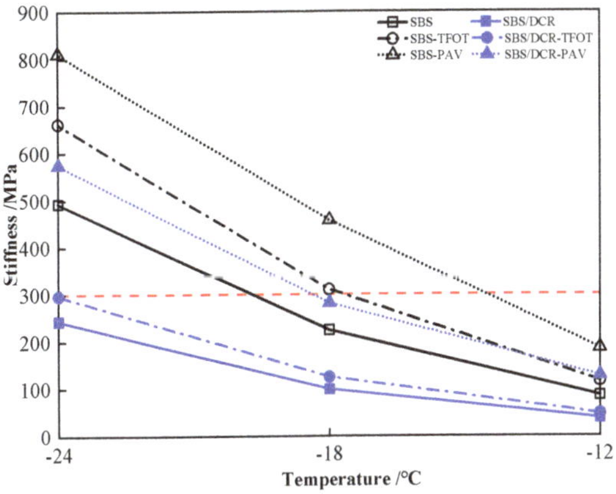

(**a**) Stiffness VS. Temperature in different aging conditions.

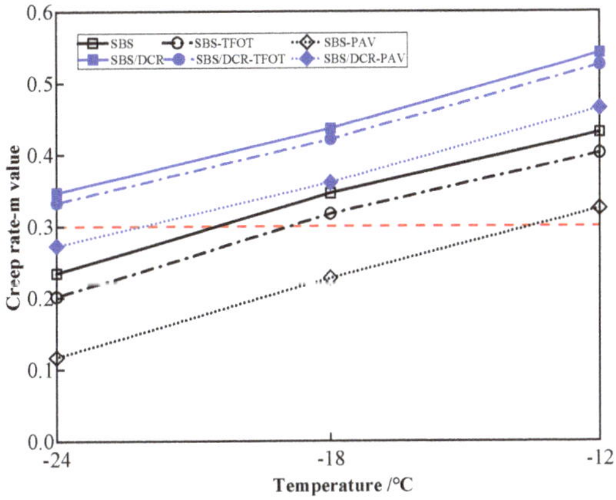

(**b**) Creep rate VS. the temperature in different aging conditions.

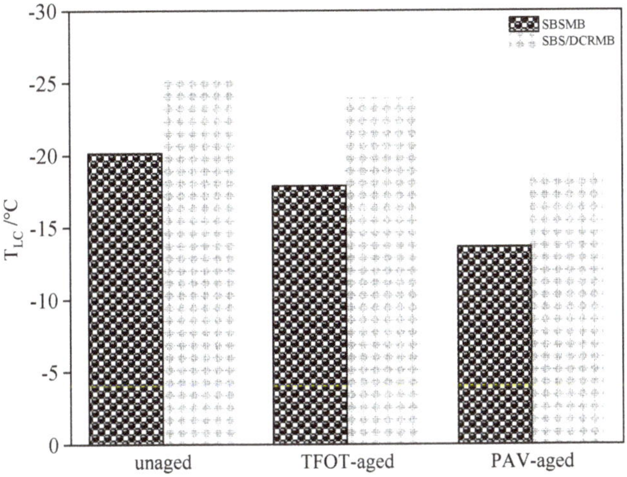

(**c**) Low service temperature T_L

Figure 8. Low-temperature performance in bending beam Rheometer.

3.6. Analysis of and Anti-aging Properties

To some extent, the sensitivity of the asphalt binder's performance affected by aging is revealed by the variation amplitude of the rheological performance and chemical composition indices of asphalt binder before and after aging, which are also important to evaluate the anti-aging performance of asphalt binder. These aging indices were presented in Equations (9)–(12) [49,50].

$$MAI = \frac{\left|\text{Unaged } \frac{\text{Mn}}{\text{Mw}} - \text{Aged } \frac{\text{Mn}}{\text{Mw}}\right|}{\text{Unaged } \frac{\text{Mn}}{\text{Mw}}} \times 100\% \tag{9}$$

$$JDAI = \frac{\left|\text{Unaged } J_{\text{nr}-diff} - \text{Aged } J_{\text{nr}-diff}\right|}{\text{Unaged } J_{\text{nr}-diff}} \times 100\% \tag{10}$$

$$N_f AI = \frac{\left|\text{Unaged } N_f - \text{Aged } N_f\right|}{\text{Unaged } N_f} \times 100\% \tag{11}$$

$$T_L AI = \frac{\left|\text{Unaged } T_L - \text{Aged } T_L\right|}{\text{Unaged } T_L} \times 100\% \tag{12}$$

According to the above definition of the aging index, the sensitivity of the SBSMB and SBS/DCRMB to aging in different aspects of various temperature rheological performance and chemical composition is listed in Table 4. It can be seen that aging sensitivity of SBSMB and SBS/DCRMB is different in the total temperature domain.

Table 4. Aging sensitivity of two modified bitumen.

Modified Asphalt	MAI, %	JDAI, %	T_LAI, %	N_fAI, % (Strain = 1%)
SBSMB-TFOT	2.63	12.13	11.67	45.7
SBS/DCRMB-TFOT	0.33	6.75	5.72	14.0
SBSMB-PAV	4.87	32.67	56.63	80.6
SBS/DCRMB-PAV	2.33	20.47	28.4	63.0

According to the above definition of the aging index, the sensitivity of the SBSMB and SBS/DCRMB to aging in different aspects of various temperature rheological performance and chemical composition are listed in Table 4. It can be seen that aging sensitivity of SBSMB and SBS/DCRMB is different in the total temperature domain, but all aging indexes of SBSMB are larger than that of SBS/DCRMB in both short-aging and long-aging at the macro and micro levels, SBSMB is approximately twice that of SBS/DCRMB. It indicated that SBS/DCRMB possesses a better anti-aging ability due to absorption of lightweight components by composite modifiers (SBS and DCR) and release of carbon black from DCR.

4. Conclusions

The objective of this study was to evaluate the modification mechanism of SBS/DCRMB and investigate the evolution of the resulting rheological properties before and after TFOT and PAV aging by using fluorescence microscopy, GPC, DSR, and BBR. These experimental methods provide multi-scale data identifying the relationship between microstructure and macroscopic rheological properties. The specific benefits of these testing methods are listed in the following.

(1) The fluorescence micrographs confirmed that the DCR and SBS modifiers in SBS/DCRMB had been vulcanized and produced a three-dimensional network structure under the action of sulfur. Moreover, using GPC to analyze the molecular distribution of SBS/DCRMB shows only two regions (residual asphalt and polymer), which further indicates that a crosslinking reaction occurred between the polymers of SBS/DCRMB.

(2) GPC testing results indicated that the introduction of DCR in SBS/DCRMB significantly improved the thermal oxygen anti-aging ability, and the results of the LAS and MSCR test also indicated the same pattern.
(3) The MSCR testing results illustrated that DCR could effectively improve the elastic properties and decrease the viscous parts of asphalt. Furthermore, the three-dimensional polymer network structure formed by DCR and SBS plays a significant role in the high-temperature properties.
(4) DCR enhanced the resistance to fatigue cracking of SBS/DCRMB and combination with the GPC test results, shows that the influence of aging on the fatigue life of SBSMB was mainly the result of the hardening of residual asphalt. In contrast, the fatigue performance loss of SBS/DCRMB was primarily caused by the condensation of the polymer part and the disintegration of the polymer network structure.
(5) According to the stiffness modulus and creep rate results of the BBR test, the temperature sensitivity and anti-aging properties of SBS/DCRMB were significantly better than those of SBSMB.

Author Contributions: Conceptualization, X.C. and G.X.; methodology, G.X.; validation, G.X., Y.Y. and P.K.; formal analysis, J.Y.; investigation, J.Y.; resources, X.C.; data curation, T.W.; writing—original draft preparation, P.K.; writing—review and editing, G.X.; visualization, Y.Y.; supervision, X.C.; project administration, X.C.; funding acquisition, X.C. All authors have read and agreed to the published version of the manuscript.

Funding: This research was funded by National Natural Science Foundation of China (No. 51778136 and No. 51778140, Beijing, China), Technology Research and Development Program of China State Railway Group Co., Ltd. (P2019G030, Beijing, China). And the APC was funded by Science Foundation of China (No. 51778136 Beijing, China). The authors also wish to thank the reviewers for their valuable comments and suggestions concerning this manuscript.

Institutional Review Board Statement: Not applicable.

Informed Consent Statement: Not applicable.

Conflicts of Interest: The authors declare no conflict of interest.

References

1. Yilmaz, M.; Yalcin, E. The effects of using different bitumen modifiers and hydrated lime together on the properties of hot mix asphalts. *Road Mater. Pavement Des.* **2016**, *17*, 499–511. [CrossRef]
2. Chen, J.S.; Liao, M.C.; Shiah, M.S. Asphalt modified by styrene-butadiene-styrene triblock copolymer: Morphology and model. *J. Mater. Civ. Eng.* **2002**, *14*, 224–229. [CrossRef]
3. Wu, S.-p.; Pang, L.; Mo, L.-t.; Chen, Y.-c.; Zhu, G.-j. Influence of aging on the evolution of structure, morphology and rheology of base and SBS modified bitumen. *Constr. Build. Mater.* **2009**, *23*, 1005–1010. [CrossRef]
4. Dong, F.; Fan, W.; Yang, G.; Wei, J.; Luo, H.; Wu, M.; Zhang, Y. Dispersion of SBS and its Influence on the Performance of SBS Modified Asphalt. *J. Test. Eval.* **2014**, *42*, 1073–1080. [CrossRef]
5. Dong, Z.J.; Zhou, T.; Luan, H.; Williams, R.C.; Wang, P.; Leng, Z. Composite modification mechanism of blended bio-asphalt combining styrene-butadiene-styrene with crumb rubber: A sustainable and environmental-friendly solution for wastes. *J. Clean Prod.* **2019**, *214*, 593–605. [CrossRef]
6. Lee, S.-J.; Akisetty, C.K.; Amirkhanian, S.N. The effect of crumb rubber modifier (CRM) on the performance properties of rubberized binders in HMA pavements. *Constr. Build. Mater.* **2008**, *22*, 1368–1376. [CrossRef]
7. Wang, H.; Liu, X.; Apostolidis, P.; Scarpas, T. Review of warm mix rubberized asphalt concrete: Towards a sustainable paving technology. *J. Clean Prod.* **2018**, *177*, 302–314. [CrossRef]
8. Lo Presti, D. Recycled Tyre Rubber Modified Bitumens for road asphalt mixtures: A literature review. *Constr. Build. Mater.* **2013**, *49*, 863–881. [CrossRef]
9. Lundy, J.R.; Hicks, R.G.; Zhou, H. Ground Rubber Tires in Asphalt-Concrete Mixtures—Three Case Histories. In *Use of Waste Materials in Hot-Mix Asphalt*; Waller, H.F., Ed.; ASTM International: West Conshohocken, PA, USA, 1993; Volume 1193, pp. 262–275.
10. Mull, M.A.; Stuart, K.; Yehia, A. Fracture resistance characterization of chemically modified crumb rubber asphalt pavement. *J. Mater. Sci.* **2002**, *37*, 557–566. [CrossRef]
11. Ye, Z.G.; Zhang, Y.Z.; Kong, X.M. Modification of bitumen with desulfurized crumb rubber in the present of reactive additives. *J. Wuhan Univ. Technol. Mater. Sci. Ed.* **2005**, *20*, 95–97.

12. He, L.; Ma, Y.; Huang, X.; Ma, T. Research on Performance and Microstructure of Desulfurized Rubber Asphalt. *J. Build. Mater.* **2012**, *15*, 227–231.
13. Subhy, A.; Lo Presti, D.; Airey, G. An investigation on using pre-treated tyre rubber as a replacement of synthetic polymers for bitumen modification. *Road Mater. Pavement Des.* **2015**, *16*, 245–264. [CrossRef]
14. Singh, B.; Kumar, L.; Gupta, M.; Chauhan, M.; Chauhan, G.S. Effect of activated crumb rubber on the properties of crumb rubber-modified bitumen. *J. Appl. Polym. Sci.* **2013**, *129*, 2821–2831. [CrossRef]
15. Liang, M.; Xin, X.; Fan, W.; Ren, S.; Shi, J.; Luo, H. Thermo-stability and aging performance of modified asphalt with crumb rubber activated by microwave and TOR. *Mater. Des.* **2017**, *127*, 84–96. [CrossRef]
16. Li, B.; Zhou, J.; Zhang, Z.; Yang, X.; Wu, Y. Effect of Short-Term Aging on Asphalt Modified Using Microwave Activation Crumb Rubber. *Mater* **2019**, *12*, 1039. [CrossRef] [PubMed]
17. Cao, D.; Zhang, H. Study on the Mechanism and Properties of Activated Crumb Rubber Modified Asphalt. *Advan Mater Resear.* **2011**, *239*, 191–196. [CrossRef]
18. Wang, H.; You, Z.; Mills-Beale, J.; Hao, P. Laboratory evaluation on high temperature viscosity and low temperature stiffness of asphalt binder with high percent scrap tire rubber. *Constr. Build. Mater.* **2012**, *26*, 583–590. [CrossRef]
19. Navarro, F.J.; Partal, P.; Martinez-Boza, F.; Gallegos, C. Influence of crumb rubber concentration on the rheological behavior of a crumb rubber modified bitumen. *Energy Fuels* **2005**, *19*, 1984–1990. [CrossRef]
20. Nejad, F.M.; Aghajani, P.; Modarres, A.; Firoozifar, H. Investigating the properties of crumb rubber modified bitumen using classic and SHRP testing methods. *Constr. Build. Mater.* **2012**, *26*, 481–489. [CrossRef]
21. Liang, M.; Ren, S.S.; Sun, C.J.; Zhang, J.Z.; Jiang, H.G.; Yao, Z.Y. Extruded Tire Crumb-Rubber Recycled Polyethylene Melt Blend as Asphalt Composite Additive for Enhancing the Performance of Binder. *J. Mater. Civ. Eng.* **2020**, *32*, 10. [CrossRef]
22. Yidirim, Y. Polymer modified asphalt binders. *Constr. Build. Mater.* **2007**, *21*, 66–72. [CrossRef]
23. Kodrat, I.; Sohn, D.; Hesp, S.A.M. Comparison of polyphosphoric acid-modified asphalt binders with straight and polymer-modified materials. *Transp. Res. Rec.* **2007**, *1998*, 47–55. [CrossRef]
24. Tan, Y.Q.; Guo, M.; Cao, L.P.; Zhang, L. Performance optimization of composite modified asphalt sealant based on rheological behavior. *Constr. Build. Mater.* **2013**, *47*, 799–805. [CrossRef]
25. Xiang, L.; Cheng, J.; Kang, S. Thermal oxidative aging mechanism of crumb rubber/SBS composite modified asphalt. *Constr. Build. Mater.* **2015**, *75*, 169–175. [CrossRef]
26. Javier Lopez-Moro, F.; Candelas Moro, M.; Hernandez-Olivares, F.; Witoszek-Schultz, B.; Alonso-Fernandez, M. Microscopic analysis of the interaction between crumb rubber and bitumen in asphalt mixtures using the dry process. *Constr. Build. Mater.* **2013**, *48*, 691–699. [CrossRef]
27. Zhang, F.; Hu, C. The research for structural characteristics and modification mechanism of crumb rubber compound modified asphalts. *Constr. Build. Mater.* **2015**, *76*, 330–342. [CrossRef]
28. Fu, Q.; Xu, G.; Chen, X.; Zhou, J.; Sun, F. Rheological properties of SBS/CR-C composite modified asphalt binders in different aging conditions. *Constr. Build. Mater.* **2019**, *215*, 1–8. [CrossRef]
29. Wang, S.; Zhao, X.; Wang, Q. Rheological and Structural Evolution of Rubberized Asphalts under Weathering. *J. Mater. Civ. Eng.* **2017**, *29*, 04017180. [CrossRef]
30. ASTM D5/D5M-20. *Standard Test Method for Penetration of Bituminous Materials*; ASTM International: West Conshohocken, PA, USA, 2020; Available online: www.astm.org (accessed on 5 September 2021).
31. ASTM D36-06. *Standard Test Method for Softening Point of Bitumen (Ring-and-Ball Apparatus)*; ASTM International: West Conshohocken, PA, USA, 2006; Available online: www.astm.org (accessed on 5 September 2021).
32. ASTM D113-17. *Standard Test Method for Ductility of Asphalt Materials*; ASTM International: West Conshohocken, PA, USA, 2017; Available online: www.astm.org (accessed on 5 September 2021).
33. ASTM D2872-97. *Standard Test Method for Effect of Heat and Air on a Moving Film of Asphalt (Rolling Thin-Film Oven Test)*; ASTM International: West Conshohocken, PA, USA, 1997; Available online: www.astm.org (accessed on 5 September 2021).
34. ASTM D92-16. *Standard Test Method for Flash and Fire Points by Cleveland Open Cup Tester*; ASTM International: West Conshohocken, PA, USA, 2016; Available online: www.astm.org (accessed on 5 September 2021).
35. ASTM D1754/D1754M-09. *Standard Test Method for Effect of Heat and Air on Asphaltic Materials (Thin-Film Oven Test)*; ASTM International: West Conshohocken, PA, USA, 2009; Available online: www.astm.org (accessed on 5 September 2021).
36. ASTM D6531-00. *Standard Test Method for Relative Tinting Strength of Aqueous Ink Systems by Instrumental Measurement*; ASTM International: West Conshohocken, PA, USA, 2019; Available online: www.astm.org (accessed on 5 September 2021).
37. Buisine, J.; Buisine, J.M.; Joly, G.; Eladlani, A.; Such, C.; FARCAS, F.; Ramond, G.; Claudy, P.; Letoffe, J.M.; King, G.N.; et al. Thermodynamic behavior and physicochemical analysis of eight SHRP bitumens. *Transp. Res. Rec.* **1993**, *1386*, 1–9.
38. Daly, W.H.; Qui Chiu, Z.; Negulescu, I. Preparation and characterization of asphalt-modified polyethylene blends. *Transp. Res. Rec.* **1993**, *1391*, 56.
39. D'Angelo, J.; Dongre, R. Practical Use of Multiple Stress Creep and Recovery Test Characterization of Styrene-Butadiene-Styrene Dispersion and Other Additives in Polymer-Modified Asphalt Binders. *Transp. Res. Rec.* **2009**, *2126*, 73–82. [CrossRef]
40. Hao, G.R.; Huang, W.D.; Yuan, J.; Tang, N.P.; Xiao, F.P. Effect of aging on chemical and rheological properties of SBS modified asphalt with different compositions. *Constr. Build. Mater.* **2017**, *156*, 902–910. [CrossRef]

41. Themeli, A.; Chailleux, E.; Farcas, F.; Chazallon, C.; Migault, B. Molecular weight distribution of asphaltic paving binders from phase-angle measurements. *Road Mater. Pavement Des.* **2015**, *16*, 228–244. [CrossRef]
42. Wang, H.; Liu, X.; Apostolidis, P.; Scarpas, T. Rheological Behavior and Its Chemical Interpretation of Crumb Rubber Modified Asphalt Containing Warm-Mix Additives. *Transp. Res. Rec.* **2018**, *2672*, 337–348. [CrossRef]
43. AASHTO T350. *Multiple Stress Creep Recovery(MSCR) Test of Asphalt Binder Using a Dynamic Shear Rheometer (DSR)*; AASHTO: Washington, DC, USA, 2016.
44. AASHTO TP101. *Standard Method of Test for Estimating Fatigue Resistance of Asphalt Binders Using the Linear Amplitude Sweep*; AASHTO: Washington, DC, USA, 2012.
45. ASTM D7643-16. *Standard Practice for Determining the Continuous Grading Temperatures and Continuous Grades for PG Graded Asphalt Binders*; ASTM International: West Conshohocken, PA, USA, 2016; Available online: www.astm.org (accessed on 5 September 2021).
46. Wekumbura, C.; Stastna, J.; Zanzotto, L. Destruction and recovery of internal structure in polymer-modified asphalts. *J. Mater. Civ. Eng.* **2007**, *19*, 227–232. [CrossRef]
47. Golalipour, A.; Bahia, H.U.; Tabatabaee, H.A. Critical Considerations toward Better Implementation of the Multiple Stress Creep and Recovery Test. *J. Mater. Civ. Eng.* **2017**, *29*, 7. [CrossRef]
48. Zhou, Z.; Gu, X.Y.; Dong, Q.; Ni, F.J.; Jiang, Y.X. Rutting and fatigue cracking performance of SBS-RAP blended binders with a rejuvenator. *Constr. Build. Mater.* **2019**, *203*, 294–303. [CrossRef]
49. Ren, S.S.; Liu, X.Y.; Fan, W.Y.; Qian, C.D.; Nan, G.Z.; Erkens, S. Investigating the effects of waste oil and styrene-butadiene rubber on restoring and improving the viscoelastic, compatibility, and aging properties of aged asphalt. *Constr. Build. Mater.* **2021**, *269*, 19. [CrossRef]
50. Wang, R.; Xu, G.; Chen, X.H.; Zhou, W.B.; Zhang, H.Y. Evaluation of aging resistance for high-performance crumb tire rubber compound modified asphalt. *Constr. Build. Mater.* **2019**, *218*, 497–505. [CrossRef]

polymers

MDPI

Article

Experimental Investigation of the High-Temperature Rheological and Aging Resistance Properties of Activated Crumb Rubber Powder/SBS Composite-Modified Asphalt

Zhizhong Zhao [1,†], Longlin Wang [2,3,*,†], Wensheng Wang [4,*] and Xuanhao Shangguan [4]

1 Guangxi Hetian Expressway Co., Ltd., Nanning 530029, China; liuzyjlu2019@163.com
2 School of Civil Engineering, Southeast University, Nanjing 211189, China
3 Bridge Engineering Research Institute, Guangxi Transportation Science and Technology Group Co., Ltd., Nanning 530007, China
4 Department of Road and Bridge, College of Transportation, Jilin University, Changchun 130025, China; sgxh1719@mails.jlu.edu.cn
* Correspondence: wll955@163.com (L.W.); wangws@jlu.edu.cn (W.W.)
† These authors contributed equally to this work.

Abstract: Crumb rubber could form the active groups on the surface by interrupting the crosslinking bond to improve the compatibility with asphalt. While styrene-butadiene-styrene block copolymer (SBS)-modified asphalt has excellent comprehensive properties, it has poor anti-aging performance and a high cost. To explore the influence of composite modification of activated crumb rubber powder (ACR) and SBS on asphalt, modified asphalt samples with different modifiers and SBS contents were prepared. Conventional physical properties tests, a dynamic shear rheometer (DSR), and the thin-film oven test (TFOT) were used to study the conventional physical properties, high-temperature rheological properties, and aging resistance of asphalt. In addition, the action forms and distribution of modifiers in asphalt were observed by an optical microscope to characterize the micro-morphology of ACR/SBS composite-modified asphalt. Test results showed that after adding SBS, the softening point, ductility, and elastic recovery of ACR/SBS asphalt could be significantly improved, but the viscosity and softening point difference were also larger. At the same time, according to the complex shear modulus, phase angle, and rutting factor, SBS can effectively improve the high-temperature deformation resistance of ACR/SBS asphalt. The modified asphalt (ACR/SBS-2) had good high- and low- temperature performances, as well as an appropriate viscosity and low softening point difference, as a research object of aging. After short-term aging, the changes in the high- and low-temperature performances and workability of ACR/SBS asphalt were reduced. Taking the softening point as the target performance, the softening point of ACR/SBS asphalt was less affected by aging time and temperature, indicating that ACR/SBS asphalt was not sensitive to aging temperature and had good stability and aging resistance. From the micrograph by microscope, it was found that ACR/SBS asphalt could maintain a relatively stable polyphase structure for aging resistance.

Keywords: asphalt; activated crumb rubber; SBS; short-term aging; physical properties; rheological property

Citation: Zhao, Z.; Wang, L.; Wang, W.; Shangguan, X. Experimental Investigation of the High-Temperature Rheological and Aging Resistance Properties of Activated Crumb Rubber Powder/SBS Composite-Modified Asphalt. *Polymers* **2022**, *14*, 1905. https://doi.org/10.3390/polym14091905

Academic Editor: Enzo Martinelli

Received: 7 April 2022
Accepted: 5 May 2022
Published: 6 May 2022

1. Introduction

With the rapid development of new solid waste utilization technology in the road industry, renewable materials-modified asphalts represented by waste crumb rubber powder have been widely used in the field of civil engineering [1–3]. Scholars have systematically studied the modification process, road performance, and application of waste crumb rubber powder-modified asphalt in hot mix asphalt mixtures [4–8]. In order to further improve the utilization efficiency of waste crumb rubber and improve the road performance of crumb rubber-modified asphalt, many scholars have studied the activation process of crumb rubber powder.

The commonly used crumb rubber powder activation methods include physical activation methods such as the microwave method based on the thermal environment and ultrasonic method [9], chemical activation methods using chemical adjuvants [9], and biological methods [10]. In addition to the above activation methods, the mechanochemical methods using chemical adjuvants combined with mechanical force are also applied to the activation of crumb rubber powder [11]. The mechanochemical method can effectively play the role of desulfurization activation of additives through rubber broken to increase the surface roughness and activity of rubber powder. The mechanochemical method has the advantages of simple equipment, high production efficiency, low cost, and weak smell, and is very suitable for large factory production, which was chosen in this study. Many researchers believe that the storage stability and workability of crumb rubber-modified asphalt prepared from rubber powder significantly improve after desulfurization activation [12,13]. Juganaru et al. observed that according to the micro-morphology of activated crumb rubber powder (ACR)-modified asphalt, the ACR was more evenly distributed in asphalt, and the viscosity of ACR-modified asphalt was significantly reduced [14]. Shatanawi et al. pointed out that the water stability and rutting resistance of the ACR-modified asphalt mixture were better than those of the ordinary crumb rubber-modified asphalt mixture [15]. Liu et al. used a wet process to prepare crumb rubber-modified asphalt with different ACR contents and optimized the preparation parameters of ACR based on viscosity and the softening point difference. The asphalt mixture containing a 60% content of ACR possessed optimal comprehensive properties in terms of high-temperature rutting resistance, cracking resistance, and moisture resistance [16]. Chen et al. studied a new treatment method of crumb rubber impact on the high-temperature rheological performances and found that ACR-modified asphalt had better rheological performances at high temperatures and short-term aging performances [17]. The above studies showed that ACR can improve some properties of crumb rubber-modified asphalt, but ACR-modified asphalt still has problems such as difficult construction and poor aging resistance.

As a widely used asphalt modifier, styrene-butadiene-styrene block copolymer (SBS) can be used with crumb rubber powder to combine the advantages of two kinds of asphalt modifier through composite modification to meet the higher performance requirements of asphalt pavement [18–21]. Zhang et al. found that adding SBS with the content of 3% (mass fraction) into crumb rubber-modified asphalt with a rubber content of 15% can improve the high- and low-temperature performance, and the toughness of composite-modified asphalt could be enhanced. Adding sulfur with the content of 0.2% to moderate the vulcanization of composite-modified asphalt can further improve its storage stability and flexibility [19]. Huang et al. studied the low-temperature properties of soluble rubber powder/SBS composite-modified asphalt by the low-temperature ductility test and bending beam rheological test. The results showed that when the SBS content is 2%, the low-temperature plastic deformation capacity and low-temperature rheological properties of composite-modified asphalt are improved. However, the higher addition of SBS or crumb rubber may weaken the improvement effect of low-temperature properties of crumb rubber-modified asphalt. When the SBS content is too high, the aromatic content in asphalt is excessively reduced, resulting in the decline in the low-temperature rheological properties of composite-modified asphalt [22,23]. Liang et al. prepared crumb rubber powder/SBS composite-modified asphalt by the high-speed shear process, and studied its linear viscoelasticity and storage stability. The results showed that compared with base asphalt, adding SBS and crumb rubber powder to asphalt can significantly improve its viscoelasticity and viscosity. When the SBS content exceeds 1%, it is feasible to use crumb rubber powder instead of some SBS as a modifier, and its modulus and viscoelasticity are significantly improved. However, the phase separation of composite-modified asphalt will occur during storage, and the storage stability still needs to be further improved [16]. Therefore, under the compound modification of crumb rubber powder and SBS, the high-temperature stability and aging resistance of composite-modified asphalt are significantly improved compared with SBS-modified asphalt. The relevant indexes of low-temperature crack resistance, viscosity toughness, and storage stability of composite-modified asphalt are higher than those of ordinary crumb rubber asphalt.

Crumb rubber powder/SBS composite-modified asphalt has been proven to improve the road performance of crumb rubber-modified asphalt and have good aging resistance. However, due to the change in physical and chemical properties before and after rubber powder activation, it is different from original crumb-rubber modified asphalt in the modification system, which will have an impact on the composite modification technology. Therefore, using ACR prepared by the mechanochemical method instead of original crumb rubber powder as a modifier, this study prepared modified asphalt samples with different modifiers and SBS contents. Conventional physical properties tests, a dynamic shear rheometer (DSR), and the thin-film oven test (TFOT) were used to study the conventional physical properties, high-temperature rheological properties, and aging resistance of asphalt. In addition, the action forms and distribution of modifiers in asphalt were observed by an optical microscope to characterize the micro-morphology of ACR/SBS composite-modified asphalt. The innovation of this study was to compare and analyze the influence of crumb rubber powder activation and the effect of SBS on crumb rubber powder-modified asphalt.

2. Materials and Methods

2.1. Experimental Raw Materials

The raw materials in this study included Maoming 70# road petroleum asphalt, ACR, and SBS, in which the ACR was obtained from crumb rubber powder with a size of 30 mesh by the mechanochemical method using a chemical adjuvant (organic disulfide, i.e., OD) combined with the mechanical force to accelerate the chemical reaction. Based on the previous studies [11,24,25], the activation process parameters of the mechanochemical method using the OD adjuvant (3% of crumb rubber powder by mass) were optimized as follows: the OD adjuvant content was 3%, the mixing temperature was 160 °C, and the mixing time was 30 min. A commercially available linear SBS was selected, and its main physical indexes are shown in Table 1.

Table 1. The basic physical properties of SBS.

Structure Type	Block Ratio	Tensile Rate (%)	Tensile Strength (MPa)	Melt Index (g/10 min)	Density (g/cm^3)
Linear	30/70	800	25	0.6	0.93

2.2. Preparation of ACR/SBS Composite-Modified Asphalt

Based on the existing literature [11], the asphalt modified by untreated or activated crumb rubber powder (25% of base asphalt by mass) could be prepared, which are labeled as UCR asphalt or ACR asphalt, respectively. According to the previous process exploration, the following steps were adopted to prepare ACR/SBS composite-modified asphalt (as shown in Figure 1):

1. Place the base asphalt in an oven at 140 °C and heat it to the flowing state, and slowly add the weighed SBS (1%, 2%, 3%, and 3.5% of base asphalt by mass) with a mixing speed of 300 r/min. During the addition process, raise the temperature to 170 °C and maintain it for 20 min.
2. Put the mixed asphalt in an oven at 175 °C and swell for 30 min.
3. Place the swelled SBS asphalt into high-speed shear equipment, gradually increase the shear rate to about 4500 r/min, control the temperature at 170–180 °C, and then take it out after shearing for 1 h.
4. Put the modified asphalt in an oven at 175 °C and develop for 40 min.
5. Heat the developed SBS asphalt to the flowing state and slowly add the weighed ACR (25% of base asphalt by mass) with a mixing speed of 300 r/min. During the addition process, raise the temperature to 175 °C and maintain it for 30 min.

6. Place the blended asphalt into high-speed shear equipment, gradually increase the shear rate to about 4500 r/min, control the temperature at 185 °C, and then take it out after shearing for 1 h.
7. Put the prepared ACR/SBS asphalt into an oven at 175 °C for 30 min.

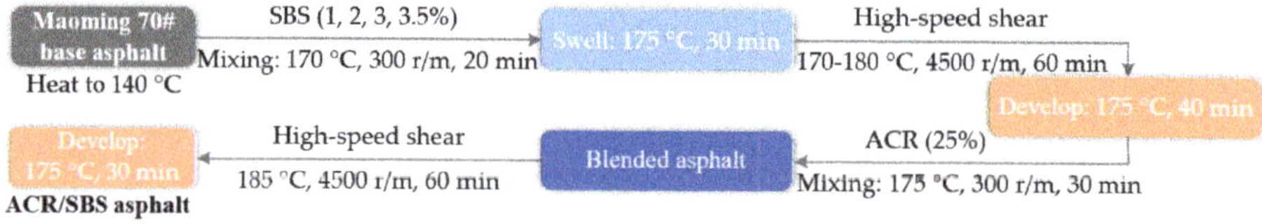

Figure 1. The preparation process of ACR/SBS-modified asphalt.

2.3. Experimental Scheme and Methods

The asphalt samples were first prepared including seven asphalt codes, i.e., base, UCR, ACR, and ACR/SBS-1/2/3/3.5. Then, the asphalt samples could be used for the conventional physical properties test, high-temperature rheological test, thin-film oven test, and micro-characterization test including a stereoscopic microscope and Fourier infrared spectroscopy. The flow chart of this study is shown in Figure 2.

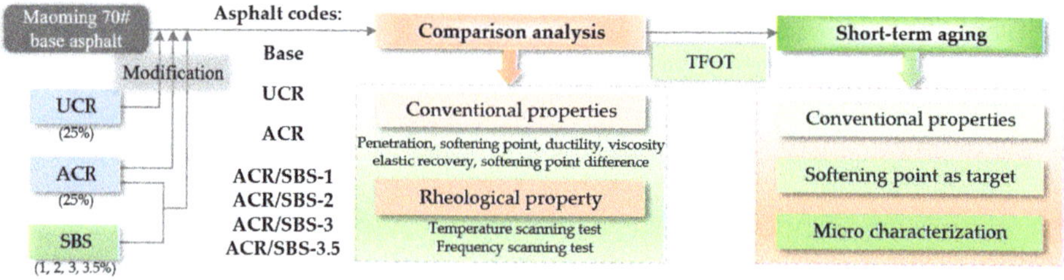

Figure 2. The flow chart of this study.

2.3.1. Conventional Physical Properties Test

In this study, according to the Chinese standard "Standard Test Methods of Bitumen and Bituminous Mixtures for Highway Engineering" (JTG E20-2011) and previous studies [26,27], the penetration, softening point, ductility at 5 °C, viscosity, elastic recovery, and softening point difference (Δ) were tested for base asphalt, UCR asphalt, ACR asphalt, and ACR/SBS-1/2/3/3.5 asphalt.

2.3.2. High-Temperature Rheological Test

The high-temperature rheological test of base asphalt, UCR asphalt, ACR asphalt, and ACR/SBS-1/2/3/3.5 asphalt was carried out by using a dynamic shear rheometer (DSR) [28]. In this study, the temperature scanning test in the temperature range of 52–82 °C was performed under the strain control mode with the strain value of γ = 12% and frequency of ω = 10 rad/s, and the frequency scanning test in the frequency range of 0.1 rad/s~100 rad/s was carried out under the strain control mode with the strain value of γ = 2% and temperature of 60 °C. By comparing and analyzing the rutting factor, complex shear modulus, and phase angle, the high-temperature properties of composite-modified asphalt could be characterized for different asphalt samples. Temperature scanning at the same frequency and different temperatures, and frequency scanning at the same temperature and different frequencies were conducted to explore the changes in asphalt viscoelastic parameters under two modes for the high-temperature performance from the perspective of temperature dependence and time dependence, respectively.

2.3.3. Thin-Film Oven Test

In this study, the thin-film oven test (TFOT) was used to simulate the short-term aging behavior of asphalt in the process of storage, transportation, mixing, and paving. The prepared UCR asphalt, ACR asphalt, and ACR/SBS-2 asphalt were insulated for 5 h in a film oven at 163 °C for the short-term aging test. Then, the physical properties of asphalt samples before and after aging were tested to characterize the corresponding aging resistance by comparison and analysis.

In addition, the prepared UCR asphalt, ACR asphalt, and ACR/SBS-2 asphalt were insulated for different aging durations (5 h, 10 h, 15 h, and 20 h) in a film oven at different aging temperatures (150 °C, 163 °C, and 180 °C) for the short-term aging test. By comparing the variation in softening point of modified asphalt before and after aging, the effects of short-term aging factors (i.e., aging time and temperature) on the aging resistance of asphalt were studied with the softening point as the target performance.

3. Results and Discussion

3.1. *Effect of ACR/SBS on the Conventional Physical Properties of Asphalt*

According to the above experimental scheme, the prepared base asphalt, UCR asphalt, ACR asphalt, and ACR/SBS-1/2/3/3.5 asphalt were tested to obtain the conventional physical properties, and the comparison results are shown in Figure 3.

From Figure 3a, it can be seen that the penetration of asphalt decreases by adding crumb rubber powder and SBS, and the penetration of ACR/SBS asphalt decreases with the increase in SBS content. Except for the SBS content being 3.5% (i.e., ACR/SBS-3.5 asphalt), the penetration of ACR/SBS asphalt is greater than that of UCR asphalt. In Figure 3b, the values of the softening point of asphalt modified by crumb rubber powder and SBS increase in comparison with base asphalt. It could also be found that with the increase in SBS content, the softening point of ACR/SBS asphalt gradually increases, and the softening point of ACR/SBS asphalt is higher than that of crumb rubber-modified asphalt.

In Figure 3c, it can be seen that with the addition of crumb rubber and SBS, the ductility of modified asphalt gradually increases, and the ductility of ACR/SBS asphalt increases with the SBS content. When the SBS content is 3.5%, the ductility of ACR/SBS-3.5 asphalt reaches more than 15 cm. As shown in Figure 3d, with the increase in SBS content, the viscosity of ACR/SBS asphalt increases gradually. When the SBS content is 2.0%, the viscosity of ACR/SBS-2 asphalt is basically the same as that of UCR asphalt.

It can be seen from Figure 3e that with the increase in SBS content, the elastic recovery rate of ACR/SBS asphalt gradually increases. When the SBS content is 3.5%, the elastic recovery rate of ACR/SBS-3.5 asphalt can reach 96%. At the same time, in Figure 3f, with the increase in SBS content, the softening point difference (Δ) of ACR/SBS asphalt also increases gradually, but it still meets the specification requirements. When the SBS content is less than 3%, the softening point difference (Δ) of ACR/SBS asphalt is less than that of UCR asphalt. According to the softening point difference (Δ) of UCR, it can be seen that one of the risks of using crumb rubber is the poor storage stability. Compared with UCR, the storage stability of ACR or ACR/SBS composite-modified asphalt improves due to the smaller softening point difference (Δ). However, when the SBS content is too high, the storage stability of ACR/SBS asphalt becomes worse.

The reason for the above phenomena is that the modification process of asphalt containing crumb rubber powder and SBS is mainly physical blending, which is also consistent with previous studies [13,29,30]. The addition of crumb rubber powder or SBS is equivalent to the increase in the reinforcing phase in the modified asphalt system. The performance variation trend of ACR/SBS asphalt is equivalent to the superposition of the performance trends of ACR-modified asphalt and SBS-modified asphalt. Thus, when the SBS content is less than 3%, the softening point and ductility of ACR/SBS asphalt improve at the same time, and the viscosity and penetration are more appropriate, enhancing the high-temperature performance. In addition, the elastic recovery performance of modified asphalt is also enhanced. However, when the SBS content is too high, the storage stability of modified asphalt becomes worse. Therefore, it is necessary to control the contents of ACR and SBS to reduce the segregation of modified asphalt.

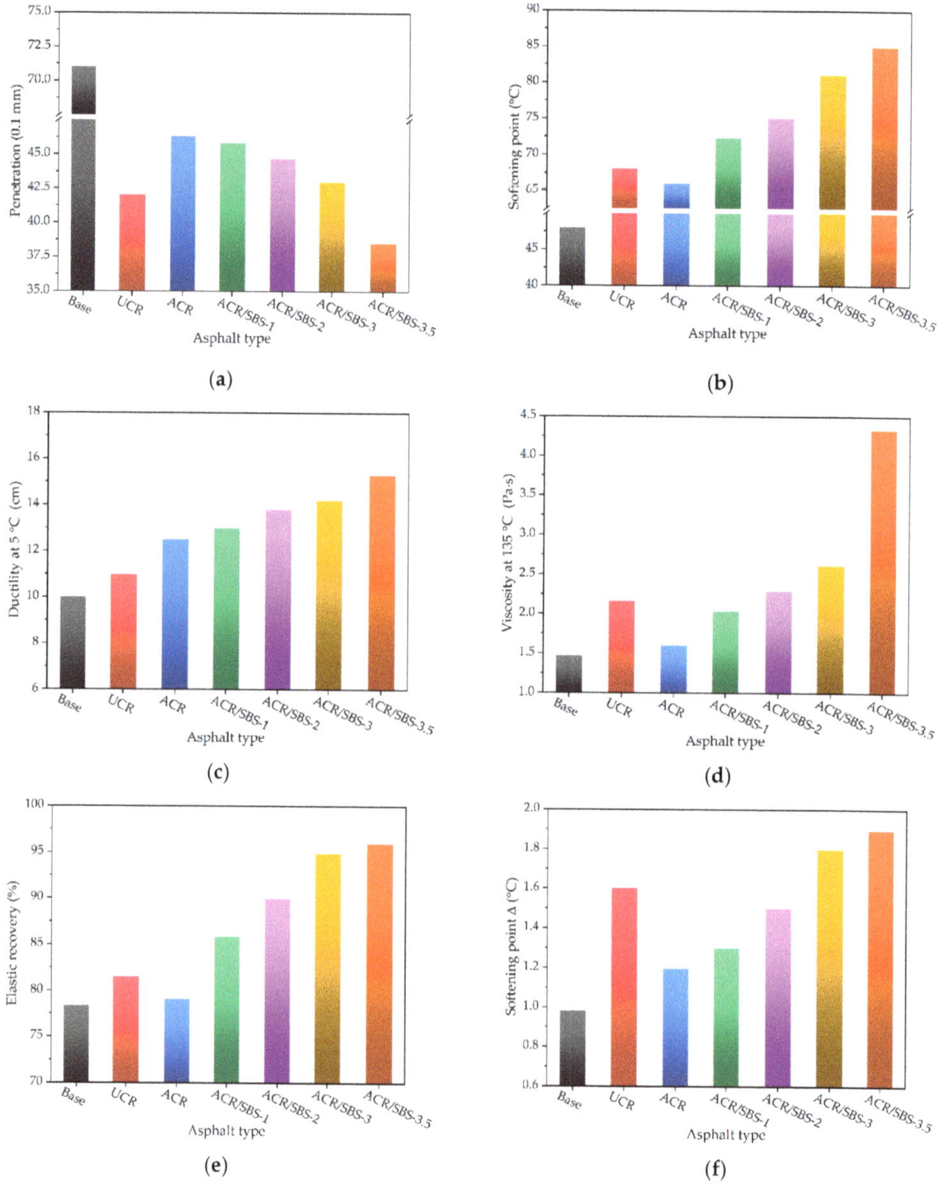

Figure 3. The conventional physical properties of different asphalt types: (**a**) penetration; (**b**) softening point; (**c**) ductility at 5 °C; (**d**) viscosity at 135 °C; (**e**) elastic recovery; (**f**) softening point Δ.

3.2. Effect of ACR/SBS on the High-Temperature Rheological Properties of Asphalt

3.2.1. Temperature Scanning Test

The temperature scanning test was performed in the temperature range of 52–82 °C under the strain control mode with the strain value of γ = 12% and frequency of ω = 10 rad/s, and the comparison results of complex shear modulus and phase angle are shown in

Figure 4. The complex shear modulus G^* is used to characterize the deformation resistance, and the phase angle δ indicates the proportion and corresponding influence degree of elastic and viscous components of materials.

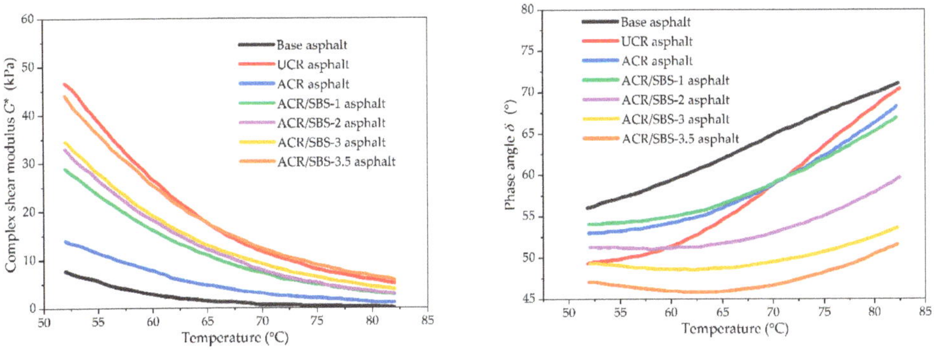

(**a**) (**b**)

Figure 4. The temperature scanning test results of different asphalt types: (**a**) complex shear modulus; (**b**) phase angle.

Figure 4a shows the relationship between the complex shear modulus of asphalt and temperature. It can be seen from Figure 4a that for the same asphalt type, the higher the test temperature, the smaller the complex shear modulus of asphalt. The complex shear modulus of asphalt modified by ACR or ACR/SBS is significantly higher than that of base asphalt, but lower than that of UCR asphalt. At the same temperature, with the increase in SBS content, the complex shear modulus of ACR/SBS asphalt gradually increases, but the difference between the complex shear modulus gradually decreases with the temperature. When the temperature is too high, the complex shear modulus of ACR/SBS asphalt is close to or higher than that of UCR asphalt. The addition of SBS greatly improves the complex shear modulus of ACR/SBS asphalt.

The relationship between the phase angle of asphalt and temperature is shown in Figure 4b. It can be seen that with the increase in temperature, the phase angle of ACR/SBS-2/3/3.5 asphalt decreases first and then increases. When the temperature is about 63 °C, the phase angle of these three ACR/SBS asphalt samples appears with the lowest value. However, the phase angle of other asphalt samples increases with the temperature. At the same temperature, the greater the SBS content, the smaller the phase angle of the corresponding ACR/SBS asphalt, and the better the high-temperature deformation resistance of ACR/SBS asphalt. When the test temperature reaches 80 °C, the phase angle of ACR/SBS asphalt is still less than 65°, which shows that the three types of asphalt have excellent high-temperature performance, and the presence of SBS has a significant effect on the phase angle of modified asphalt.

In general, the rheological properties of asphalt cannot be fully reflected only by a single basic index of rheological properties (i.e., G^* or δ). The rutting factor ($G^*/\sin\delta$) is a new index derived from the basic index of rheological properties, which can characterize the ability of asphalt materials to resist permanent deformation at high temperature. Figure 5 shows the relationship between the rutting factor and temperature for asphalt samples. It can be seen from Figure 5 that the variation trends of the rutting factor with temperature and SBS content are consistent with the complex shear modulus for different asphalt types. Compared with ACR asphalt, the addition of SBS greatly improves the rutting factor of ACR/SBS asphalt. As an elastomer, SBS will form a network structure in asphalt, limiting the movement of crumb rubber powder particles. Thus, SBS plays the role of fixing crumb rubber powder particles, to significantly improve the high-temperature stability of modified asphalt.

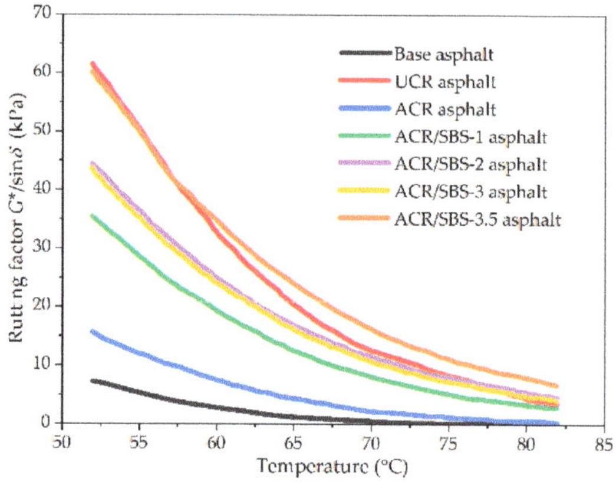

Figure 5. The rutting factor results of different asphalt types in the temperature scanning test.

3.2.2. Frequency Scanning Test

The frequency scanning test was carried out in the frequency range of 0.1 rad/s~100 rad/s under the strain control mode with the strain value of γ = 2% and temperature of 60 °C. The comparison results of complex shear modulus and phase angle for different asphalt samples in the frequency scanning test are shown in Figure 6.

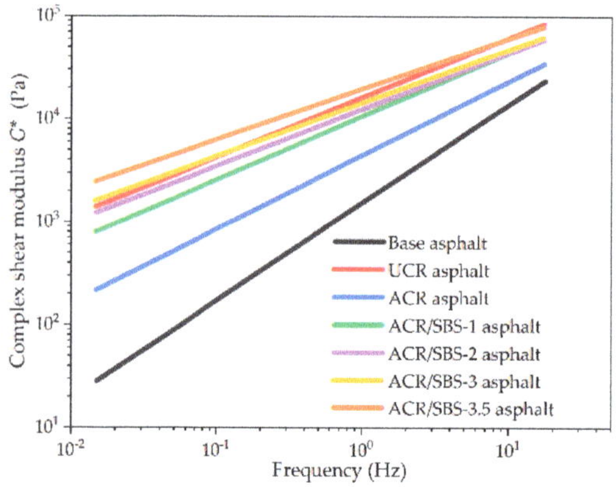

(**a**)

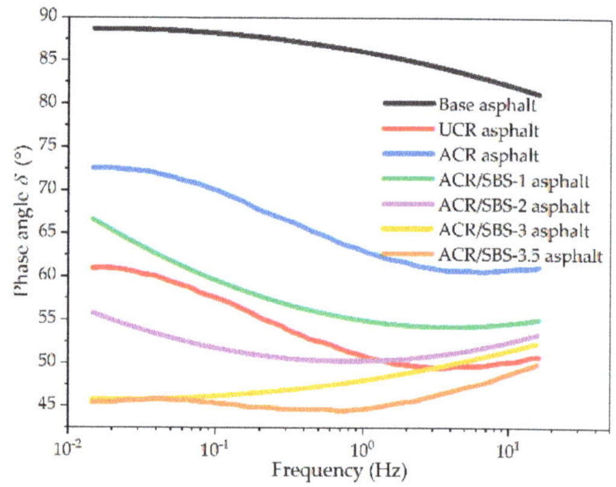

(**b**)

Figure 6. The frequency scanning test results of different asphalt types: (**a**) complex shear modulus; (**b**) phase angle.

In Figure 6a, there are linear relationships between the complex shear modulus and frequency for different asphalt samples in double logarithmic coordinates. It can be seen from Figure 6a that compared with ACR asphalt, the complex shear modulus of ACR/SBS asphalt is greatly improved, and the complex shear modulus of ACR/SBS asphalt increases with the frequency. At the same frequency, with the increase in SBS content, the complex shear modulus of ACR/SBS asphalt gradually increases. The complex shear modulus of asphalt modified by ACR or ACR/SBS is also significantly higher than that of base asphalt. When the SBS content increases, the complex shear modulus of ACR/SBS asphalt is close to or higher than that of UCR asphalt.

The relationship between the phase angle and frequency for different asphalt samples in single-logarithmic coordinates is shown in Figure 6b. It can be seen that when the temperature is 60 °C, the phase angle of different asphalt samples presents a certain plateau period under low-frequency loading. With the increase in frequency, the phase angle of base asphalt and asphalt-modified crumb rubber particles decreases; however, the phase angle of ACR/SBS composite-modified asphalt first decreases and then increases. At the same frequency, the greater the SBS content, the smaller the phase angle of the corresponding ACR/SBS asphalt, and the better the high-temperature deformation resistance of ACR/SBS asphalt. When the test frequency reaches 10 Hz, the phase angle of ACR/SBS asphalt is less than 60°, which shows that ACR/SBS asphalt has excellent high-temperature performance. As a typical elastomer, SBS can effectively fill the lack of elastic components caused by desulfurization activation, and the phase angle of ACR/SBS composite-modified asphalt is greatly reduced. Thus, the presence of SBS has a significant effect on the phase angle of modified asphalt.

The relationship results between the rutting factor and frequency for asphalt samples in the frequency scanning test are shown in Figure 7. As seen in Figure 7, similar to the complex shear modulus, the variation trend of the rutting factor with frequency and SBS content is consistent with that of the complex shear modulus for various asphalt types, and the variation relationship between the complex shear modulus and frequency is also linear in double logarithmic coordinates. The rutting factor of different asphalt samples increases with the frequency. The addition of SBS greatly improves the rutting factor of ACR/SBS asphalt compared with ACR asphalt, which could be beneficial to improve its high-temperature performance.

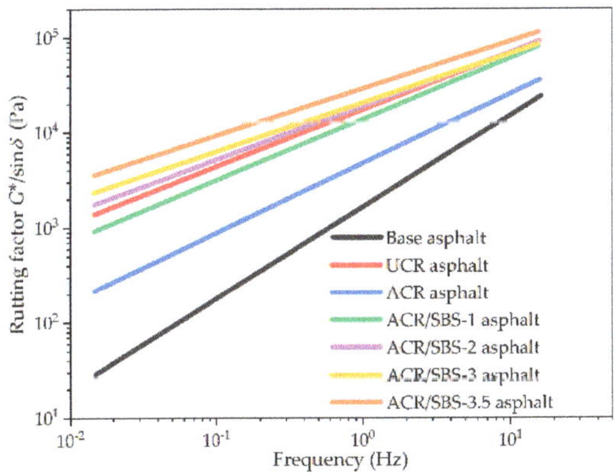

Figure 7. The rutting factor results of different asphalt types in the frequency scanning test.

3.3. *Effect of ACR/SBS on the Aging Resistance of Asphalt*

According to the comparison results including penetration, softening point, ductility at 5 °C, viscosity at 135 °C, elastic recovery, softening point Δ, and rutting factor in the radar chart (Figure 8), the addition of SBS improves the high- and low-temperature performances and elastic recovery performance of modified asphalt. However, when the SBS content is higher, the viscosity and storage stability of ACR/SBS asphalt become worse. Considering the indexes of high- and low-temperature performances, viscosity, and construction workability, ACR/SBS-2 asphalt has the best comprehensive performance, which not only has good high- and low-temperature performance, but also appropriate viscosity and low softening point difference. Then, the ACR (25% of base asphalt by mass) and SBS (2% of base asphalt by mass) composite-modified asphalt was selected as the research object used for the follow-up study.

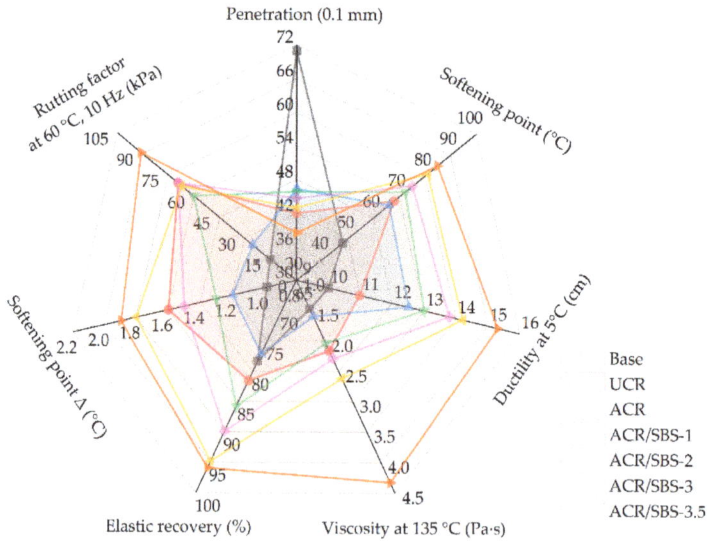

Figure 8. Radar chart based on conventional physical properties and high-temperature rheological properties of different asphalt types.

3.3.1. Conventional Physical Properties of Modified Asphalt under Short-Term Aging

According to the penetration test of modified asphalt (i.e., UCR asphalt, ACR asphalt, and ACR/SBS-2 asphalt) before and after short-term aging, the effect of short-term aging on the viscosity of modified asphalt was analyzed, as shown in Figure 9a. It can be seen from Figure 9a that the penetration of the three modified asphalt samples after the TFOT decreases, among which the penetration reduction of ACR/SBS-2 asphalt is the least, and the penetration reduction of ACR asphalt is the most. This may be because under the action of thermal oxygen aging, the light components in asphalt volatilize, oxidize, and harden the intermolecular structure, which leads to the hardening of asphalt. Thus, the addition of SBS can reduce the hardening degree of modified asphalt.

The softening point variation of modified asphalt before and after short-term aging is used to analyze the effect of short-term aging on the high-temperature stability of modified asphalt, and Figure 9b shows the softening point test results. It can be seen that the softening point of the three modified asphalt samples increases in varying degrees after the TFOT, among which the softening point of ACR asphalt increases the most, and the softening point of ACR/SBS asphalt increases the least. After thermal oxygen aging, the light components in asphalt volatilize, oxidize, and harden the intermolecular structure, leading to a larger softening point value of asphalt. However, the aging process of crumb rubber-modified asphalt is more complex, in which, in addition to the aging of asphalt and crumb rubber powder, the interaction between crumb rubber powder and asphalt in the process of thermal oxygen aging should also be considered. The particle structure of ordinary crumb rubber powder is complete, so the light components that enter the internal network structure are not easy to volatilize, indicating that the thermal stability of ordinary crumb rubber-modified asphalt (i.e., UCR asphalt) is good. After desulfurization and activation of crumb rubber powder, the internal network structure of crumb rubber powder is destroyed. In the process of short-term aging, the light components of asphalt are easier to volatilize, resulting in a large increase in the softening point of ACR asphalt. On the other hand, the addition of SBS into ACR asphalt can more fully absorb the light components and reconstruct the network structure. Therefore, the thermal stability of ACR/SBS asphalt significantly improves.

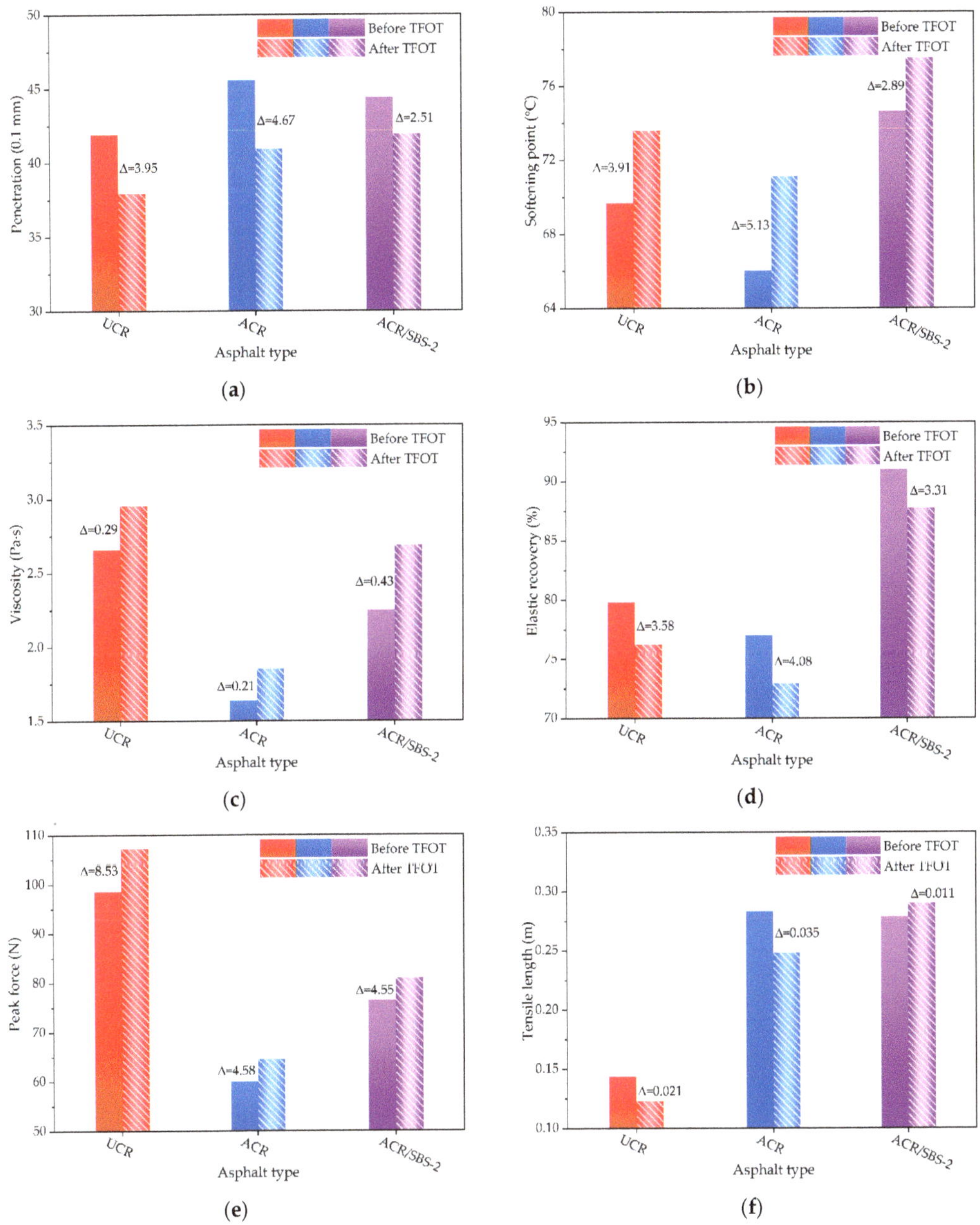

Figure 9. The conventional physical properties of different asphalt types: (**a**) penetration; (**b**) softening point; (**c**) viscosity; (**d**) elastic recovery; (**e**) peak force; (**f**) tensile length.

The rotary viscosity test of modified asphalt before and after short-term aging was carried out to analyze the effect of short-term aging on the construction workability of modified asphalt, as shown in Figure 9c. As seen in Figure 9c, the viscosity of the three modified asphalts increases in varying degrees after the TFOT, in which the increase variation in viscosity of ACR asphalt is the least, and the increase variation in viscosity of

ACR/SBS asphalt is larger. The viscosity values of three kinds of modified asphalt after the TFOT are still less than 3.5 Pa·s, which meets the specification requirements. Similarly, the light components in asphalt volatilize, oxidize, and harden the intermolecular structure, leading to the viscosity increasing of modified asphalt. Through the viscosity between UCR asphalt and ACR asphalt, the internal network structure of crumb rubber powder is damaged after activation treatment, resulting in the viscosity of ACR asphalt being lower than that of UCR asphalt. With the progression of short-term aging, the viscosity of ACR asphalt still increases slightly. In addition, due to the larger average particle size of the SBS modifier and smaller average particle size of ACR, the viscosity of ACR asphalt before and after aging is smaller, while the viscosity of ACR/SBS asphalt is larger.

The elastic recovery test of modified asphalt before and after short-term aging was conducted to analyze the effect of short-term aging on the elastic recovery performance of modified asphalt, as shown in Figure 9d. The elastic recovery rate of ACR asphalt is lower than that of UCR asphalt, but the addition of SBS greatly improves the elastic recovery rate of modified asphalt. After short-term aging, the elastic recovery rate of modified asphalt decreases in varying degrees, among which the elastic recovery rate of ACR asphalt decreases the most, and the elastic recovery rate of ACR/SBS asphalt decreases the least. The elastic recovery ability of crumb rubber-modified asphalt mainly depends on the elasticity of crumb rubber powder particles. Therefore, the swelling and degradation degree of crumb rubber powder in asphalt play a major role in its elastic recovery ability. After desulfurization and activation, the internal network structure of crumb rubber powder is damaged and becomes smaller, and crumb rubber powder further loses elasticity and reduces its elastic recovery rate. After adding SBS, the internal network structure of modified asphalt is reconstructed, and the elastic recovery performance of ACR/SBS asphalt improves.

The force ductility test of modified asphalt before and after short-term aging was performed to analyze the effect of short-term aging on the low-temperature crack resistance of modified asphalt, as shown in Figure 9e,f. The peak force of the three modified asphalts increases after short-term aging. The light components in crumb rubber-modified asphalt volatilize, resulting in the hardening of modified asphalt and the increase in peak force. The tensile length of UCR asphalt and ACR asphalt decreases, and the tensile length of ACR/SBS asphalt increases slightly. During short-term aging, asphalt hardens by oxidation, and the modifier continues to swell and crack slowly. The pyrolysis products of crumb rubber powder and SBS can effectively make up for the adverse effects caused by the volatilization and oxidation of light components in asphalt. Therefore, the tensile length and toughness of ACR/SBS asphalt improve.

3.3.2. Short-Term Aging Factors with Softening Point as Target Performance

The short-term aging test was carried out under different aging durations (5 h, 10 h, 15 h, and 20 h) in a film oven at different aging temperatures (150 °C, 163 °C, and 180 °C) for UCR asphalt, ACR asphalt, and ACR/SBS-2 asphalt. By comparing the variation in softening point of modified asphalt under different aging conditions, the effect of aging time and temperature on the aging resistance of modified asphalt was studied with the softening point as the target performance, as shown in Figure 10.

It can be seen that the softening points of modified asphalt increase linearly with aging time at different aging temperatures. Meanwhile, at the same aging temperature, the growth range of the softening point of different modified asphalt samples is significantly different. When the aging temperature is 150 °C, the linear curve slope of the softening point versus aging time for ACR/SBS-2 asphalt is the smallest, and the slope of ACR asphalt is the largest, indicating that ACR/SBS-2 asphalt has the best aging resistance at the aging temperature of 150 °C. At the same time, there is a small difference between the curve slopes of these three modified asphalts at the aging temperature of 150 °C.

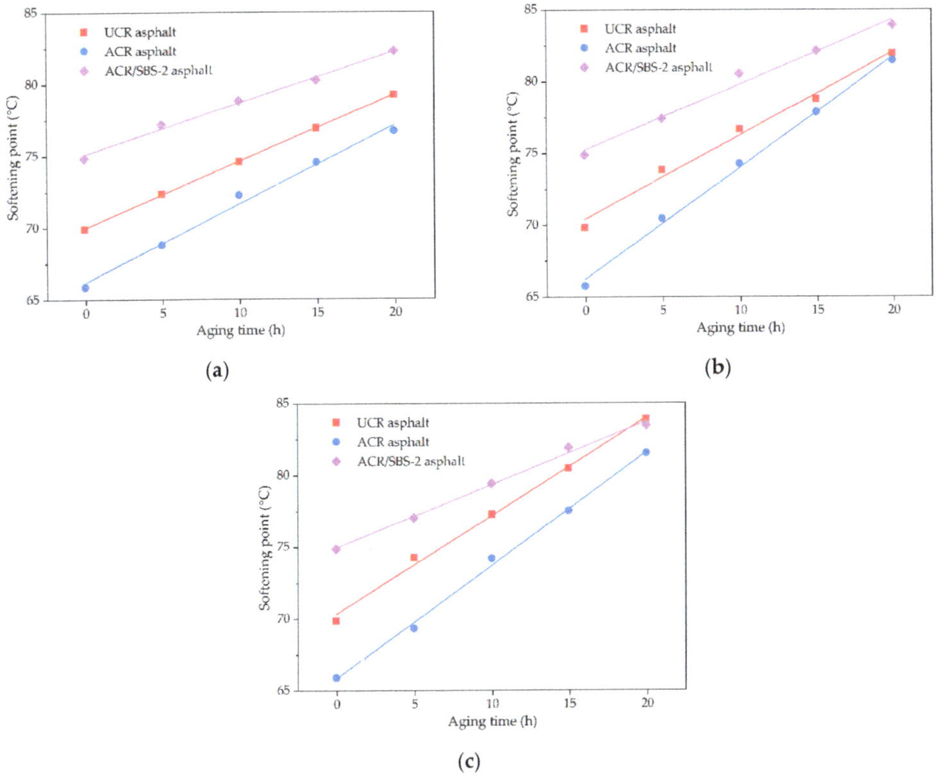

Figure 10. The softening point results of different asphalt types under different aging conditions: (**a**) 150 °C; (**b**) 163 °C; (**c**) 180 °C.

For the same modified asphalt, the growth range of the softening point versus aging time is different at different aging temperatures. The linear curve slope of the softening point versus aging time for ACR/SBS-2 asphalt changes little with the aging temperature, while the slope of ACR asphalt changes the most, indicating that ACR/SBS-2 asphalt is not sensitive to the aging temperature and has good short-term aging resistance. When the aging temperature is lower, crumb rubber powder and SBS in modified asphalt basically do not change, while the light components in base asphalt volatilize and oxidize. Therefore, the softening point of modified asphalt increases with the aging time, and the growth range difference of the softening point is less for these three kinds of modified asphalt. When the aging temperature is higher, the modifier in modified asphalt degrades and desulfurizes. These depolymerized and desulfurized components can make up for the adverse effects of asphalt aging, and the degradation of SBS is more obvious. Therefore, the softening point of ACR/SBS-2 asphalt increases slightly with aging time, and its aging resistance is better.

3.4. Micro-Characterization of ACR/SBS Composite-Modified Asphalt

The asphalt sample sections of ACR asphalt and ACR/SBS-2 asphalt before and after short-term aging were magnified by 80 times with an optical microscope, and are shown in Figure 11.

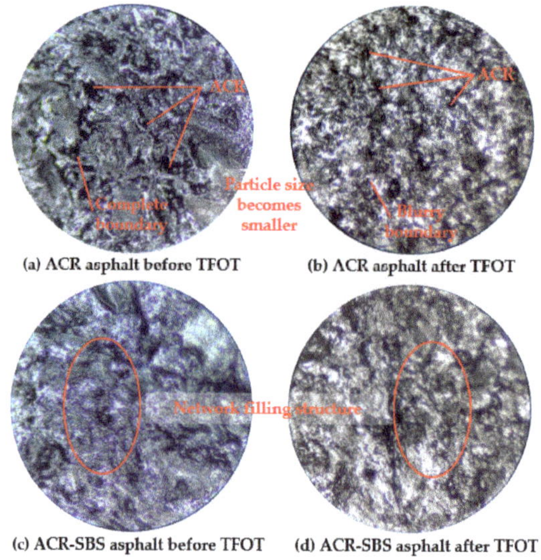

Figure 11. The 80× micrograph of ACR asphalt and ACR/SBS-12 asphalt before and after TFOT: (**a**) ACR asphalt before TFOT; (**b**) ACR asphalt after TFOT; (**c**) ACR/SBS-12 asphalt before TFOT; (**d**) ACR/SBS-12 asphalt after TFOT.

It can be seen from Figure 11a that the particle sizes of ACR are different, which are more evenly dispersed in asphalt, and there is an obvious boundary between ACR and asphalt. ACR particles with smaller particle size could form a discontinuous network structure with asphalt, and ACR particles with larger particle size would disperse in asphalt and play a filling role. It can be seen from Figure 11b that after short-term aging of ACR asphalt, the boundary between ACR particles and asphalt becomes blurred, and the ACR particle size becomes smaller, in which their dispersion is more uniform, and discontinuous network structures would be formed in asphalt. During the short-term aging process, ACR particles are partially degraded, and the degradation products are distributed in asphalt in the form close to the original components of asphalt such as resin and asphaltene, making them easier for ACR and asphalt to form a network structure.

In Figure 11c, after the composite modification of ACR and SBS, SBS particles and ACR with smaller particle size could form an interpenetrating network, and ACR with larger particle size are filled in the network. A network filling structure composed of asphalt, SBS, and ACR with different particle sizes is formed inside ACR/SBS-2 asphalt, and the existence of the network structure improves the compatibility and stability of the heterogeneous structure. As can be seen from Figure 11d, the structure of ACR/SBS-2 asphalt remains relatively complete, and the boundary between ACR and asphalt becomes blurred. However, due to the network structure formed by SBS, some ACR with large particle size does not undergo desulfurization and degradation, and their dispersion in the network structure would play a filling role. During the short-term aging process, the light components in base asphalt undergo condensation reaction to form resin and asphaltene. The activity of resin and asphaltene obtained by condensation reaction is higher than that of the original components. Because ACR and SBS will undergo desulfurization and degradation reaction under heating and oxygen conditions, the desulfurization degradation products could react with resin and asphaltene with strong activity to produce resin, asphaltene, and toluene-insoluble substances different from the original components of asphalt. Therefore, ACR/SBS-2 asphalt has excellent anti-aging properties, and ACR and SBS could play an anti-aging role.

4. Conclusions

In this study, the conventional physical properties, high-temperature rheological properties, and aging resistance of ACR/SBS composite-modified asphalt were investigated based on conventional physical properties tests, DSR, and TFOT. In addition, the action forms and distribution of modifiers in asphalt were observed by an optical microscope to characterize the micro-morphology. From the test results, the following conclusion could be drawn:

(1) After adding SBS, the softening point, ductility, and elastic recovery of ACR/SBS asphalt could be significantly improved, but the viscosity and softening point difference were also larger with the increase in SBS content. The complex shear modulus and rutting factor of ACR/SBS asphalt were greatly improved, while the phase angle was significantly reduced, indicating that SBS can effectively improve the high-temperature deformation resistance.

(2) The changes in high- and low-temperature performances and workability of ACR/SBS asphalt were reduced after short-term aging, and the addition of SBS could improve the aging resistance of modified asphalt. The softening point as the target performance of ACR/SBS asphalt was less affected by aging time and temperature, indicating that ACR/SBS asphalt was not sensitive to aging temperature with good stability and aging resistance.

(3) From the micrograph by microscope, ACR/SBS asphalt could maintain a relatively stable polyphase structure in the short-term aging process, which helped the degradation products of crumb rubber powder and SBS supplement the components in asphalt for aging resistance.

Author Contributions: Conceptualization, Z.Z. and L.W.; methodology, Z.Z. and L.W.; validation, L.W. and X.S.; formal analysis, Z.Z. and W.W.; investigation, Z.Z., L.W., W.W. and X.S.; writing—original draft preparation, Z.Z. and W.W.; writing—review and editing, L.W. and X.S.; project administration, W.W.; funding acquisition, W.W. All authors have read and agreed to the published version of the manuscript.

Funding: This research was supported by the Scientific and Technological Project of Science and Technology Department of Jilin Province (grant number: 20210508028RQ), Scientific Research Project of Department of Education of Jilin Province (grant number: JJKH20221019KJ), China Postdoctoral Science Foundation (grant number: 2021T140262), and Innovation and Entrepreneurship Training Fund of Jilin University (grant number: S202110183260).

Institutional Review Board Statement: Not applicable.

Informed Consent Statement: Not applicable.

Data Availability Statement: The data presented in this study are available on request from the corresponding author.

Conflicts of Interest: The authors declare no conflict of interest.

References

1. Picado-Santos, L.G.; Capitao, S.D.; Neves, J.M.C. Crumb rubber asphalt mixtures: A literature review. *Constr. Build. Mater.* **2020**, *247*, 118577. [CrossRef]
2. Ma, T.; Wang, H.; He, L.; Zhao, Y.L.; Huang, X.M.; Chen, J. Property characterization of asphalt binders and mixtures modified by different crumb rubbers. *J. Mater. Civ. Eng.* **2017**, *29*, 04017036. [CrossRef]
3. Saberi, K.F.; Fakhri, M.; Azami, A. Evaluation of warm mix asphalt mixtures containing reclaimed asphalt pavement and crumb rubber. *J. Clean. Prod.* **2017**, *165*, 1125–1132. [CrossRef]
4. Riekstins, A.; Baumanis, J.; Barbars, J. Laboratory investigation of crumb rubber in dense graded asphalt by wet and dry processes. *Constr. Build. Mater.* **2021**, *292*, 123459. [CrossRef]
5. Ameli, A.; Babagoli, R.; Asadi, S.; Norouzi, N. Investigation of the performance properties of asphalt binders and mixtures modified by crumb rubber and gilsonite. *Constr. Build. Mater.* **2021**, *279*, 122424. [CrossRef]
6. Wang, Q.Z.; Wang, N.N.; Tseng, M.L.; Huang, Y.M.; Li, N.L. Waste tire recycling assessment: Road application potential and carbon emissions reduction analysis of crumb rubber modified asphalt in china. *J. Clean. Prod.* **2020**, *249*, 119411. [CrossRef]

7. Gao, J.; Yang, J.G.; Yu, D.; Jiang, Y.; Ruan, K.G.; Tao, W.J.; Sun, C.; Luo, L.H. Reducing the variability of multi-source reclaimed asphalt pavement materials: A practice in china. *Constr. Build. Mater.* **2021**, *278*, 122389. [CrossRef]
8. Gao, J.; Yao, Y.; Song, L.; Xu, J.; Yang, J. Determining the maximum permissible content of recycled asphalt pavement stockpile in plant hot-mix recycled asphalt mixtures considering homogeneity: A case study in china. *Case Stud. Constr. Mater.* **2022**, *16*, e00961. [CrossRef]
9. Xu, M.Z.; Liu, J.J.; Li, W.Z.; Duan, W.F. Novel method to prepare activated crumb rubber used for synthesis of activated crumb rubber modified asphalt. *J. Mater. Civ. Eng.* **2015**, *27*, 04014173. [CrossRef]
10. Yu, R.B.; Gong, Z.H.; Guo, W.H.; Zhang, H.B.; Liu, C.L. A novel grafting-modified waste rubber powder as filler in natural rubber vulcanizates. *J. Appl. Polym. Sci.* **2016**, *133*, 42993. [CrossRef]
11. Zhang, H.G.; Zhang, Y.P.; Chen, J.; Liu, W.C.; Wang, W.S. Effect of desulfurization process variables on the properties of crumb rubber modified asphalt. *Polymers* **2022**, *14*, 1365. [CrossRef] [PubMed]
12. Kedarisetty, S.; Biligiri, K.P.; Sousa, J.B. Advanced rheological characterization of reacted and activated rubber (rar) modified asphalt binders. *Constr. Build. Mater.* **2016**, *122*, 12–22. [CrossRef]
13. Lv, S.T.; Ma, W.B.; Zhao, Z.G.; Guo, S.C. Improvement on the high-temperature stability and anti-aging performance of the rubberized asphalt binder with the lucobit additive. *Constr. Build. Mater.* **2021**, *299*, 124304. [CrossRef]
14. Juganaru, T.; Bombos, M.; Vasilievici, G.; Bombos, D. Devulcanized rubber for bitumen modification. *Mater. Plast.* **2015**, *52*, 336–339.
15. Shatanawi, K.M.; Biro, S.; Naser, M.; Amirkhanian, S.N. Improving the rheological properties of crumb rubber modified binder using hydrogen peroxide. *Road Mater. Pavement Des.* **2013**, *14*, 723–734. [CrossRef]
16. Liu, Q.; Liu, J.Z.; Yu, B.; Zhang, J.P.; Pei, J.Z. Evaluation and optimization of asphalt binder and mixture modified with high activated crumb rubber content. *Constr. Build. Mater.* **2022**, *314*, 125676. [CrossRef]
17. Chen, Z.X.; Pei, J.Z.; Wang, T.; Amirkhanian, S. High temperature rheological characteristics of activated crumb rubber modified asphalts. *Constr. Build. Mater.* **2019**, *194*, 122–131. [CrossRef]
18. Rasool, R.T.; Song, P.; Wang, S.F. Thermal analysis on the interactions among asphalt modified with sbs and different degraded tire rubber. *Constr. Build. Mater.* **2018**, *182*, 134–143. [CrossRef]
19. Zhang, F.; Hu, C.B. The research for structural characteristics and modification mechanism of crumb rubber compound modified asphalts. *Constr. Build. Mater.* **2015**, *76*, 330–342. [CrossRef]
20. Wang, W.S.; Tan, G.J.; Liang, C.Y.; Wang, Y.; Cheng, Y.C. Study on viscoelastic properties of asphalt mixtures incorporating sbs polymer and basalt fiber under freeze-thaw cycles. *Polymers* **2020**, *12*, 1804. [CrossRef]
21. Song, L.; Zhang, G.Q.; Xie, H.F.; Gao, J. Laboratory study on cr/sbs modified asphalt: Preparation and performance characterization. *J. Renew. Mater.* **2022**, *10*, 1659–1674. [CrossRef]
22. Tang, N.P.; Lv, Q.; Huang, W.D.; Lin, P.; Yan, C.Q. Chemical and rheological evaluation of aging characteristics of terminal blend rubberized asphalt binder. *Constr. Build. Mater.* **2019**, *205*, 87–96. [CrossRef]
23. Yan, C.; Lv, Q.; Zhang, A.A.; Ai, C.; Huang, W.; Ren, D. Modeling the modulus of bitumen/sbs composite at different temperatures based on kinetic models. *Compos. Sci. Technol.* **2022**, *218*, 109146. [CrossRef]
24. Zhang, B.; Chen, H.X.; Zhang, H.G.; Kuang, D.L.; Wu, J.Y.; Zhang, X.L. A study on physical and rheological properties of rubberized bitumen modified by different methods. *Materials* **2019**, *12*, 3538. [CrossRef]
25. Zhang, B.; Chen, H.X.; Zhang, H.G.; Wu, Y.C.; Kuang, D.L.; Guo, F.J. Laboratory investigation of aging resistance for rubberized bitumen modified by using microwave activation crumb rubber and different modifiers. *Materials* **2020**, *13*, 4230. [CrossRef]
26. Wang, W.S.; Cheng, Y.C.; Chen, H.P.; Tan, G.J.; Lv, Z.H.; Bai, Y.S. Study on the performances of waste crumb rubber modified asphalt mixture with eco-friendly diatomite and basalt fiber. *Sustainability* **2019**, *11*, 5282. [CrossRef]
27. Wang, W.S.; Cheng, Y.C.; Tan, G.J.; Liu, Z.Y.; Shi, C.L. Laboratory investigation on high- and low-temperature performances of asphalt mastics modified by waste oil shale ash. *J. Mater. Cycles Waste* **2018**, *20*, 1710–1723. [CrossRef]
28. Zhang, L.M.; Gao, X.K.; Wang, W.S.; Wang, H.; Zheng, K.K. Laboratory evaluation of rheological properties of asphalt binder modified by nano-tio2/caco3. *Adv. Mater. Sci. Eng.* **2021**, *2021*, 5522025. [CrossRef]
29. Chen, T.; Ma, T.; Huang, X.M.; Guan, Y.S.; Zhang, Z.X.; Tang, F.L. The performance of hot-recycling asphalt binder containing crumb rubber modified asphalt based on physiochemical and rheological measurements. *Constr. Build. Mater.* **2019**, *226*, 83–93. [CrossRef]
30. Zhang, J.W.; Chen, M.Z.; Wu, S.P.; Zhou, X.X.; Zhao, G.Y.; Zhao, Y.C.; Cheng, M. Evaluation of vocs inhibited effects and rheological properties of asphalt with high-content waste rubber powder. *Constr. Build. Mater.* **2021**, *300*, 124320. [CrossRef]

polymers

Article

Investigation of the High-Temperature and Rheological Properties for Asphalt Sealant Modified by SBS and Rubber Crumb

Yafeng Gong, Shuzheng Wu, Yuwei Zhang *, Yunze Pang and Yulin Ma

College of Transportation, Jilin University, Changchun 130025, China; gongyf@jlu.edu.cn (Y.G.); wusz20@mails.jlu.edu.cn (S.W.); pangyz19@mails.jlu.edu.cn (Y.P.); ylma18@mails.jlu.edu.cn (Y.M.)
* Correspondence: ywzhang@mails.jlu.edu.cn; Tel.: +86-135-0089-8555

Abstract: Crack sealing is an important measure for pavement maintenance. Hot-poured crack sealant is the most utilized material for crack sealing. However, its poor high-temperature and rheological properties seriously weaken the mechanical properties of repaired pavement. Thus, to overcome the disadvantage of the poor high-temperature and rheological properties of sealant, styrene–butadiene–styrene (SBS) and rubber crumb (CR) were utilized for modifying the asphalt-based sealants. Softening point tests, temperature tests, frequency scan tests, and multiple stress creep recovery tests (MSCR) were conducted to evaluate the high-temperature and rheological properties of the modified sealant. Additionally, the influence of SBS and CR on the high-temperature performance of the modified sealant was quantitatively analyzed by the grey relational analysis method. The results reveal that the SBS has a greater enhancement effect on the high-temperature performance of sealant than CR. Increasing the SBS and CR content in the sealant could enhance the sealant's high-temperature performance, stiffness, and elasticity. Compared with asphalt-based sealant and one-component modified asphalt-based sealant, SBS/CR-modified asphalt sealant has greater viscosity and higher temperature deformation resistance. Additionally, SBS can increase the stress level of the sealant, thereby enhancing the resistance of the sealant to permanent deformation.

Keywords: sealant; SBS; rubber crumb; high-temperature performance; permanent deformation resistance

Citation: Gong, Y.; Wu, S.; Zhang, Y.; Pang, Y.; Ma, Y. Investigation of the High-Temperature and Rheological Properties for Asphalt Sealant Modified by SBS and Rubber Crumb. *Polymers* **2022**, *14*, 2558. https://doi.org/10.3390/polym14132558

Academic Editor: Pablo Marcelo Stefani

Received: 25 May 2022
Accepted: 17 June 2022
Published: 23 June 2022

1. Introduction

Asphalt pavement is widely used in road engineering for its advantageous properties, which include short construction period, low noise, safety, comfortable driving, recyclability, etc. In recent years, with the rapid increase in traffic volume and vehicle load, serious asphalt pavement damage has occurred, and road maintenance measures have gradually become key to guaranteeing service performance [1,2]. Among many types of pavement damage, pavement cracking is one of the most common forms of damage, not only destroying the integrity of the pavement, but also leading to the infiltration of water, softening the base, weakening the bearing capacity of the base, and causing structural damage to the road [3–5]. Most of the pollution caused by asphalt pavement is secondary environmental pollution caused by pavement damage resulting from poor asphalt durability. Using asphalt to cover the earth as a road surface was once considered to be contrary to environmental friendliness. However, the environmental pollution caused by refurbishment due to road damage will undoubtedly lead to more serious environmental degradation. Engineering practice shows that crack sealing is an effective way of dealing with cracking-type damage to pavement [6,7]. Therefore, increasing attention is being paid to research into sealing materials, and various types of sealants have emerged in accordance with the requirements of the times [8]. Among these, heated asphalt-based sealant has been widely used in road maintenance projects for its outstanding crack sealing effect and low price [9,10].

However, the sealant material left on the surface after filling the road cracks will be directly exposed to the natural environment and constantly exposed to sunlight. On-site investigations have shown that the maximum temperature of asphalt pavement exposed to direct sunlight can exceed 60 °C. Continuous high temperature is likely to cause sealant aging, premature decline in mechanical properties, and a reduction in the bonding force with the road, meaning that the asphalt can be easily carried away by vehicles [11]. Therefore, it is important to improve the high-temperature stability of the sealant material [12,13].

Asphalt-based sealant materials are mainly composed of matrix asphalt, rubber powder, various polymers, softeners, fillers, etc., and improve the performance of asphalt sealants/mastics by introducing different waste/virgin polymers. There are many kinds of polymer-modified asphalt, the most common of which are rubber-modified asphalt using rubber powder as modifier. Thermoplastic-rubber-modified asphalt includes modifications with styrene-butadiene-styrene (SBS), styrene-isoprene (SIS), styrene-polyethylene-butadiene-styrene (SEBS), and other block copolymers. Resin-based asphalt modifications include the addition of polyethylene (PE) and ethylene-vinyl acetate copolymer (EVA) [14]. For example, Rosa Veropalumbo and Russo, F. et al. [15–17] have been working on the incorporation of polymer-functional plastic waste into asphalt mixtures to improve the performance of pavement layers. Since the asphalt-based sealant material prepared with a single modifier has disadvantages in terms of performance, and the single modifier is expensive, some low-cost modifiers need to be used to reduce costs; therefore, composite modification has been developed and applied [14]. With advances in technology, SBS has gradually replaced other types of polymers due to its superior high- and low-temperature properties. In view of the excellent anti-deformation and anti-cracking properties of rubber-modified asphalt, this paper uses CR/SBS to modify asphalt in order to study the properties of the sealant material.

A huge amount of waste rubber tires is generated globally every year. The reasonable disposal of the ever-increasing number of waste rubber tires is a global concern, and the tires could cause serious harm to the ecological environment if disposed of improperly. Some of these waste tires are eventually landfilled or incinerated, causing pollution to either land resources or air, and posing a severe threat to the environment. In recent years, scholars have begun to focus on the addition of waste rubber tires, physically ground into a granular state, into construction materials to prepare modified asphalt or concrete, thus realizing the recycling of waste rubber and opening a novel route for the treatment of waste rubber products. In the field of pavement material preparation, rubber crumb (CR)-modified asphalt presents good anti-deformation and anti-cracking ability [18–20]. However, the compatibility between CR and matrix asphalt is poor, and the high-temperature resistance of CR is not stable, which shortens the service life of CR-modified asphalt [21,22]. In addition, in the preparation process of CR-modified asphalt, CR expands due to the absorbance the light components (aromatic and saturated) of asphalt, increasing the viscosity of the asphalt and therefore affecting the fluidity of the sealant material made from CR-modified asphalt [23–25]. Therefore, in the production of CR-modified asphalt sealant, the proportion of added rubber, mixing temperature, and the performance of the matrix asphalt have a crucial impact on its high-temperature resistance and other properties [26,27].

Styrene-butadiene-styrene (SBS) is currently the most widely used asphalt modifier. It possesses a multiphase structure and is mainly composed of a butadiene segment and a styrene segment. SBS melts when the ambient temperature is between the glass transition temperatures of styrene and butadiene [28,29]. SBS-modified asphalt exhibits both high-temperature resistance and high elasticity, which is due to the cross-linking of SBS inside the asphalt to create a stable three-dimensional network structure [30]. Therefore, utilizing SBS-modified asphalt as the sealant for filling pavement cracks will significantly improve its high-temperature resistance. However, the cost of SBS-modified asphalt is relatively high, so the proportion of SBS must be controlled when preparing sealant in order to reduce the engineering cost, while at the same time guaranteeing good service performance (such as high-temperature properties) [31]. Thus, some researchers have thought of combining

SBS and CR to modify asphalt materials so as to improve their high-temperature resistance while reducing their cost [32].

In summary, although much literature has reported the high-temperature resistance properties of CR/SBS-modified asphalt, as a sealant material for cracked pavements, the evaluation indices for the high-temperature performance of CR/SBS-modified asphalt sealant are incomplete. Further performance evaluation is needed to verify whether it meets the requirements of the Chinese national standard, making it suitable for being adopted in actual road maintenance projects [33,34]. Meanwhile, the asphalt sealant is often subjected to continuously repeating vehicle load; therefore, its real mechanical response needs to be further evaluated on the basis of the rheological performance of the sealant under dynamic load [35,36].

This study is dedicated to preparing an eco-friendly asphalt sealant material for pavement maintenance engineering that possesses excellent high-temperature stability and anti-deformation ability through the addition of waste rubber and SBS, which is expected to expand the application of waste rubber in pavement maintenance engineering, representing a more sustainable utilization of waste rubber products, thus protecting the ecological environment. In addition, softening point, complex shear modulus, phase angle, mean non-recoverable creep modulus, and other indicators are used to systematically evaluate the high-temperature properties and rheological properties of the modified asphalt sealants. Firstly, the enhancement effect of SBS and CR on the high-temperature stability of asphalt was evaluated separately. Secondly, the differences in the high-temperature performances of the SBS/CR-modified asphalt sealant, unmodified asphalt sealant (UAS), SBS-modified asphalt sealant, and CR-modified asphalt sealant were evaluated. Subsequently, the high-temperature stability and anti-deformation ability of the SBS/CR asphalt sealant was evaluated on the basis of softening point tests, temperature scan tests, frequency scan tests, MSCR, and dynamic shear rheology tests (DSR). The master curve of modulus for the SBS/CR sealant was established, and the viscoelastic properties of the SBS/CR sealant were evaluated over a wide frequency range.

2. Materials and Methods

2.1. Raw Materials

Donghai Brand 90# road petroleum asphalt, produced by China Petrochemical Corporation in Beijing, China, was used as the matrix asphalt for the pavement sealant. The test process was determined with reference to the standard test methods of Asphalt and asphalt mixture test regulations for highway engineering (JTG E20-2011) [37]. The technical indices of the matrix asphalt are provided in Table 1. The selected SBS (grade 4402), produced by China Petrochemical Corporation, was a star-shaped structure, with a styrene content of 30 wt%. CR was purchased from Shaanxi Hongrui Rubber Co., Ltd. in Xi'an, China, with a particle size of 0.3 mm; the technical indices are listed in Table 2.

Table 1. Properties of 90# matrix asphalt.

Index	Standard Value	Measured Results
Softening Point (Global Method)	≥45 (°C)	46.0 (°C)
Penetration (25 °C, 100 g, s)	80~100 (0.01 mm)	84 (0.01 mm)
Ductility (15 °C)	≥100 (cm)	>100 (cm)
TFOT (Thin Film Oven Test) Residue		
Quality Change	±0.8 (%)	−0.112 (%)
Residual Penetration Ratio	≥57 (%)	62.4 (%)
Residual Ductility (10 °C)	≥8 (cm)	11.9 (cm)

Table 2. Technical indices of CR.

Index	Standard Value	Measured Value
Sieving rate	≥45%	91%
Ash content	≤10%	4.5%
Moisture content	<1%	0.6%
Fiber content	<1%	0.5%
Rubber content	≥48%	51%
Burn Participation	<38%	37.5%

2.2. Mixture and Preparation for the CR- and SBS-Compound-Modified Asphalt Sealant

The preparation technique for the modified asphalt sealant is as follows: (1) heat 500 g of 90# matrix asphalt to a fluid state at 140 °C; (2) add SBS to the sample and stir at a speed of 800 r/min for 30 min on a high-speed mixer at 170 °C; (3) cut the sample at a speed of 5000 r/min for 45 min on a high-speed shearing machine at 170 °C; and (4) add CR to the sample and stir at a speed of 800 r/min for 1 h on a high-speed mixer at 170 °C to produce the modified asphalt sealant.

Mixing excessive CR into the asphalt will lead to an increase in the viscosity of the modified asphalt sealant, which will result in asphalt segregation [38]. This is because the swelling rate of CR does not increase infinitely, and reaches a peak when the content is around 20% [39]. The softening point of the modified asphalt sealant depends mainly on the amount of dispersed SBS and the perfection of the star-shaped structure. When the content of SBS is about 4% to 6%, a relatively perfect network structure will be formed in asphalt. Therefore, the use of 4% to 6% SBS in modified asphalt production is most appropriate, in accordance with Chinese Standard (JT/T 740-2015) [40]. The research presented in this paper also take the content of star-shaped SBS as a variable for conducting research. Hence, 12 groups of sealants based on CR- and SBS-compound-modified asphalt sealants were prepared in this study; 3 samples were prepared for each group, and experimental error was eliminated by averaging the results. The mix proportions are listed in Table 3.

Table 3. The mix proportions for the CR- and SBS-compound-modified sealants.

Group	CR Contents (%)	SBS Contents (%)
1	10	1
2	10	3
3	10	5
4	10	7
5	15	1
6	15	3
7	15	5
8	15	7
9	20	1
10	20	3
11	20	5
12	20	7

2.3. Test Methods

2.3.1. Softening Point Test

Asphalt softening point is an important indicator of asphalt sealant performance, and is usually used to measure the temperature sensitivity of asphalt sealant. In this study, the softening point values of all groups of the sealant samples were tested according to the test procedure of the Chinese Standard (JTG E20-2011) [37]. The softening point experiments were performed using a DF-10 computer automatic softening point instrument produced by Nanjing Dongyong Shenfu Technology Co., Ltd. in Nanjing, China. Note that the softening point tests for the sealant samples were carried out using the ring and ball softening point method. The sealant samples were cured in the specified curing environment, in line with the test requirements. Then, the sealant samples were heated at a rate of 5 °C/min in the cup, and the heating and cooling temperatures were recorded. The average value of the temperature is taken as the softening point test result for each sealant sample.

2.3.2. Temperature Sweep Test

A Dynamic Shear Rheometer (DSR) produced by Anton Paar Co., Ltd. in Graz, Austria (Model: SmartPave 102) was utilized in this study to conduct temperature sweep experiments on modified asphalt sealant. The test results were used to reflect the rheological behavior of the asphalt-based material at medium and high temperature. The temperature range of the test was 34~94 °C, and the temperature gradient was 6 °C, as well as having an angular frequency of 10 rad/s. The complex shear modulus (G^*) and phase angle (δ) were tested, and the change rule of the viscoelastic parameters was evaluated, which was used to determine the temperature dependency of the samples.

2.3.3. Muti-Stress Creep and Recovery Test

The muti-stress creep and recovery (MSCR) tests were used to measure the deformation value of asphalt-based material under different constant stress and the creep deformation recovery values after the disappearance of stress. The stress control approach was used in the repeated stress creep recovery experiments. The test process can be divided into the creeping stage, with a constant stress loading time of 1 s, and the recovery stage, with a zero-stress no-load time of 9 s under each cycle at 64 °C. For each experiment, 30 stress cycles were carried out. The adopted stress for cycles 0~20 was 0.1 kPa, and for cycles 21~30 it was 3.2 kPa. Nevertheless, the test results for cycles 0~10 were only used to adjust the properties of the asphalt sealant, and were not used for analysis. During the test, stress and strain were recorded at a frequency of 0.1 s during the creep phase and at a frequency of at least 0.45 s during the recovery phase.

2.3.4. Frequency Sweep Test

To study the mechanical properties of modified asphalt sealant materials under different loading frequencies, frequency sweep tests for different sealant samples were performed at a test frequency of 0.1~100 Hz at different temperatures (30 °C, 40 °C, 50 °C, 60 °C, and 70 °C). The parallel plate's diameter was 8 mm, and the thickness values of the sealant samples were 2 mm at 30 °C and 40 °C, while the parallel plate's diameter was 25 mm and the thickness values of the sealant samples were 1 mm at 50 °C, 60 °C, and 70 °C (as shown in Figure 1). Before the frequency sweep test, the parallel plates were heated to 70 °C in order to avoid damage to the sealant samples when removing excess asphalt and achieve a good connection between the asphalt samples and the parallel plates.

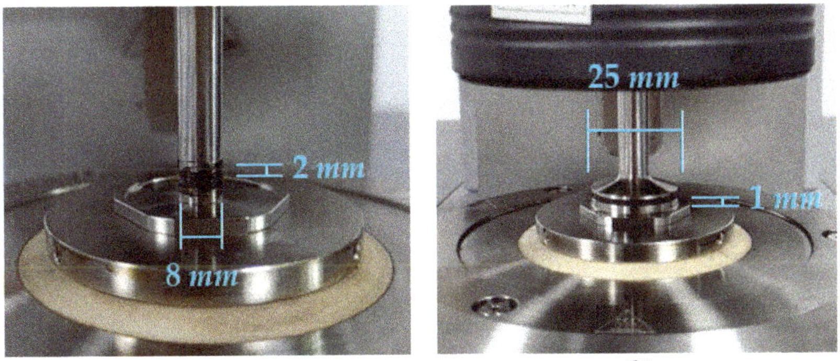

(a) (b)

Figure 1. Parallel plate size and sealant sample size in DSR at different temperatures. (**a**) 30 °C and 40 °C; (**b**) 50 °C, 60 °C and 70 °C.

3. Results and Discussion

3.1. Softening Point Results and Grey Correlation Analysis

This study used softening point values to assess the high-temperature performance of modified asphalt sealant samples (as shown in Figure 2), and the effect of CR and SBS on high-temperature performance for the modified asphalt sealant was investigated.

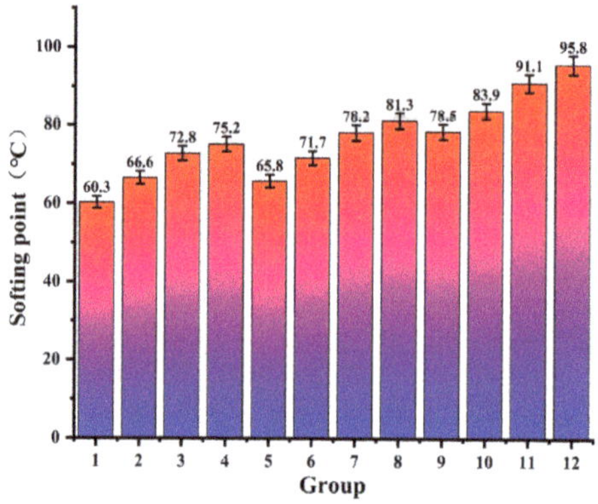

Figure 2. The results of the softening point test.

As shown in Figure 2, the softening point of modified asphalt sealant increases with increasing CR and SBS content, which indicates that CR and SBS can significantly enhance the high-temperature performance of asphalt sealant [39]. With reference to the softening point regulations for hot-poured sealants for pavement (JT/T 740-2015) [40], the required minimum softening point for sealant is 70 °C. Compared with the matrix asphalt, the softening points of Group 1, Group 2, and Group 5 were significantly improved, but were still not able to meet the requirements of the sealant. This shows that the sealants these contents were not able to achieve the best performance, which could be achieved by increasing the modifier content in order to meet performance requirements [41]. When the SBS content was 20% and the CR content was 7% (Group 12), the softening point of the modified asphalt sealant reached 95.8 °C, which is 58.9% higher than that of Group 1. When the SBS content in the modified asphalt sealant was constant, the softening point value increased by 12.9% on average with a 5% increase in CR content. Similarly, when the CR content was constant, the softening point value increased by 7.3% on average with a 2% increase in SBS content [42]. Furthermore, it can be seen from the test results that the modification effect of SBS per unit content is about 1.4 times that of CR. Intuitively, the increase in unit SBS content had a greater impact on the softening point of the asphalt sealant than CR.

Nevertheless, the analysis of the basic data cannot be used to accurately measure the degree of correlation between the modifier and the high-temperature performance of the modified asphalt sealant. Therefore, the grey correlation degree was used to further analyze the basic data. Grey correlation analysis is an academic analysis method for multi-factor degrees of correlation and the comparison of data development trends [43,44], and has been successfully applied in agronomy, medicine, management, and business [45,46]. In this paper, the softening point was selected as the reference sequence, and the correlation degree between CR, SBS content, and high-temperature performance in a single-component modified asphalt sealant was analyzed using the grey correlation analysis method.

According to the theory of grey relational analysis, it is first necessary to determine the comparison sequence and the reference sequence:

The comparison sequence is written as Equation (1):

$$X_1', X_2', \ldots X_n' = \begin{pmatrix} X_1'(1) & X_2'(1) & \cdots & X_n'(1) \\ X_1'(2) & X_2'(2) & \cdots & X_n'(2) \\ \vdots & \vdots & \vdots & \vdots \\ X_1'(m) & X_2'(m) & \cdots & X_n'(m) \end{pmatrix} \quad (1)$$

The reference sequence is an ideal comparison standard in theory. The optimal value (or worst value) of each index can be used to form the reference sequence, and other reference values can also be selected according to the purposes of the evaluation. The reference sequence is written as: $X_0' = X_0'(1), X_0'(2), \ldots X_0'(m)$.

The index data are dimensionless for easy comparison. In this paper, the difference method is used to obtain the data sequence matrix. The absolute differences between each evaluated object index sequence (comparison sequence) and the corresponding element of the reference sequence are calculated one by one. Then, the correlation coefficient between each comparison sequence and the corresponding element of the reference sequence is calculated according to Equation (2):

$$\zeta_i(k) = \frac{\underset{i}{min}\,\underset{k}{min}|X_0(k) - X_1(k)| + \rho\cdot\underset{i}{min}\,\underset{k}{min}|X_0(k) - X_1(k)|}{|X_0(k) - X_1(k)| + \rho\cdot\underset{i}{min}\,\underset{k}{min}|X_0(k) - X_1(k)|} \quad k = 1, 2, \ldots m \quad (2)$$

where ρ is the resolution coefficient, $0 < \rho < 1$. Smaller values of ρ indicate a larger difference between the correlation coefficients, and stronger discrimination ability. Usually, ρ is taken as 0.5.

For each evaluation object, the mean value of the correlation coefficient between each index and the corresponding element of the reference sequence is calculated to reflect the relationship between each evaluation object and the reference sequence; this is referred to as the correlation sequence, and is determined using Equation (3) [44]:

$$R = \frac{1}{m}\sum_{k}^{m}\zeta_i(k) \quad (3)$$

The difference sequences for each factor were calculated, and the results are listed in Table 4. The grey correlation coefficients under different contents were calculated, and the results are listed in Table 5.

Table 4. The difference sequences of CR and SBS.

Index	Sample											
	1	**2**	**3**	**4**	**5**	**6**	**7**	**8**	**9**	**10**	**11**	**12**
CR	10	10	10	10	15	15	15	15	20	20	20	20
SBS	1	3	5	7	1	3	5	7	1	3	5	7
Difference sequence	35.5	29.2	23	20.6	30	24.1	17.6	14.5	17.3	11.9	4.7	0

Table 5. The grey correlation coefficients for different contents of modifiers.

Factor	k												Grey Relation Coefficient
	1	**2**	**3**	**4**	**5**	**6**	**7**	**8**	**9**	**10**	**11**	**12**	
R1	0.64	0.64	0.64	0.64	0.54	0.54	0.54	0.54	0.47	0.47	0.47	0.47	0.55
R2	0.94	0.85	0.78	0.72	0.94	0.85	0.78	0.72	0.94	0.85	0.78	0.72	0.82

The quantitative characterization of the effect of SBS and CR on high-temperature performance was realized through correlation analysis. It can be seen from Table 5 from the correlation coefficients between the modifiers (CR and SBS), as well as the order of the grey relation coefficients from high to low, that R2 > R1. The correlation between SBS and asphalt sealant softening point is 1.5 times that of CR, which also confirms the previous conclusions. The correlation sequence results show that, compared with CR, the content of SBS has a greater correlation with the high-temperature performance of the modified asphalt sealant.

Most previous studies have concluded that both CR and SBS are able to improve the softening point of asphalt sealants to a certain extent [4,20]. However, there have been few studies on the softening point improvement efficiency of SBS and CR in CR/SBS-composite-modified asphalt. This paper confirms the dominant role of SBS in affecting the softening point of sealants through grey correlation. SBS is the most significant factor affecting the softening point in the CR/SBS-modified asphalt sealant. Therefore, when preparing environmental protection sealants, we should not only consider the addition of waste/virgin polymers, but also ensure an adequate content of SBS, in order to meet the high-temperature performance requirements.

Both CR and SBS have a significant effect on the softening point of the modified asphalt sealant. The changing trend of the softening point shows that 20% CR generally has higher conventional high-temperature performance. Hu M et al. [39] also reported that once the content of rubber powder has reached 20%, the viscosity of the dispersion medium is no longer significantly improved by increasing content and mesh number. To improve creep recovery ability and rheological properties under dynamic load, once the requirements of conventional high-temperature performance had been satisfied, SBS was added. The grey correlation analysis results show that SBS has a greater impact on the softening point of the modified asphalt sealant (conventional high-temperature performance). However, it is still necessary to investigate whether higher SBS content results in more beneficial rheological performance in the modified asphalt sealant under high-temperature conditions in order to determine the optimal amount of added SBS. In the experiments that follow in this paper, the rheological properties are studied by controlling the gradient of SBS content for the asphalt sealant. Under the premise of satisfying a certain high-temperature performance (20% CR), the rheology of the modified asphalt sealant can be optimized to the greatest extent so that the performance can be improved more comprehensively while taking the project cost into account.

The three groups with the best softening point performance in the orthogonal experiment were selected: Group 10, Group 11, and Group 12. Therefore, the temperature sweep test was carried out on Group 10, Group 11, and Group 12. Three control groups were determined: unmodified asphalt sealant (marked as Group a), single-component modified asphalt with 7% SBS content (marked as Group b), and single-component modified asphalt with 20% CR content (marked as Group c). The MSCR test and frequency sweep test were carried out together for Group 10, Group 11, and Group 12. The high-temperature performance of MSCR and the viscoelastic mechanical properties of composite-modified asphalt sealant were evaluated.

3.2. Anti-Deformation Ability

It is well known that the viscoelastic properties of modified asphalt sealant are sensitive to temperature. Therefore, it is necessary to evaluate the sensitivity of the viscoelastic properties of SBS/CR-modified asphalt sealant to verify whether this new asphalt sealant is able to meet the mechanical performance requirements of future road maintenance engineering applications. Hence, a temperature scanning test was carried out on the modified asphalt sealant (Group 10, Group 11, and Group 12) to evaluate their temperature sensitivity. In the temperature sweep test, the complex shear modulus (G^*) can be used to evaluate the anti-deformation ability of asphalt sealant. The higher the composite shear modulus, the greater the high-temperature deformation ability of the asphalt sealant. In addition, the viscoelastic characteristics of asphalt sealant could be assessed by the phase angle (δ). The asphalt sealant becomes more pliable with decreasing phase angle, which means that the asphalt sealant has stronger high-temperature resistance [47,48]. In this study, the variation in the parameters G^* and δ with modifier content (CR and SBS) was analyzed at different temperatures. Parallel experiments were performed for three samples in this experiment, and the maximum standard deviation was calculated to be 2.7345 by filtering and integrating the data. The average value of the data points was calculated, and the test results are shown in Figure 3.

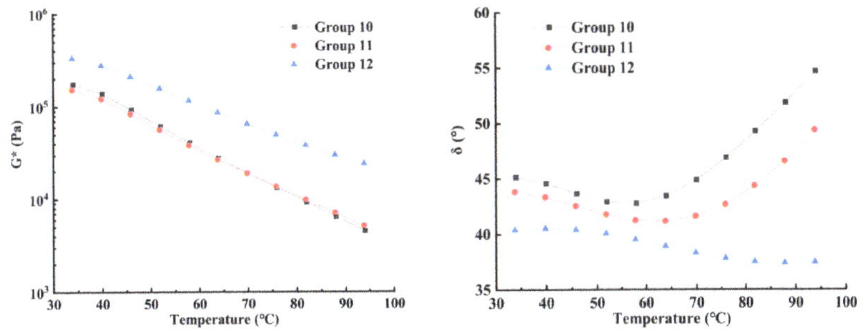

(a) (b)

Figure 3. The temperature scanning test results. (**a**) The complex shear modulus; (**b**) the phase angle.

On the basis of the test results presented in Figure 3a, the complex shear modulus G^* of the modified asphalt sealant shows a decreasing trend with increasing temperature. In the semi-logarithmic coordinates, the regularity of the change of G^* with temperature approximates a linear functional relationship. Furthermore, the curves of Group 10 and Group 11 have similar trends and values, indicating that the two sealants have the same anti-deformation ability within the temperature range 34~94 °C. The complex shear modulus of the three sealants is about 1.75×10^5 Pa at 34 °C, while the complex shear modulus of the Group 12 sealant decreases to 2.4×10^4 Pa, and the modulus values of the Group 10 and Group 11 sealants decrease to 4.5×10^3~5.2×10^3 Pa at 94 °C. Compared with Group 10 and Group 11, the complex modulus of Group 12 has a smaller variation in amplitude with angular frequency, which also means that its temperature sensitivity is lower. These experimental results are consistent with the results reported by Liang M. [22].

In addition, the phase angle of the Group 10 and Group 11 modified asphalt sealants first decreased and then increased with increasing temperature, which indicates that the viscous and elastic components of the two sealants change continuously at different temperatures. When the temperature is lower than 60 °C, the proportion of elastic components in the modified asphalt sealants increases with increasing temperature, which implies that the deformation resistance of the modified material gradually increases. When the temperature is higher than 60 °C, the proportion of viscous components in the modified asphalt sealants increases gradually with increasing temperature, revealing that the deformation resistance of the modified asphalt sealant decreases. Moreover, the change in the phase angle of the Group 12 sealant is small, and the phase angle is less than 45° within the test temperature range, meaning that modified asphalt sealants with high modifier contents (CR and SBS) have high elasticity and low temperature sensitivity. The temperature change has a small impact on the proportion of the viscoelastic components.

From the complex modulus and phase angle obtained from the temperature sweep test, it can be seen that the complex modulus of Group 12 is generally higher than that of Group 10 and Group 11, although this shows that its stiffness and resistance to deformation are higher. However, its phase angle is always lower than 45° in the test temperature range, and even decreases with increasing temperature, meaning that its viscoelasticity fails to reach equilibrium. After repairing the crack, the adhesion to the original crack wall is insufficient, and the toughness is insufficient due to its excessive hardness. The modulus values of Group 10 and Group 11 are similar. In the high-temperature range of 60~94 °C, Group 11 shows better elastic behavior than Group 10 and has stronger resistance to foreign body embedding. Therefore, when the content of CR is 20%, when preparing SBS/CR-modified asphalt sealant, the addition of 5% SBS provides the best deformation resistance and elastic toughness for the modified asphalt sealant.

3.3. High-Temperature Performance on the Basis of MSCR

The deformation of asphalt sealant is a dynamic process caused by the combined effect of dynamic load and temperature. The MSCR test is able to accuratley describe the high-temperature deformation resistance of asphalt sealant under the synergistic action of dynamic load and temperature [49]. MSCR tests were conducted on unmodified asphalt sealant, SBS-modified asphalt sealant, CR-modified asphalt sealant, and SBS/CR-modified asphalt sealant. The average recovery rate ($\overline{R}$) and the average non-recoverable creep modulus ($\overline{J_{nr}}$) at different creep stress levels were used to evaluate the capacity of asphalt sealant to withstand deformation on a permanent basis under various creep stress levels. The calculation method was as shown in Equations (4) and (5):

$$R = \frac{\varepsilon_c - \varepsilon_0}{\varepsilon_r - \varepsilon_0}; \overline{R} = \sum_{N=1}^{10} \frac{R(\tau, N)}{10} \tag{4}$$

$$J_{nr} = \frac{\varepsilon_r - \varepsilon_0}{\tau}; \overline{J_{nr}} = \sum_{N=1}^{10} \frac{J_{nr}(\tau, N)}{10} \tag{5}$$

where τ (kPa) is the creep stress corresponding to each loading period; $\overline{R}$ is the average recovery rate; $\overline{J_{nr}}$ (kPa^{-1}) is the average unrecoverable creep modulus.

The average recovery rate represents the elastic response of asphalt sealant, while the average unrecoverable creep modulus represents the viscous deformation of asphalt sealant. As shown in Figure 4, the average unrecoverable creep modulus of SBS/CR-modified asphalt sealants was lower than that of the CR-modified asphalt sealant and the SBS-modified asphalt sealant when the degree of stress was 0.1 kPa and the test temperature was 64 °C, while the SBS/CR-modified asphalt sealant's average recovery rate was comparable to that of the CR-modified asphalt sealant and significantly superior to that of the SBS-modified asphalt sealant. This indicates that modification with SBS and CR enhanced the stiffness and elasticity of the asphalt sealant, thus increasing its deformation resistance.

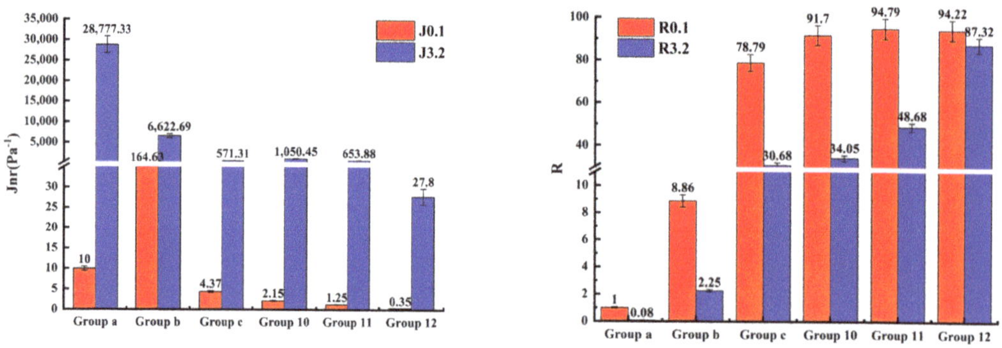

(a) (b)

Figure 4. MSCR test results (test temperature: 64 °C). (**a**) The average non-recoverable creep modulus; (**b**) the average recovery rate.

The average unrecoverable creep modulus of asphalt increased and the average recovery rate dropped when the degree of stress was increased from 0.1 kPa to 3.2 kPa at 64 °C, indicating that the deformation resistance of the asphalt sealant decreases under high load. Hence, high traffic volumes will increase the risk of the asphalt pavement breaking. However, when the percent of SBS in the modified asphalt sealant was increased, the growth rate of the unrecoverable creep modulus decreased significantly, and the reduction range of the average recovery rate slowed down, indicating that the ability of the modified asphalt sealant to withstand permanent deformation at high stress levels increases with increasing SBS content.

Ren S.S. et al. [50] calculated the average non-recoverable creep modulus and the average recovery rate. Accurate quantitative evaluation of the effect on rutting resistance of modifying asphalt with CR/SBS was performed. Low J_{nr} with high R-value CR/SBS-modified asphalt outperformed the unmodified asphalt at both stress levels, indicating that the presence of elastic response in the modified asphalt makes it less sensitive to rutting and permanent deformation. This is due to the interaction of rubber powder and SBS mesh in asphalt, which is beneficial for hindering the accumulation of permanent strain. In this section, not only the above conclusions are verified, but 20% CR and 7% SBS-modified asphalt sealant were demonstrated to offer the best stiffness and elasticity, as well as excellent resistance to deformation.

3.4. Viscoelastic Mechanical Properties and CAM Model Fitting

As a viscoelastic material, the properties of asphalt vary with temperature and loading time, which has a great impact on its uses in road engineering [1,9,27]. The softening point index described in Section 3.1 can only be used to carry out a macro evaluation of the high-temperature resistance of asphalt sealant, and cannot achieve an accurate description of the microstructure of the asphalt sealant [22]. Consequently, the frequency scanning test was used to evaluate the viscoelastic behavior of asphalt sealant from a microscopic point of view. Similarly, parallel experiments were performed on three samples in this experiment, and the average value of the data points was calculated. Figure 5 shows the average experimental results of the frequency scanning of Group a, Group b, Group c, Group 10, Group 11, and Group 12. The composite shear modulus of the six sealants increased with increasing frequency and decreased with increasing temperature. This phenomenon is more obvious in the unmodified asphalt sealant, which demonstrated poor deformation resistance at high temperature [34].

As a kind of viscoelastic material, the change in the mechanical properties of asphalt abides by the principle of time-temperature equivalence; therefore, valid data that are outside the test frequency and temperature range can be obtained by means of calculation [51,52]. The time-temperature equivalence principle indicates that the frequency at a given test temperature can be altered along the logarithmic frequency axis by a displacement factor. Of these, the shift factor is only related to temperature, and can be calculated by the WLF equation, as shown in Equation (6):

$$\log_{10} \alpha_T = \frac{-C_1(T - T_0)}{C_2 + (T - T_0)} \tag{6}$$

where $\log_{10} \alpha_T$—shift factor; C_1,C_2—constants; T—test temperature, °C; T_0—reference temperature, °C.

Based on the time-temperature equivalence principle, the reference temperature was set as 30 °C, and the shift factors of six asphalt sealants at different temperatures were calculated according to Equation (5) (listed in Table 6), and the main curves of complex shear modulus of composites were drawn (shown in Figure 6).

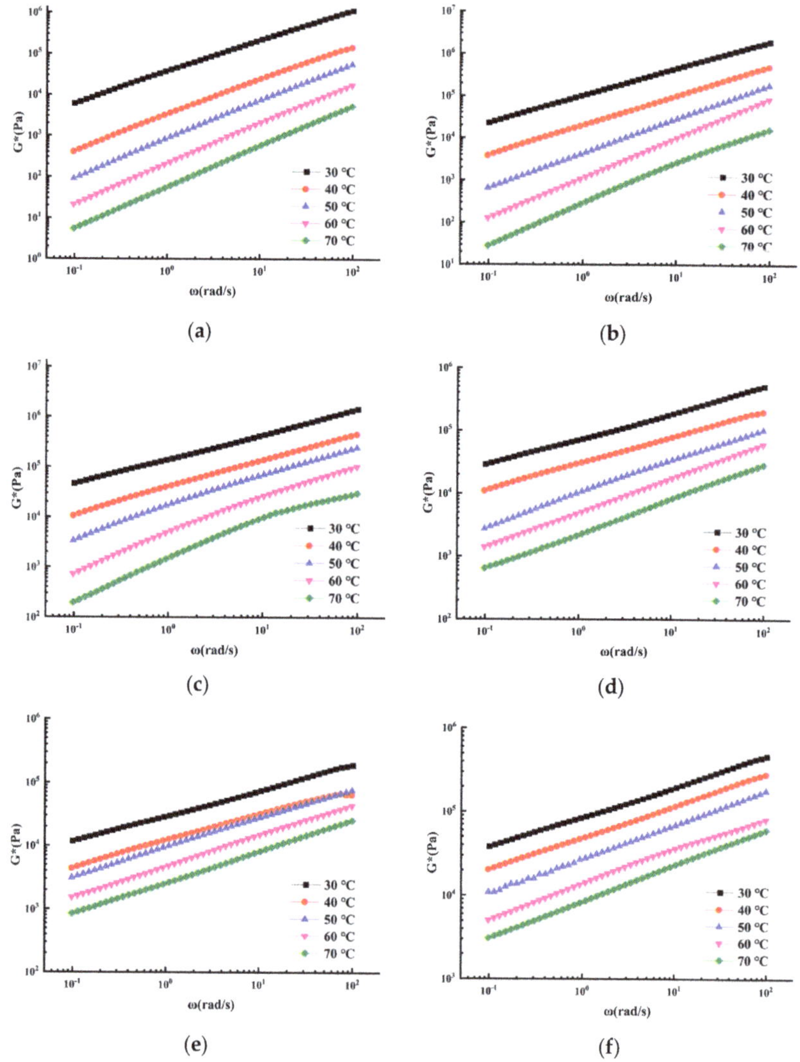

Figure 5. Frequency scanning test results. (**a**) Group a, (**b**) Group b, (**c**) Group c, (**d**) Group 10, (**e**) Group 11, (**f**) Group 12.

Table 6. Shift factors and C1 and C2 parameter values of materials at different test temperatures (reference temperature: 30 °C).

Asphalt	Parameter		Shift Factor					R^2
	C1	C2	30	40	50	60	70	
Group a	6.349	46.509	0	−1.217	−1.844	−2.433	−2.990	0.9952
Group b	10.508	86.7	0	−1.045	−1.988	−2.732	−3.295	0.9993
Group c	9.329	87.64	0	−1.078	−1.642	−2.345	−2.964	0.9933
Group 10	10.109	100.127	0	−0.905	−1.718	−2.295	−2.899	0.9993
Group 11	6.188	42.558	0	−1.117	−2.002	−2.633	−2.944	0.9971
Group 12	17.545	238.033	0	−0.710	−1.301	−2.046	−2.491	0.9962

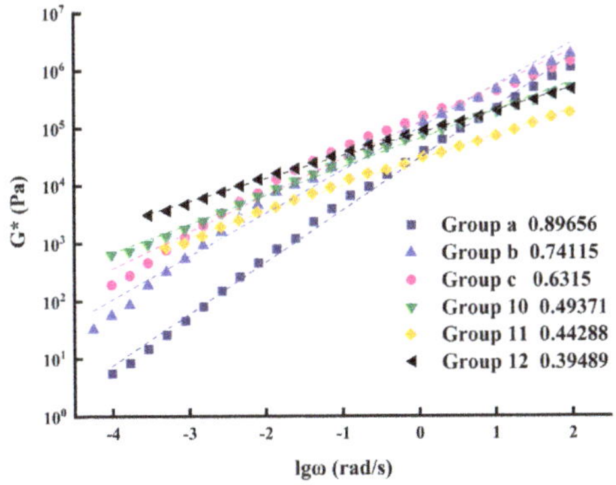

Figure 6. The linear fit curve of complex shear modulus (reference temperature: 30 °C).

The slope of the curve obtained by linear fitting is able to reflect the variation trend of complex shear modulus with frequency. Therefore, this slope can be used to characterize the ratio of complex shear modulus to angular frequency. In Figure 6, it can be seen that the ratios of complex shear modulus and angular frequency of the six materials, in decreasing value, are in the following order: Group a > Group b > Group c > Group 10 > Group 11 > Group 12. This indicates that the susceptibility of the modified asphalt sealant to temperature following modification with a composite of SBS and CR was less than that of the modified asphalt sealant modified with a single additive (either SBS or CR), and the five modified asphalt sealant materials had greater temperature sensitivity than the unmodified asphalt sealant.

Simultaneously, in the low-frequency range, the complex shear modulus of the three kinds of composite-modified asphalt sealant (Group 10, Group 11, and Group 12) was greater than that of the unmodified asphalt sealant and the asphalt sealant modified with a single additive (Group b and Group c). However, in the high-frequency range, the complex shear modulus of the three kinds of composite-modified asphalt sealant was lower than that of the unmodified asphalt sealant and the asphalt sealant modified with a single additive. The low frequency of the primary curve corresponds to asphalt's performance at high temperatures, while the high frequency refers to the asphalt's low-temperature performance. Consequently, it can be inferred that the composite-modified asphalt sealant has greater resistance to permanent deformation than unmodified asphalt sealant and asphalt sealant modified with a single additive at elevated temperatures. Furthermore, the viscosity of the composite-modified asphalt sealant at low temperature was also superior to that of the unmodified asphalt sealant and the asphalt sealant modified with a single additive at low temperature.

In addition, in order to more clearly describe the change rule of the master curve, the CAM model was used to fit the master curves, and is expressed as shown in Equation (7):

$$|G*| = \frac{|G_g*|}{\left[1 + f_c/(\alpha_T f)^K\right]^{M/K}} \tag{7}$$

where $|G_g*|$ is the glassy shear modulus of the asphalt sealant; K and M are the shape-fitting parameters of the master curve; f_c is the position-fitting parameter of the master curve; f is the loading frequency; and α_T is the shift factor.

The fitting curves are shown in Figure 7. It can be observed that the CAM model has a high degree of fitting to the sample, indicating that the CAM model is suitable for fitting the dynamic modulus master curves of various sealants. In this way, the main curve of dynamic shear modulus was also drawn using the CAM model in order to verify that the CAM model is able to fit the main curves of various types of asphalt sealant, predict the changing trend of asphalt sealant shear modulus, and perform a reasonable evaluation of the rheological properties of modified asphalt sealants.

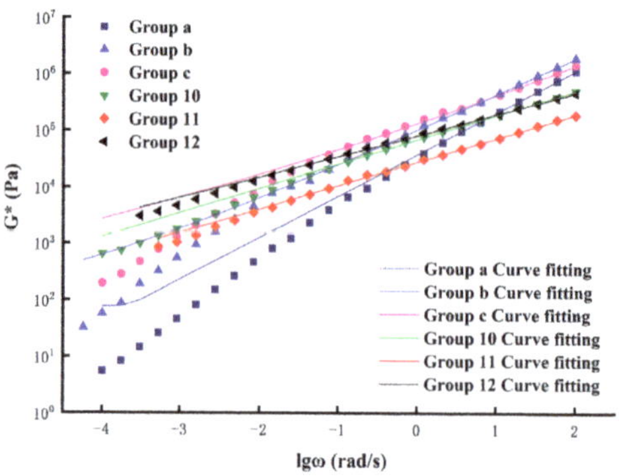

Figure 7. The main curve of complex shear modulus (reference temperature: 30 °C).

In summary, the mechanical properties of asphalt sealant materials are very sensitive to the frequency of load action. When the load frequency is low, the viscous properties of the asphalt sealant material are more obvious; when the load frequency in increased, the elastic properties of the asphalt sealant material become more obvious [51]. On the basis of the frequency sweep experiment, it can be seen that the modified asphalt sealant had lower temperature sensitivity, among which Group 12 had the lowest temperature sensitivity. Compared with the other groups of modified asphalt sealants, Group 12 still had better elasticity at high temperatures and better viscosity at low temperatures. This gives asphalt sealant modified with 20% CR and 7% SBS broad application prospects in cold regions.

4. Conclusions

With the aim of addressing the issue of the low resistance of conventional sealants to high-temperature deformation, this paper utilized waste rubber and SBS to prepare an eco-friendly asphalt sealant material for pavement maintenance engineering that possessed excellent high-temperature stability and anti-deformation ability, expanding the application of waste rubber in pavement maintenance engineering. Additionally, the high-temperature and rheological properties of this new type of SBS/CR-composite-modified asphalt sealant were estimated by performing a softening point test, a temperature scan test, a frequency scan test, and an MSCR and frequency sweep test. The following conclusions can be drawn:

(1) SBS and CR can improve the high-temperature performance of sealants, and the softening point of the modified sealant was able to meet the Chinese national standard (JT/T 740-2015). On the basis of grey correlation analysis, the correlation between SBS and the asphalt sealant softening point was 1.5 times that between CR and the asphalt sealant softening point.

(2) The temperature sweep test revealed that both SBS and CR can increase the complex shear modulus while simultaneously decreasing the phase angle, thereby improving the performance of the modified asphalts in high-temperature environments. In all groups, asphalt sealant modified with 20% CR and 5% SBS demonstrated superior deformation resistance and high-temperature resistance. The temperature sweep test revealed that SBS and CR improved the elasticity of the modified asphalt sealants in high-temperature environments. With comprehensive consideration of the composite shear modulus and phase angle of the sealants, the viscoelastic balance of the 20% CR and 5% SBS-modified asphalt sealant was more suitable for crack repair in all groups.

(3) The MSCR test showed that the composite modification procedure greatly improved the stiffness and elasticity of the modified asphalt sealants, among which the 20% CR

and 7% SBS-modified asphalt sealants had the lowest $\overline{J_{nr}}$ and the highest $\overline{R}$, implying that it had good deformation resistance.

(4) The complex main curve of shear modulus for different asphalt sealant materials at 30 °C showed that the SBS/CR-composite-modified asphalt sealant possessed greater persistent deformation resistance at elevated temperatures and lower temperature sensitivity than unmodified asphalt sealant and asphalt sealant modified with a single modifier. Additionally, the CAM model demonstrated a good ability to fit to the dynamic model curve of the modified asphalt sealant and was able to be used to reasonably evaluate the rheological properties of the sealant under load conditions that are difficult to test.

This paper systematically evaluated the high-temperature rheological properties of CR/SBS-modified asphalt sealant. As for determining the actual environmental benefits and actual road performance of this new kind of asphalt sealant, this is a topic on which the authors will deliberate in the future.

Author Contributions: Conceptualization: Y.Z.; Methodology: Y.G.; Formal analysis and investigation: Y.G., S.W.; Writing—original draft preparation: Y.G., S.W.; Writing—review and editing: Y.Z.; Funding acquisition: Y.G.; Resources: Y.P.; Supervision: Y.M. All authors have read and agreed to the published version of the manuscript.

Funding: The research was funded by the Science and Technology Development Program of Jilin Province (20200403157SF), the Transportation Science and Technology Project of Jilin Province (2021-1-1), the Science and Technology Research and Planning Project of Jilin Education Department (JJKH20211113KJ), the Special Funding for Basic Scientific Research Operation Fees of Central Universities and the Scientific (Grant No. 2020-JCXK-03), the Transportation Science and Technology Project of Jilin Province (2022 1 8), and the Major Emphasis Project of Department of Transportation of Heilongjiang Province (2022-02).

Institutional Review Board Statement: Not applicable.

Informed Consent Statement: Not applicable.

Data Availability Statement: The data presented in this study are available on request from the corresponding author.

Acknowledgments: The authors would like to thank Xueliang Wei for his guidance during manuscript revision. The authors would like to thank the anonymous reviewers for their constructive suggestions and comments that improved the quality of the paper.

Conflicts of Interest: The authors declare no conflict of interest.

References

1. Zhang, H.S.; Sheng, X.H.; Wang, S.W.; Xu, T. Effects of different modifiers on thermal stability, constituents and microstructures of asphalt-based sealant. *J. Therm. Anal. Calorim.* **2020**, *142*, 1183–1192. [CrossRef]
2. Zhou, L.; Huang, W.D.; Xiao, F.P.; Lv, Q. Shear adhesion evaluation of various modified asphalt binders by an innovative testing method. *Constr. Build. Mater.* **2018**, *183*, 253–263. [CrossRef]
3. Mazumder, M.; Sriraman, V.; Kim, H.H.; Lee, S. Quantifying the environmental impacts of crack sealing and filling treatment in hot mix asphalt pavement. *Innov. Infrastruct. Solut.* **2018**, *3*, 61. [CrossRef]
4. Kim, H.H.; Mazumder, M.; Lee, M.; Lee, S. Laboratory Evaluation of SBS Modified Asphalt Binder Containing GTR, SIS, and PE. *Adv. Civ. Eng.* **2020**, *2020*, 8830622. [CrossRef]
5. Gong, Y.; Xu, J.; Chang, R.; Yan, E. Effect of water diffusion and thermal coupling condition on SBS modified asphalts' surface micro properties. *Constr. Build. Mater.* **2021**, *273*, 121758. [CrossRef]
6. Munch, J.; Arepalli, U.M.; Barman, M. Decision Trees for Selecting Asphalt Pavement Crack Sealing Method. *Transp. Res. Rec.* **2020**, *2675*, 172–183. [CrossRef]
7. Yildirim, Y. Field performance comparison of asphalt crack-filling materials: Hot pour versus cold pour. *Can. J. Civ. Eng.* **2007**, *34*, 505–512. [CrossRef]
8. Cao, L.P.; Yang, C.; Dong, Z.J.; Wang, W.J.; Yin, H.M. Aging mechanism of hot-poured sealants for asphalt pavement under natural environmental exposure. *Int. J. Pavement Eng.* **2020**, *23*, 197–206. [CrossRef]

9. Tan, X.Y.; Zhang, J.P.; Guo, D.; Sun, G.Q.; Zhou, Y.Y.; Zhang, W.W.; Guan, Y.S. Preparation, characterization and repeated repair ability evaluation of asphalt-based crack sealant containing microencapsulated epoxy resin and curing agent. *Constr. Build. Mater.* **2020**, *256*, 119443. [CrossRef]
10. Liu, S.; Mo, L.T.; Wang, K.; Xie, Y.J.; Woldekidan, M.F. Preparation, microstructure and rheological properties of asphalt sealants for bridge expansion joints. *Constr. Build. Mater.* **2016**, *105*, 1–13. [CrossRef]
11. Zhang, G.; Zhang, H.L.; Bu, X.D.; Yang, H.Z. Laboratory study on performances of bimaleimide/unsaturated polyester resin modified asphalt. *Constr. Build. Mater.* **2018**, *179*, 576–586. [CrossRef]
12. Tan, Y.Q.; Guo, M.; Cao, L.P.; Zhang, L. Performance optimization of composite modified asphalt sealant based on rheological behavior. *Constr. Build. Mater.* **2013**, *47*, 799–805. [CrossRef]
13. Munera, J.C.; Erick, A.O. Polymer modified bitumen: Optimization and selection. *Mater. Des.* **2014**, *62*, 91–97. [CrossRef]
14. Wang, L.; Bi, F.; Zhao, P. Research progress of evaluation of rheological index properties of polymer modified asphalt. *J. Shandong Jianzhu Univ.* **2018**, *33*, 56–62. [CrossRef]
15. Veropalumbo, R.; Russo, F.; Oreto, C.; Buonocore, G.G.; Verdolotti, L.; Muiambo, H.; Biancardo, S.A.; Viscione, N. Chemical, Thermal, and Rheological Performance of Asphalt Binder Containing Plastic Waste. *Sustainability* **2021**, *13*, 13887. [CrossRef]
16. Veropalumbo, R.; Oreto, C.; Viscione, N.; Biancardo, S.A.; Russo, F. Environmental assessment of asphalt mastics containing plastic bottles and jet grouting waste. *Environ. Impact Assess. Rev.* **2022**, *93*, 106736. [CrossRef]
17. Russo, F.; Veropalumbo, R.; Viscione, N.; Oreto, C.; Biancardo, S.A. Rheological performance of soft and rigid waste plastic-modified bitumen and mastics. *Plast. Waste Sustain. Asph. Roads* **2022**, *4*, 61–83.
18. Fang, Y.; Zhan, M.S.; Wang, Y. The status of recycling of waste rubber. *Mater. Des.* **2000**, *22*, 123–127. [CrossRef]
19. Nejad, F.M.; Aghajani, P.; Modarres, A.; Firoozifar, H. Investigating the properties of crumb rubber modified bitumen using classic and shrp testing methods. *Constr. Build. Mater.* **2012**, *26*, 481–489. [CrossRef]
20. Kaloush, K.E. Asphalt rubber: Performance tests and pavement design issues. *Constr. Build. Mater.* **2014**, *67*, 258–264. [CrossRef]
21. Zhang, B.; Chen, H.X.; Zhang, H.G.; Wu, Y.C.; Kuang, D.L.; Guo, F.J. Laboratory investigation of aging resistance for rubberized bitumen modified by using microwave activation crumb rubber and different modifiers. *Materials* **2020**, *13*, 4230. [CrossRef] [PubMed]
22. Liang, M.; Xin, X.; Fan, W.Y.; Luo, H.; Wang, X.B.; Xing, B.D. Investigation of the rheological properties and storage stability of CR/SBS modified asphalt. *Constr. Build. Mater.* **2015**, *74*, 235–240. [CrossRef]
23. Cong, P.L.; Xun, P.J.; Xing, M.L.; Chen, S.F. Investigation of asphalt binder containing various crumb rubbers and asphalts. *Constr. Build. Mater.* **2013**, *40*, 632–641. [CrossRef]
24. Liu, H.Y.; Chen, Z.J.; Wang, W.; Wang, H.N.; Hao, P.W. Investigation of the rheological modification mechanism of crumb rubber modified asphalt (CRMA) containing tor additive. *Constr. Build. Mater.* **2014**, *67*, 225–233. [CrossRef]
25. Subhy, A.; Pires, G.M.; Lo Presti, D.; Airey, G. The effects of laboratory ageing on rheological and fracture characteristics of different rubberised bitumens. *Constr. Build. Mater.* **2018**, *180*, 188–198. [CrossRef]
26. Nanjegowda, V.H.; Biligiri, K.P. Recyclability of rubber in asphalt roadway systems: A review of applied research and advancement in technology. *Resour. Conserv. Recycl.* **2020**, *155*, 104655. [CrossRef]
27. Meng, Y.J.; Yan, T.Y.; Muhammad, Y.; Li, J.; Qin, P.F.; Ling, L.S.; Rong, H.L.; Yang, X.L. Study on the performance and sustainability of modified waste crumb rubber and steel slag powder/SBS composite modified asphalt mastic. *J. Clean. Prod.* **2022**, *338*, 130563. [CrossRef]
28. Becker, Y.M.; Muller, A.J.; Rodriguez, Y. Use of rheological compatibility criteria to study SBS modified asphalts. *J. Appl. Polym. Sci.* **2003**, *90*, 1772–1782. [CrossRef]
29. Liang, M.; Xin, X.; Fan, W.Y.; Ren, S.S.; Shi, J.T.; Luo, H. Thermo-stability and aging performance of modified asphalt with crumb rubber activated by microwave and tor. *Mater. Des.* **2017**, *127*, 84–96. [CrossRef]
30. Airey, G.D. Rheological properties of styrene butadiene styrene polymer modified road bitumens. *Fuel* **2003**, *82*, 1709–1719. [CrossRef]
31. Behnood, A.; Olek, J. Rheological properties of asphalt binders modified with styrene-butadiene-styrene (SBS), ground tire rubber (GTR), or polyphosphoric acid (PPA). *Constr. Build. Mater.* **2017**, *151*, 464–478. [CrossRef]
32. Huang, W.T.; Liu, X.; Zhang, S.W.; Zheng, Y.; Ding, Q.; Tong, B. Performance-guided design of permeable asphalt concrete with modified asphalt binder using crumb rubber and SBS modifier for sponge cities. *Materials* **2021**, *14*, 1266. [CrossRef] [PubMed]
33. Ma, F.; Li, C.; Fu, Z.; Huang, Y.; Dai, J.S.; Feng, Q. Evaluation of high temperature rheological performance of polyphosphoric acid-SBS and polyphosphoric acid-crumb rubber modified asphalt. *Constr. Build. Mater.* **2021**, *306*, 124926. [CrossRef]
34. Qian, C.D.; Fan, W.Y.; Yang, G.M. Influence of crumb rubber particle size and SBS structure on properties of CR/SBS composite modified asphalt. *Constr. Build. Mater.* **2020**, *235*, 117517. [CrossRef]
35. Kołodziej, K.; Bichajło, L.; Siwowski, T. Effects of aging on the physical and rheological properties of trinidad lake asphalt modified bitumen. *Materials* **2021**, *14*, 2532. [CrossRef]
36. Cheng, P.F.; Li, Y.M.; Zhang, Z.M. Effect of phenolic resin on the rheological, chemical, and aging properties of SBR-modified asphalt. *Int. J. Pavement Res. Technol.* **2021**, *14*, 421–427. [CrossRef]
37. Ministry of Transport of the People's Republic of China. *Standard Test Methods of Bitumen and Bituminous Mixtures for Highway Engineering*; Ministry of Transport of the People's Republic of China: Beijing, China, 2011. (In Chinese)
38. Yuan, D.M.; Su, B.; Liao, K.J.; Yan, F. Preparation of rubber crumb modified asphalt and its influencing factors. *Hecheng Xiangjiao Gongye* **2008**, *30*, 382–386.

39. Hu, M.; Mi, H.C.; Zhang, L. Mechanism Study of Impaction of Rubber Powder Content and the Mesh on Rubber Modified Asphalt Viscosity. *Pet. Asph.* **2011**, *2*, 41–43.
40. Ministry of Transport of the People's Republic of China. *Hot-Poured Sealants for Pavement*; Ministry of Transport of the People's Republic of China: Beijing, China, 2015. (In Chinese)
41. Zhu, M.L.; Zhang, Q.S. Applied research on SBS modified bitumen and SMA mixture. *China J. Highw. Transp.* **2001**, *14*, 13–20.
42. Liu, Y. *Study on Mechanism and Performance Evaluation of SBS and Crumb Rubber Composite Modified Asphalt*; Southeast University: Nanjing, China, 2018.
43. Shan, C.Y.; Wei, Y.G.; Zhang, Y.J.; Li, X.H.; Nie, L.; Dong, X.L.; Liu, C. Grey correlation degree analysis on main traits of soybean varieties in Heilongjiang Province. *Soybean Sci.* **2009**, *28*, 945–948.
44. Wang, K.F.; Ren, S.Y.; Qian, S.Y.; Liu, P. Grey relational analysis of benefit of surgical management for abdominal aortic aneurysm. *Int. Surg.* **2014**, *99*, 189–194. [CrossRef] [PubMed]
45. Lin, S.M. Associating DEA with grey relational analysis for performance assessment of management competencies in logistic industry. *J. Stat. Manag. Syst.* **2010**, *13*, 1117–1130. [CrossRef]
46. Hamzacebi, C.; Pekkaya, M. Determining of stock investments with grey relational analysis. *Expert Syst. Appl.* **2011**, *38*, 9186–9195. [CrossRef]
47. Xu, O.M.; Xiao, F.P.; Han, S.; Amirkhanian, S.N.; Wang, Z.J. High temperature rheological properties of crumb rubber modified asphalt binders with various modifiers. *Constr. Build. Mater.* **2016**, *112*, 49–58. [CrossRef]
48. Qian, C.D.; Fan, W.Y.; Liang, M.; He, Y.; Ren, S.S.; Lv, X.B.; Nan, G.Z.; Luo, H. Rheological properties, storage stability and morphology of CR/SBS composite modified asphalt by high-cured method. *Constr. Build. Mater.* **2018**, *193*, 312–322. [CrossRef]
49. Xiao, B.Q.; Xiao, M.L.; Ding, B.W. Research on Rubber/SBS modified asphalt by MSCR test. *Adv. Mater. Res.* **2012**, *557–559*, 1066–1069. [CrossRef]
50. Ren, S.S.; Liu, X.Y.; Li, M.L.; Fan, W.Y.; Xu, J.; Erkens, S. Experimental characterization of viscoelastic behaviors, microstructure and thermal stability of CR/SBS modified asphalt with TOR. *Constr. Build. Mater.* **2020**, *261*, 120524. [CrossRef]
51. Li, C.M.; Ning, F.B.; Li, Y.Y. Effect of carbon black on the dynamic moduli of asphalt mixtures and its master curves. *Front. Struct. Civ. Eng.* **2012**, *13*, 918–925. [CrossRef]
52. Tan, G.J.; Wang, W.S.; Cheng, Y.C.; Wang, Y.; Zhu, Z.Q. Master curve establishment and complex modulus evaluation of sbs-modified asphalt mixture reinforced with basalt fiber based on generalized sigmoidal model. *Polymers* **2020**, *12*, 1586. [CrossRef]

polymers

MDPI

Article

Investigation on Fatigue Performance of Diatomite/Basalt Fiber Composite Modified Asphalt Mixture

Chunfeng Zhu [1], Huijin Luo [1], Wei Tian [1,*], Binbin Teng [1], Yongmei Qian [1], Huaxue Ai [1] and Bo Xiao [2]

[1] College of Civil Engineering, Jilin Jianzhu University, Changchun 130118, China; zhuchunfeng@jlju.edu.cn (C.Z.); luohuijin@jlju.edu.cn (H.L.); tengbinbin@jlju.edu.cn (B.T.); qianyongmei@jlju.edu.cn (Y.Q.); aihuaxue@jlju.edu.cn (H.A.)

[2] Changchun Municipal Engineering Design & Research Institute, Changchun 130033, China; xiaobo_1980@sina.com

* Correspondence: tianwei@jlju.edu.cn; Tel.: +86-0431-8456-6178

Abstract: The fatigue resistance of asphalt mixture is an important indicator to evaluate the durability of asphalt pavement. In order to improve the fatigue properties of asphalt mixture, diatomite and environmental basalt fiber were added. Four types of asphalt mixtures, ordinary asphalt mixture (AM), diatomite modified asphalt mixture (DAM), basalt fiber modified asphalt mixture (BFAM) and diatomite/basalt fiber composite modified asphalt mixture (DBFAM), were chosen, whose optimum asphalt–aggregate ratio, optimum content of diatomite and optimum content of basalt fiber could be determined by Marshall test and response surface methodology (RSM). The multi-functional pneumatic servo Cooper test machine was carried out by a four-point bending fatigue test. Through the comparative analysis of flexural-tensile stiffness modulus (S), initial stiffness modulus(S_0), residual stiffness modulus ratio, lag angle (ϕ) and cumulative dissipation energy (E_{CD}), the fatigue resistance of asphalt mixture can be effectively improved by adding diatomite and basalt fiber. Grey correlation analysis was also used to analyze the degree of correlation between the fatigue life and the influencing factors such as VV, VMA, VFA, OAC, S, and E_{CD}. The analysis results indicate that E_{CD} has the greatest impact on the fatigue life of the asphalt mixture.

Keywords: fatigue performance; diatomite; basalt fiber; asphalt mixture; four-point bending fatigue test; grey correlation analysis

Citation: Zhu, C.; Luo, H.; Tian, W.; Teng, B.; Qian, Y.; Ai, H.; Xiao, B. Investigation on Fatigue Performance of Diatomite/Basalt Fiber Composite Modified Asphalt Mixture. *Polymers* **2022**, *14*, 414. https://doi.org/10.3390/polym14030414

Academic Editor: Marco Morreale

Received: 26 November 2021
Accepted: 18 January 2022
Published: 20 January 2022

1. Introduction

Asphalt is a mixture of black-brown polymers composed of hydrocarbons with different molecular weights and their non-metallic derivatives. Asphalt is mainly used in coatings, plastics, rubber and other industries and paving pavements. Asphalt is a widely used pavement structural cementitious material in road engineering. It can be mixed with different mineral materials in proportion to build asphalt pavement with different structures. Improving the performance of asphalt or asphalt mixture is an important means to prolong the service life of pavements. In recent years, many researchers have improved the performance of asphalt or asphalt mixture by adding modified substances. The anti-fatigue characteristic of the asphalt mixture is one of the important indicators of road performance.

Fatigue refers to a phenomenon caused by the accumulation of unrecoverable strength attenuation of asphalt mixture pavement under repeated loads [1–3]. In the service process, asphalt pavement is subjected to the repeated action of vehicle load and temperature stress for a long time. With the increase of load times, defects and microcracks will occur in the asphalt mixture [4,5]. Moreover, these defects and micro-cracks will continue to extend under dynamic load, the strength of the pavement structure will gradually decay and finally, fatigue failure and cracks will occur in the pavement. In order to ensure that the asphalt pavement has good usability and durability, the asphalt mixture must have good fatigue resistance. Fatigue life is an important indicator to evaluate the service life of asphalt pavement [6–8].

In order to improve the fatigue of asphalt pavement, many researchers add different types of reinforcement materials in asphalt mixtures, such as cellulose, lignin, glass fiber, polyester fiber, inorganic fiber and so on [9–13]. Zheng et al. [14] studied the fatigue property of basalt fiber-modified asphalt mixtures under complicated environments. Zhao et al. [15] investigated the effect of basalt fiber on the road performance of asphalt mixture. Riekstins et al. [16] evaluated the anti-fatigue property of brucite fiber reinforced asphalt mixture under sulfate and dry-wet circle corrosion environments. Lou et al. [17] evaluated the enhancement effect of basalt fiber on the fatigue performance of the mixtures.

Considering the climatic characteristics of the seasonal freezing region in Jilin Province, China, this study mixed diatomite and basalt fiber into asphalt mixtures and evaluated their improvement effect on the fatigue resistance of asphalt mixture through fatigue test. A fatigue test mainly employs phenomenon mechanics and mechanics approximation methods [18,19], but there are more experimental methods for using phenomenon mechanics at present. Due to the low cost and short cycle of laboratory tests, they are widely used. An indoor fatigue test mainly includes a repeated bending test, support bending test, uniaxial test, triaxial test, indirect tensile test and so on. SHRPA-003A indicates that the repeated bending fatigue test (especially the four-point bending test) is most conforming to the actual pavement stress, and it is more reasonable to use the four-point bending fatigue test of the rectangular beam as the standard test for the fatigue performance of asphalt mixture. Yu et al. [20,21] studied three asphalt mixture fatigue testing machines, James Cox & Sons (USA), Cooper μμ(UK) and IPC (Australia). The results showed that the test results of the above three four-point bending fatigue testing machines were consistent. Cooper and IPC fatigue testing machines can obtain lag angle and dissipated energy, which is beneficial to fatigue test research and analysis. Di Benedetto, Cardona DR, Sébastien Lamothe, Zhiyong Wu, Yang Panpan and other scholars studied the fatigue performance of asphalt mixture using dissipated energy, lag angle, cumulative dissipation energy and the ratio of dissipated energy change methods [22–28].

After careful consideration, this paper uses the multi-functional pneumatic servo Cooper test machine (NU-14) produced by Cooper Research Technology Limited in the UK to carry out a four-point bending fatigue test, testing the initial stiffness modulus, stress value, strain value, lag angle, cumulative dissipation energy and fatigue life, etc. The factors affecting fatigue life are analyzed by grey correlation theory.

2. Experiments and Methods

2.1. Raw Materials

The asphalt AH-90 in this paper was supplied by Panjin Petrochemical Industry, located in Liaoning Province of China. The physical properties of the asphalt were measured according to Chinese specification JTG E20-2011 [29]; the measured results are presented in Table 1. The ecofriendly diatomite was obtained from Changbai Mountain in Jilin Province. The average size of diatomite particles is approximately 10 μm and its shape is a disc under the microscope, as shown in Figure 1. The basalt fiber was supplied from Jiuxin Basalt Industry Co., Ltd. in Jilin Province, as shown in Figure 2, and the physical properties of these materials are presented in Tables 2–5 [30]. The laboratory temperature was 20 ± 2 °C; the humidity was 60%.

Table 1. Physical properties of AH-90 asphalt.

Properties	Values	Standard Values
Density (15 °C, g/cm^3)	1.016	—
Penetration (25 °C, 0.1 mm)	91.8	80–100
Softening point $T_{R\&B}$ (°C)	46.9	≥45
Ductility (25 °C, cm)	>150	≥100
Viscosity (135 °C, Pa·s)	306.9	—
	After TFOT	
Mass loss (%)	0.38	≤+0.8
Residual penetration ratio (25 °C, %)	73.3	≥57
Softening point $T_{R\&B}$ (°C)	49.6	—
Ductility (15 °C, cm)	>120	≥20
Viscosity (135 °C, Pa·s)	432.5	—

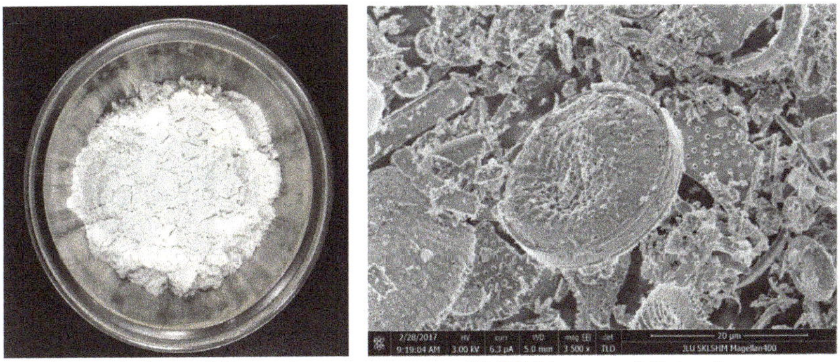

(**a**) Morphology of diatomite (**b**) SEM images of diatomite

Figure 1. Diatomite.

(**a**) Morphology of basalt fiber (**b**) ESEM images of basalt fiber

Figure 2. Basalt fiber.

Table 2. Properties of diatomite.

Properties	Color	pH	Specific Gravity (g/cm^3)	Bulk Density (g/cm^3)
Value	Off-white	7–8	2.0–2.2	0.34–0.41

Table 3. The mineral composition of diatomite (provided by manufacturer).

Mineral Composition	SiO_2	Al_2O_3	Fe_2O_3	CaO	MgO	TiO_2	K_2O	Loss on Ignition
Content (%)	85.60	4.50	1.50	0.52	0.45	0.30	0.67	4.61

Table 4. The mineral composition of basalt fiber (provided by manufacturer).

Mineral Composition	SiO_2	Al_2O_3	Fe_2O_3	CaO	MgO	Ti_2O	Na_2O	Others
Content (%)	50	15	12	10	5	1.5	4	2.5

Table 5. Properties of basalt fiber (provided by manufacturer).

Properties	Values	Standard Values
Diameter (μm)	10–13	—
Length (mm)	6	—
Water content (%)	0.030	≤0.2
Combustible content (%)	0.56	—
Linear density (Tex)	2398	2400 ± 120
Breaking strength (N/Tex)	0.55	≥0.40
Tensile strength (MPa)	2320	≥2000
Tensile modulus of elasticity (GPa)	86.3	≥85
Elongation at break (%)	2.84	≥2.5

The used aggregates were basalt produced in Liaoyuan, Jilin Province. The properties of aggregates were tested in accordance with the Chinese specification JTG E42-2005 [31]. The technical properties of aggregates are listed in Tables 6 and 7, which meet the requirements of the Chinese specification JTG F40-2004 [32].

Table 6. Aggregate test results and specifications index.

Properties	Value	
Diameter mm	13.2~16	9.5~13.2
Bulk specific gravity (g/cm^3)	2.810	2.805
Gross volume relative density (g/cm^3)	2.784	2.771
Water absorption (%)	0.33	0.44

Table 7. Coarse aggregate test results and specifications index.

Properties	Value	Standard Value
Crushing value (%)	8.9	≤26
Percentage of flat-elongated particles (≥9.5 mm) (%)	3.6	≤8
Percentage of flat-elongated particles (<9.5 mm) (%)	4.8	≤12
Abrasion value (%)	12	≤28

2.2. Gradation Design of Asphalt Mixture

The gradation type of asphalt mixture used in this study was AC-13, and the selected gradation curve is shown in Figure 3 according to Chinese specification JTG F40-2004 [32].

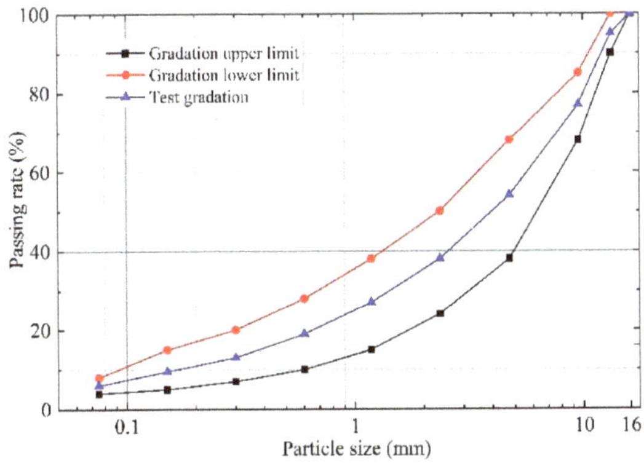

Figure 3. Aggregate gradation curve of AC-13 in this study.

According to the compaction method in the Chinese specification JTG E20-2011 [29], the asphalt mixture was developed into a standard Marshall test piece. The optimum asphalt–aggregate ratio (OAC) of the asphalt mixture is 4.78%, determined by testing the test indexes including VV, VMA, VFA, MS and FL. On this basis, the range of optimum asphalt content of diatomite/basalt fiber composite modified asphalt mixture (DBFAM) could be selected. According to a large number of research data [14,33–36], it is determined that the content range of basalt fiber is 0.2% to 0.4% and the dosage range of diatomite instead of mineral powder is 5% to 10%. Using the central composite design (CCD) of response surface methodology (RSM) to determine the optimal proportion of the DBFAM [9], the optimum asphalt–aggregate ratio is 5.22%, the optimal content of diatomite is 6.3% and basalt fiber is 0.25%.

In order to ensure the test data are comparable, basalt fiber and diatomite were separately added into AC-13, and the optimum asphalt–aggregate ratios of basalt fiber modified asphalt mixture (BFAM) and diatomite modified asphalt mixture (DAM) were determined to be 5.09% and 5.12%, respectively, based on the Chinese specification JTG F40-2004 [32]. The physical properties of asphalt mixtures are shown in Table 8. The laboratory temperature was 25 ± 2 °C and the humidity was 60% when the test was carried out.

Table 8. Physical properties of asphalt mixtures.

Indicators	AM	BFAM	DAM	DBFAM
OAC (%)	4.78	5.09	5.12	5.22
VV (%)	4.098	4.065	4.065	4.20
VMA (%)	15.6	16.2	15.4	16
VFA (%)	73.7	74.9	73.6	73.8

2.3. *Four-Point Bending Fatigue Test*

2.3.1. Preparation of Asphalt Mixture Beam

The preparation process of the four-point bending fatigue trabecular specimens mainly includes mixture mixing, specimen rolling and forming and trabecular cutting. The mixture mixing was carried out in accordance with the standard Marshall mixture specimen preparation process. The evenly mixed mixture was loaded into the test mold (400 mm × 300 mm × 75 mm), and then formed by high pressure vibration through the asphalt mixture vibratory roller compacting machine produced in Britain. The specimen was formed by controlling the height of the wheel grinding plate, and the air pump pressure was set to 2 bar, with two plates for each component type. The laboratory temperature was 20 ± 2 °C. The molding test equipment is shown in Figure 4.

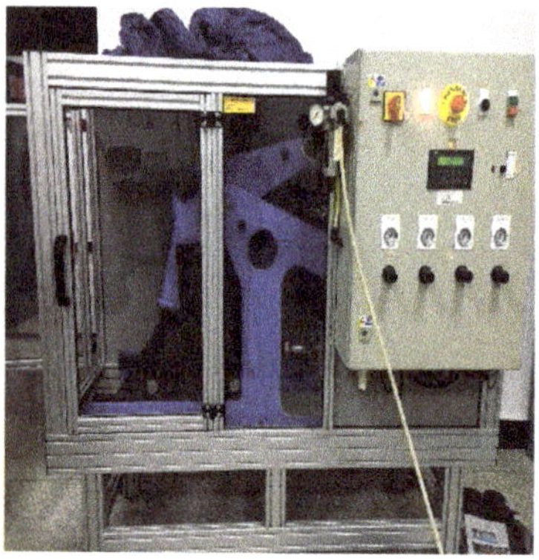

Figure 4. Vibratory roller compacting machine.

After the test specimen was formed for 24 h, the formwork was removed, the test plate was cut and the infrared control cutting machine was used. The slab was cut into small beams with dimensions of 380 mm long, 65 mm wide and 50 mm high. Each slab can be cut into four trabecular test beams. In order to ensure that the trabecular is dry enough, the fatigue test was carried out after the test beams were cut for 24 h.

2.3.2. Four-Point Bending Fatigue Test

The fatigue test was carried out using the Cooper testing machine, and 800 strain level was adopted during loading. The test device is shown in Figure 5.

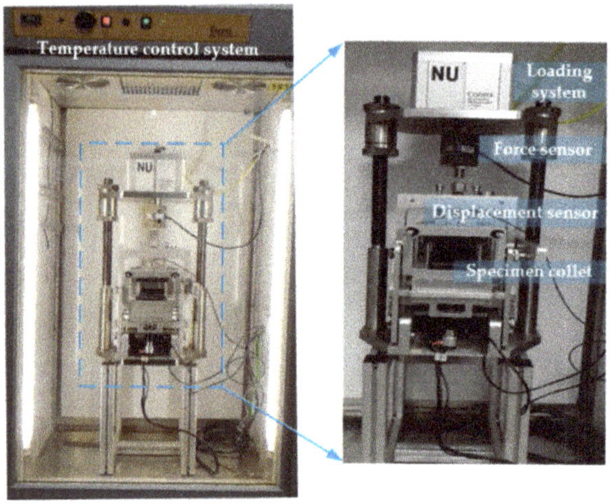

Figure 5. Fatigue test device for four-point bending beam.

The test waveform was a sine wave, the frequency was 10 Hz, the test temperature was 20 °C, the relative humidity was 60%, the test failure criterion was that the stiffness modulus of the specimen decreased to 50% of its initial modulus during loading and the cyclic loading times at this time were used as the fatigue life of the asphalt mixture to evaluate the durability of the asphalt material under the action of alternating loads.

The stiffness modulus of the 100th loading cycle was set as the initial stiffness modulus of the test specimen. Before the fatigue test, a vernier caliper was used to accurately measure the length, width and height of the beam, and the data were input into the computer software system (the distance between four points remains the default value:118 mm), setting the corresponding parameters in the software system. Before the test, the specimen was put into the test temperature for at least 4 h, then the beam was installed into the fatigue loading control system, clamped and fixed, and the test parameters were set according to the software interface. The four-point bending fatigue test would obtain the following test indexes: initial stiffness modulus, stiffness modulus attenuation percentage, stress value, strain value, lag angle, cumulative dissipation energy and fatigue life (number of cyclic loads).

2.3.3. Calculation Procedure of Fatigue Performance

- The calculation equation of maximum tensile stress is listed in Equation (1):

$$\sigma_t = \frac{L \times P}{\omega \times h^2}, \tag{1}$$

where σ_t is the maximum tensile stress (Pa); L is the beam span (the distance between two clamps at the outer end); P is a peak load (N); ω is beam width (m); h the beam height (m).

- The maximum tensile strain can be calculated by Equation (2):

$$\varepsilon_t = \frac{12 \times \delta \times h}{3 \times L^2 - 4 \times a^2} \tag{2}$$

where ε_t is the maximum tensile strain (m/m); δ is the maximum strain of beam center (m); a is the spacing between adjacent collets (i.e., $L/3$) (m).

- The flexural-tensile stiffness modulus is obtained by Equation (3):

$$S = \frac{\sigma_t}{s_t}, \tag{3}$$

where S is the flexural-tensile stiffness modulus (Pa).

- The lag angle is calculated according to Equation (4):

$$\phi = 360 \times f \times t, \tag{4}$$

where ϕ is the lag angle (°); f is the loading frequency (Hz); t is the lag time of strain peak to stress peak (s).

- The single-cycle dissipation energy is calculated by Equation (5):

$$E_D = \pi \times \sigma_t \times \varepsilon_t \times \sin\phi, \tag{5}$$

where E_D is the dissipated energy of a single cycle (J/m^3).

- The cumulative dissipation energy is calculated according to Equation (6):

$$E_{CD} = \sum_{i=1}^{n} E_{Di}, \tag{6}$$

where E_{CD} is the accumulated dissipated energy (J/m^3); E_{Di} is the dissipated energy of a single cycle for the i-th loading.

- The ratio of dissipated energy change (*RDEC*) is calculated according to Equation (7):

$$RDEC = \frac{E_{Di} - E_{Dj}}{E_{Di}(j - i)}, \tag{7}$$

where *RDEC* is the ratio of dissipated energy change; E_{Di} and E_{Dj} are the dissipated energy when the loading cycles are i and j, respectively.

2.4. Grey Correlation Analysis

Grey system theory puts forward a new analysis method, called the systematic correlation degree analysis method, which measures the degree of correlation between factors according to the degree of similarity or dissimilarity. Because the correlation analysis method is analyzed according to the development trend and can avoid the interaction of various factors, there is no excessive requirement on the sample size, and no typical distribution law is required. The amount of calculation is small, and the quantitative results of the correlation degree will not be inconsistent with the qualitative analysis. This method can analyze the influence of various factors on the fatigue life of asphalt mixture. Relevance is essentially the difference in geometry between curves. The closer the geometry is, the closer the development trend is, and the greater the degree of relevance is [37,38].

The fatigue performance of asphalt mixture is affected by a variety of factors, among which the main factors are s stiffness modulus, asphalt type, asphalt content, porosity, mineral type, gradation type and test conditions. The research shows that in the fatigue test under strain control, the lower the stiffness modulus of the mixture, the smaller the stress required to maintain the same strain, and the crack propagation will continue for a long time. The smaller the stiffness modulus, the longer the fatigue life of the material. Asphalt species, asphalt content, etc., also affect the fatigue life of the asphalt mixture by affecting the stiffness modulus. Porosity, mineral aggregate type and gradation will affect the fatigue life of asphalt mixture by affecting the composition, structure and internal defects of asphalt mixture. In this paper, grey correlation analysis is used to analyze the correlation degree between the relevant influencing factors and the fatigue life of the asphalt mixture.

The reference sequence and comparison sequence should be specified before correlation analysis. In this paper, the fatigue life of the asphalt mixture is taken as a reference sequence, and the factors influencing the fatigue life asphalt mixture are taken as a comparison sequence. The reference number is listed as x_0, $x_0 = (x_0\ (1), x_0\ (2), \ldots, x_0\ (n))$, the comparison is listed as x_1, $x_1 = (x_1\ (1), x_1\ (2), \ldots, x_1\ (n))$.

The correlation coefficient between the comparison curve and the reference curve at time k is:

$$\xi_i(k) = \frac{\underset{i}{\min}\underset{k}{\min}|x_0(k) - x_i(k)| + 0.5\underset{i}{\max}\underset{k}{\max}|x_0(k) - x_i(k)|}{|x_0(k) - x_i(k)| + 0.5\underset{i}{\max}\underset{k}{\max}|x_0(k) - x_i(k)|}, \tag{8}$$

where $\xi_i(k)$ is the relative difference between the comparison curve and the reference curve at the kth moment, and this form of relative difference is the correlation coefficient of x_i to x_0 at time k. $\underset{i}{\min}\underset{k}{\min}|x_0(k) - x_i(k)|$ is the minimum difference between two hierarchies; $\underset{i}{\max}\underset{k}{\max}|x_0(k) - x_i(k)|$ is the maximum difference between the two levels.

The general expression of grey correlation degree is expressed in Equation (9):

$$r_i = \frac{1}{N}\sum_{i=1}^{N} \xi_i(k), \tag{9}$$

where r_i is the correlation degree between the curve and the reference curve.

3. Results and Discussion

3.1. Variation Law of Flexural-Tensile Stiffness Modulus

The variation trend of flexural tensile stiffness modulus (S) of four types of asphalt mixtures with cyclic loading times is shown in Figure 6. It is evident from Figure 6 that the fatigue process of the asphalt mixture mainly goes through two stages before 50% of the initial stiffness modulus. In the first stage, S decreased sharply, and in the second stage, S decreased slowly and close to linearity. The sharp decrease of S in the first stage is mainly caused by the reorganization of the internal structural materials of the test specimen under the action of alternating load. The second stage is the main stage of material fatigue damage. The test specimen produces fatigue damage under the action of alternating load, and this stage is the initiation and development process of microcracks [39].

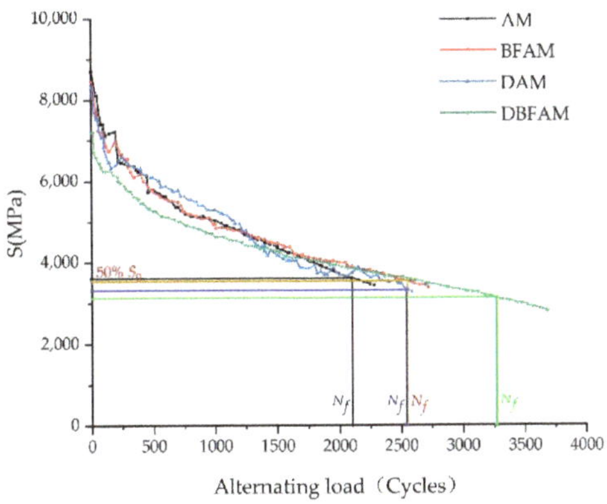

Figure 6. Attenuation law of flexural-tensile stiffness modulus of asphalt mixtures.

The fatigue test results indicated that the fatigue life order of four types of asphalt mixtures is: DBFAM > BFAM ≈ DAM > AM. The addition of diatomite and basalt fiber can significantly improve the fatigue performance of asphalt mixture, and the improvement effect of 0.25% basalt fiber and 6.3% diatomite on the fatigue performance of asphalt mixture is similar. The fatigue life of the asphalt mixture can be greatly improved by adding diatomite and basalt fiber. The test results show that the fatigue life of DBFAM is 55.7% higher than that of AM.

3.2. Analysis of Initial Stiffness Modulus

The initial flexural tensile stiffness modulus (S_0) is closely related to the determination of the fatigue life of asphalt mixture specimens. In this paper, the stiffness modulus of the 100th loading cycle is uniformly set as the initial stiffness modulus of the test specimens [40]. The experimental results of the initial flexural tensile stiffness modulus of four types of asphalt mixtures are shown in Figure 7. It can be seen from the figure that under the action of 800 strain level, the order of the initial stiffness modulus is AM > BFAM > DAM > DBFAM. The addition of basalt fiber and diatomite increases the optimum bitumen–aggregate ratio of asphalt mixture, and the increase of asphalt content is the main reason for the decrease of initial flexural tensile stiffness modulus.

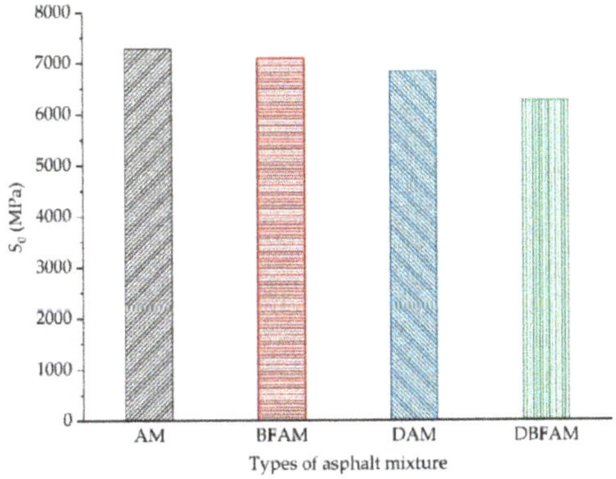

Figure 7. Initial flexural-tensile stiffness modulus test results of asphalt mixture.

3.3. Analysis of Residual Stiffness Modulus Ratio

The residual stiffness modulus ratio is defined as the percentage of flexural tensile stiffness modulus corresponding to each alternating load in the initial flexural tensile stiffness modulus, that is, the attenuation rate of flexural tensile stiffness modulus. The relationship between the attenuation rate of flexural-tensile stiffness modulus and fatigue life is shown in Figure 8.

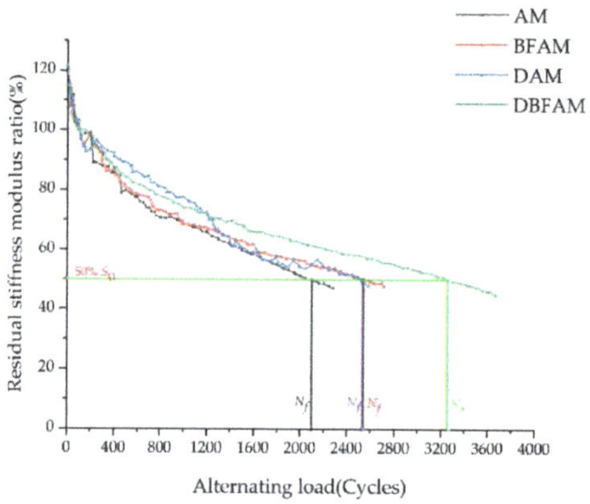

Figure 8. Residual flexural-tensile stiffness modulus ratio of asphalt mixture.

It can be seen from the test results in Figure 7 that the variation trend of the flexural tensile stiffness modulus attenuation rate of four types of asphalt mixtures under alternating load is the same as that of the above flexural tensile stiffness modulus, which is divided into two stages before 50% of the initial stiffness modulus. The first stage is a steep decline, and the second stage drops slowly and nearly linearly.

3.4. Analysis of Lag Angle

The lag angle (ϕ) is the phase difference between the strain rotation vector and the stress rotation vector [41]. The lag angle (ϕ) can reflect the viscoelasticity of the asphalt mixture. The larger ϕ is, the more viscous the mixture tends to be. Otherwise, the mixture tends to be elastic [42]. For an ideal elastomer, the lag angle ϕ is 0, for purely viscous bodies, ϕ is $\pi/2$, for viscoelastic materials, ϕ is between 0 and $\pi/2$ [43].

The variation trend of lag angle with loading times is shown in Figure 9. It can be seen from Figure 9 that the lag angle of four types of asphalt mixtures is increasing, but the increase is relatively slow. It shows that the viscoelastic properties of the four types of asphalt mixtures are constantly changing in during the process of the fatigue test. With the increase of loading times, the asphalt mixture continues to develop in the direction of viscosity. This is because, in the process of the fatigue test, the alternating load continuously works on the specimen, resulting in a certain increase in the temperature of the test piece, the lag angle gradually increases, and the asphalt binder changes from elasticity to viscosity. It can be seen from Figure 9 that after the mineral powder is replaced by diatomite, the lag angle becomes smaller and the elasticity of the asphalt mixture increases. The influence of basalt fiber on the lag angle is not very obvious; the viscoelastic change of asphalt mixture is small. Adding diatomite and basalt fiber increases the elasticity of the asphalt mixture. In addition, the fatigue tests of the four kinds of asphalt mixtures are all carried out under the action of high strain level and the time when the materials reach fatigue failure is short. The number of cycles is few, leading to the trend of a slow increase of lag angle.

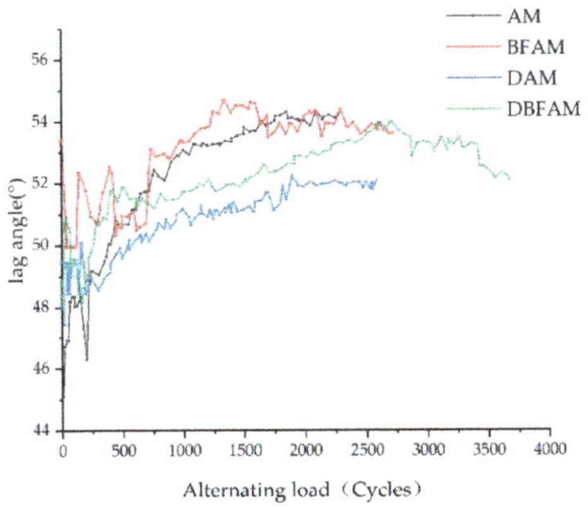

Figure 9. Variation of lag angle of asphalt mixture.

3.5. Analysis of Dissipation Energy

1. Dissipation energy under each load

Asphalt mixture is a type of viscoelastic material. Under the action of alternating load, the stress–strain curves during loading and unloading will not coincide, resulting in a hysteretic curve. The dissipated energy can be determined by the area of this stress–strain hysteretic line and the calculation equation is shown in Equation (5). The variation law of dissipated energy can indirectly reflect the fatigue damage evolution process of asphalt mixture under alternating load.

Figure 10 show the changing trend of energy consumption for four types of asphalt mixtures with loading times. It can be seen from the figure that the energy consumption of each loading of four types of asphalt mixtures decreases with the increase of loading times; compared to AM, BFAM and DAM, DBFAM has less dissipation energy per load; therefore, the addition of basalt fiber and diatomite can significantly improve the toughness of asphalt mixture, increase the elastic recovery capacity of asphalt mixture and reduce the dissipated energy under each load.

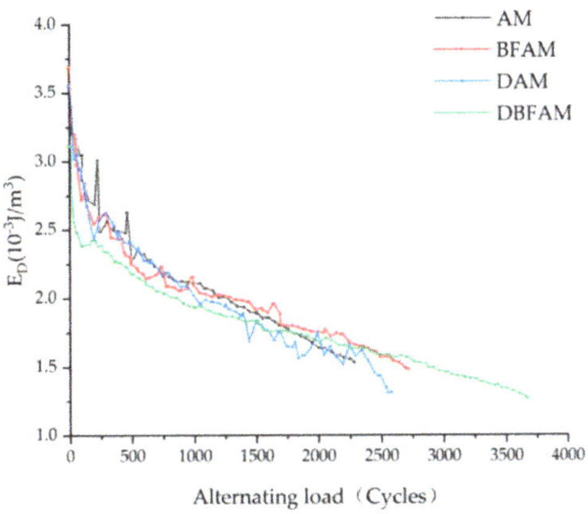

Figure 10. Dissipation energy of asphalt mixture under each loading.

2. Analysis of cumulative dissipation energy results

The fatigue failure of asphalt mixture is a process of constant energy consumption (E_{CD}) under an alternating load. According to the theory of fracture mechanics, the fatigue failure of materials is the process of continuous crack propagation, and the energy lost in each loading and unloading cycle is accumulated and transformed into the surface energy of crack propagation to form a new surface. When the accumulated dissipated energy reaches a certain limit during the action of alternating load, the material will undergo fatigue failure.

The cumulative dissipation energy is calculated according to Equation (6). As shown in Figure 11, the changing trend of cumulative dissipation energy of four types of asphalt mixtures is given. It can be seen from Figure 12 that the cumulative energy consumed by the test piece gradually increases at a certain rate with the increase of the action times of alternating load. The cumulative dissipated energy of basalt fiber modified asphalt mixture and matrix asphalt mixture increases rapidly, followed by diatomite modified asphalt mixture, and the cumulative dissipated energy of composite modified asphalt mixture increases slowly.

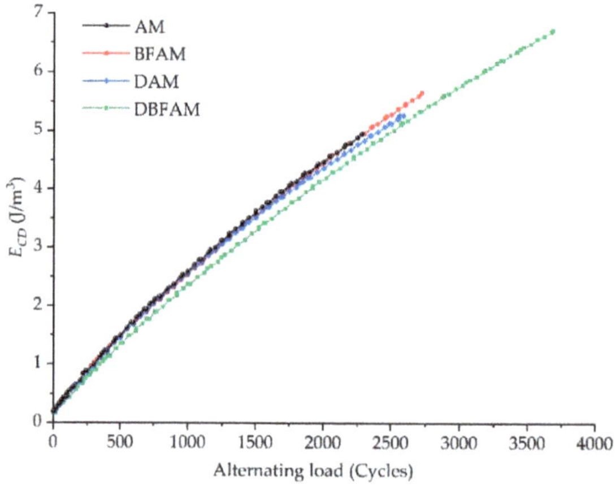

Figure 11. The changing trend of cumulative dissipation energy.

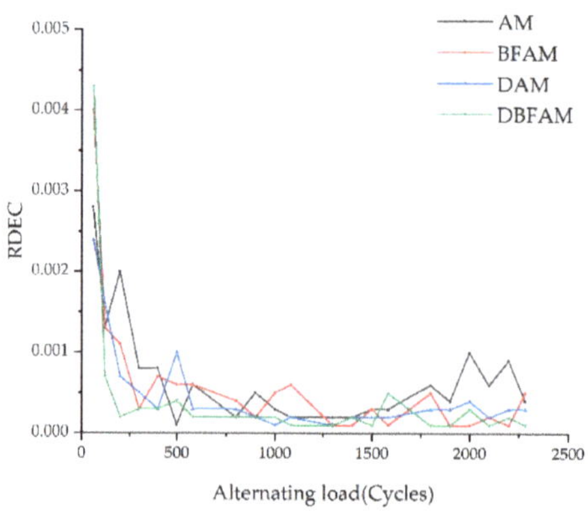

Figure 12. The ratio of dissipated energy change.

Adding basalt fiber and diatomite can improve the elastic recovery ability of asphalt mixture and reduce the energy loss of asphalt mixture caused by work, heat and material damage. The cumulative dissipation energy of DAM is smaller than that of BFAM and AM, which can improve the fatigue performance of the asphalt mixture to a certain extent. Although the cumulative dissipation energy of BFAM is roughly the same as that of AM, the addition of basalt fiber can enhance and prevent cracks. After the addition of fiber, it can prevent the propagation of original defects (microcracks) in the asphalt mixture and effectively delay the initiation of new cracks [14,17,28,44].

3. Analysis of the ratio of dissipated energy change

The variation of the rate of dissipated energy change (*RDEC*) is plotted in Figure 12. It is found that the rate dissipated energy change of asphalt mixture has only two stages, and the third stage does not appear. That is, when the modulus of the asphalt mixture reaches 50%, the failure stage of the asphalt mixture has not been reached [17]. However, *RDEC* in the second stage is in a stable state, so the plateau value (PV) is still calculated by using the average value of the change rate of dissipative energy in the second stage (600–1700 Cycles). The results are shown in Table 9.

Table 9. The Plateau Value.

Asphalt Mixture Type	Strain Level (με)	PV
AM	800	2.41×10^{-4}
BFAM	800	2.47×10^{-4}
DAM	800	2.11×10^{-4}
DBFAM	800	1.96×10^{-4}

It can be seen from Table 9 that the PV value increases slightly after adding basalt fiber. It means that the energy loss of the asphalt mixture could well increase after adding basalt fiber [17,28]. The PV value decreased after adding diatomite and two materials, indicating that the fatigue resistance of DAM and DBFAM is better than AM.

3.6. Grey Correlation Analysis of Factors Affecting Fatigue Life

3.6.1. Grey Correlation Analysis Process

When using grey correlation analysis, first select the data column and then analyze the reference number and comparison series in the data. The comparison series and reference series are dimensionless. In order to avoid the different dimensions of each influencing factor, each factor should be dimensionless in the grey correlation analysis; the comparison sequence and reference sequence are engaged in dimensionless processing. Use Equation (8) to calculate the correlation coefficient between the comparison curve and the reference curve at each time and calculate the grey correlation degree. The correlation coefficient is the correlation degree between the comparison sequence and the reference sequence at each time. Therefore, there is more than one correlation coefficient calculated by Equation (8). For analysis, the average value of the calculated correlation coefficient is calculated by Equation (9).

The analysis process of grey correlation degree is as follows:

- The data of the fatigue test and relevant influencing factors are listed in Table 10 to generate the analysis data column; in order to keep the polarity of each influencing factor consistent, the data are processed accordingly. For example, the fatigue performance decreases with the increase of void fraction. In order to keep the polarity consistent, the reciprocal of a void fraction is taken for analysis; the fatigue life decreases with the increase of the initial stiffness modulus. When analyzing the data, take the reciprocal of the initial stiffness modulus for analysis.
- The test data in Table 10 are listed in Table 8 after dimensionless initialization.
- Using the data in Table 11 to calculate the difference sequence, the difference sequence is shown in Table 12.
- According to Table 12, the maximum difference between the two levels is 0.581, and the minimum difference between the two levels is 0. Calculate the grey correlation coefficient according to Equation (8). See Table 13 for the grey correlation coefficient.

Table 10. Raw data.

Influence Factor	AM	BFAM	DAM	DBFAM
OAC	4.78	5.09	5.12	5.22
1/VV	0.244	0.246	0.246	0.238
VMA	15.6	16.2	15.4	16
VFA	73.7	74.9	73.6	73.8
$1/S_0$	0.000137	0.00014	0.000146	0.00016
E_{CD}	4.635	5.375	5.212	6.154
Fatigue life	2099	2549	2541	3269

Table 11. Data initialization results.

Influence Factor	AM	BFAM	DAM	DBFAM
OAC	1.000	1.065	1.071	1.092
1/VV	1.000	1.007	1.010	0.976
VMA	1.000	1.038	0.987	1.026
VFA	1.000	1.016	0.999	1.001
$1/S_0$	1.000	1.027	1.065	1.163
E_{CD}	1.000	1.160	1.124	1.328
Fatigue life	1.000	1.214	1.211	1.557

Table 12. Difference sequence.

Influence Factor	AM	BFAM	DAM	DBFAM
OAC	0.000	0.150	0.139	0.465
1/VV	0.000	0.207	0.201	0.581
VMA	0.000	0.176	0.223	0.532
VFA	0.000	0.198	0.212	0.556
$1/S_0$	0.000	0.187	0.146	0.394
E_{CD}	0.000	0.055	0.086	0.230

Table 13. Grey correlation coefficient results.

Influence Factor	AM	BFAM	DAM	DBFAM
OAC	1.000	0.660	0.676	0.384
1/VV	1.000	0.584	0.591	0.333
VMA	1.000	0.623	0.565	0.353
VFA	1.000	0.595	0.578	0.343
$1/S_0$	1.000	0.608	0.666	0.424
E_{CD}	1.000	0.842	0.771	0.559

3.6.2. Grey Correlation Degree Comparison of Influencing Factors

The relation degree between each factor and the fatigue life of the asphalt mixture can be calculated by Equation (9), as shown in Figure 13. From the gray correlation degree ranking in the figure, it can be seen intuitively that the degree of influence is as follows: E_{CD} > OAC > S_0 > VMA > VFA > VV. Compared with E_{CD}, OAC, S_0, VMA, VFA and VV have less influence.

In this study, the fatigue test of the asphalt mixture was used to study the effect of diatomite and basalt fiber on the fatigue life of asphalt mixture. The gradation and the specimen forming method of the four types of asphalt mixture are the same. The slight difference in the void ratio is mainly due to the increase of asphalt consent caused by the addition of diatomite and basalt fiber. The addition of basalt fiber and diatomite can affect the fatigue life of the asphalt mixture by affecting the amount of asphalt, the initial stiffness modulus and the cumulated dissipated energy. The mixture of the two has the greatest influence on the cumulative dissipative energy of the asphalt mixture fatigue test, which can greatly improve the fatigue life of asphalt mixture.

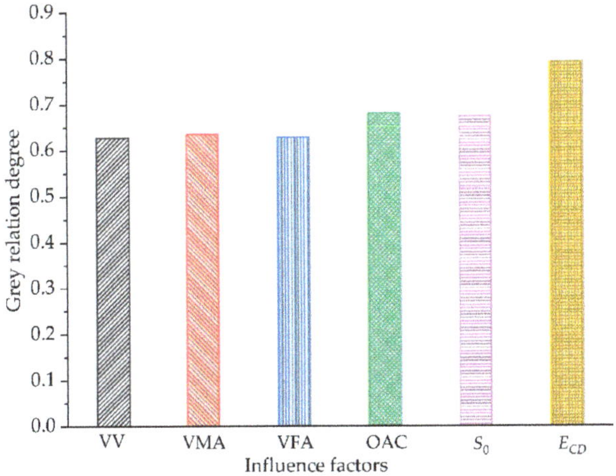

Figure 13. Grey correlation analysis results.

4. Conclusions

In this study, a Cooper universal pneumatic servo testing machine was used to carry out fatigue tests on four kinds of asphalt mixtures: DBFAM, DAM, BFAM and AM. Through the fatigue test indexes, the influence of basalt fiber and diatomite on the performance of asphalt mixture was analyzed. The effects of OAC, VV, VMA, VFA, initial stiffness modulus and cumulative dissipation energy on the fatigue life of asphalt mixture were analyzed by using the grey correlation theory. The following conclusions can be drawn:

(1) The changing trend in the flexural tensile stiffness modulus of four types of asphalt mixtures is the same as that of flexural tensile stiffness modulus attenuation rate under the alternating load. It can be seen from the flexural tensile stiffness modulus index that the fatigue life of asphalt mixture increases obviously after adding diatomite and basalt fiber, and the improvement effect of adding basalt fiber and diatomite is the best. The fatigue life of DBFAM is 55.7% higher than that of AM.

(2) There is an increasing trend in the lag angle of the four types of asphalt mixture under alternating load, but the increase is slow. The viscoelastic properties of the four types of asphalt mix materials during the fatigue test are constantly changing. With the increase of loading times, the asphalt mixture continues to develop in the direction of viscosity.

(3) The energy consumption of each loading decreases with the increase of loading times. Compared with AM, the energy consumption of DAM, BFAM, DBFAM is small under each loading. The cumulative dissipation energy and the plateau value of DBFAM are the smallest. The addition of diatomite and basalt fiber can significantly improve the toughness of asphalt mixture so that it increases the elastic recovery capacity of the asphalt mixture. When basalt fiber and diatomite are added at the same time, the improvement effect of elastic recovery ability of asphalt mixture is the most obvious.

(4) Through the grey correlation analysis, it can be seen that the cumulative dissipation energy, initial stiffness modulus and OAC have a great impact on the fatigue life of asphalt mixture, and the cumulative dissipation energy is the most significant. After adding diatomite and basalt fiber, the cumulative dissipation energy is greatly improved, and then the fatigue life of the asphalt mixture is increased.

Author Contributions: Conceptualization, C.Z. and H.L.; methodology, C.Z., H.L., W.T. and Y.Q.; validation, W.T., B.T. and Y.Q.; formal analysis, C.Z., H.L., W.T., H.A. and B.X.; investigation, H.L., W.T., B.T., H.A. and B.X.; writing—original draft preparation, C.Z., H.L., W.T. and H.A.; writing—

review and editing, B.T., Y.Q. and B.X.; supervision, C.Z.; project administration, C.Z.; funding acquisition, C.Z. All authors have read and agreed to the published version of the manuscript.

Funding: This research was funded by the Education Department of Jilin Province (grant number: JJKH20200286KJ) and the Science Technology Development Program of Jilin Province (grant number: 20160204008SF).

Institutional Review Board Statement: Not applicable.

Informed Consent Statement: Not applicable.

Data Availability Statement: The testing and analysis data used to support the findings of this study are included in the article.

Conflicts of Interest: The authors declare no conflict of interest.

References

1. Das, B.P.; Das, S.; Siddagangaiah, A.K. Probabilistic modeling of fatigue damage in asphalt mixture. *Constr. Build. Mater.* **2021**, *269*, 121300. [CrossRef]
2. Wen, Y.; Guo, N.; Wang, L.; Jin, X.; Li, W.; Wen, H. Assessment of various fatigue life indicators and fatigue properties of rock asphalt composite. *Constr. Build. Mater.* **2021**, *289*, 123147. [CrossRef]
3. Cheng, H.; Liu, J.; Sun, L.; Liu, L.; Zhang, Y. Fatigue behaviours of asphalt mixture at different temperatures in four-point bending and indirect tensile fatigue tests. *Constr. Build. Mater.* **2021**, *273*, 121675. [CrossRef]
4. Shi, C.; Cai, X.; Wang, T.; Yi, X.; Liu, S.; Yang, J.; Leng, Z. Energy-based characterization of the fatigue crack density evolution of asphalt binders through controlled-stress fatigue testing. *Constr. Build. Mater.* **2021**, *300*, 124275. [CrossRef]
5. Jin, G.-L.; Zhang, Z.-X.; Huang, X.-M. Fatigue Cracking Analysis of Asphalt Concrete Based on Coupled XFEM-Continuum Damage Mechanics Method. *J. Mater. Civ. Eng.* **2021**, *33*, 04020425. [CrossRef]
6. Ragni, D.; Sudarsanan, N.; Canestrari, F.; Kim, Y.R. Investigation into fatigue life of interface bond between asphalt concrete layers. *Int. J. Pavement Eng.* **2021**, 1–15. [CrossRef]
7. Yuan, J.; Lv, S.; Peng, X.; You, L.; Cabrera, M.B. Investigation of Strength and Fatigue Life of Rubber Asphalt Mixture. *Materials* **2020**, *13*, 3325. [CrossRef] [PubMed]
8. Złotowska, M.; Nagórski, R.; Błażejowski, K. Concept of Similarity Method for Prediction of Fatigue Life of Pavement Structures with HiMA Binder in Asphalt Layers. *Materials* **2021**, *14*, 480. [CrossRef] [PubMed]
9. Wang, W.; Cheng, Y.; Tan, G. Design Optimization of SBS-Modified Asphalt Mixture Reinforced with Eco-Friendly Basalt Fiber Based on Response Surface Methodology. *Materials* **2018**, *11*, 1311. [CrossRef]
10. Cheng, Y.; Wang, W.; Gong, Y.; Wang, S.; Yang, S.; Sun, X. Comparative Study on the Damage Characteristics of Asphalt Mixtures Reinforced with an Eco-Friendly Basalt Fiber under Freeze-thaw Cycles. *Materials* **2018**, *11*, 2488. [CrossRef]
11. Guo, Q.; Li, L.; Cheng, Y.; Jiao, Y.; Xu, C. Laboratory evaluation on performance of diatomite and glass fiber compound modified asphalt mixture. *Mater. Des.* **2015**, *66*, 51–59. [CrossRef]
12. Shu, B.; Wu, S.; Dong, L.; Norambuena-Contreras, J.; Li, Y.; Li, C.; Yang, X.; Liu, Q.; Wang, Q.; Wang, F.; et al. Self-healing capability of asphalt mixture containing polymeric composite fibers under acid and saline-alkali water solutions. *J. Clean. Prod.* **2020**, *268*, 122387. [CrossRef]
13. Zhang, X.; Xu, L.; Lv, J. Investigation on the Flexural–Tensile Rheological Behavior and Its Influence Factors of Fiber-reinforced Asphalt Mortar. *Polymers* **2020**, *12*, 1970. [CrossRef]
14. Zheng, Y.; Cai, Y.; Zhang, G.; Fang, H. Fatigue property of basalt fiber-modified asphalt mixture under complicated environment. *J. Wuhan Univ. Technol. Sci. Ed.* **2014**, *29*, 996–1004. [CrossRef]
15. Zhao, H.; Guan, B.; Xiong, R.; Zhang, A. Investigation of the Performance of Basalt Fiber Reinforced Asphalt Mixture. *Appl. Sci.* **2020**, *10*, 1561. [CrossRef]
16. Riekstins, A.; Baumanis, J.; Barbars, J. Laboratory investigation of crumb rubber in dense graded asphalt by wet and dry processes. *Constr. Build. Mater.* **2021**, *292*, 123459. [CrossRef]
17. Lou, K.; Wu, X.; Xiao, P.; Zhang, C. Investigation on Fatigue Performance of Asphalt Mixture Reinforced by Basalt Fiber. *Materials* **2021**, *14*, 5596. [CrossRef] [PubMed]
18. Zhang, H.; Shen, K.; Xu, G.; Tong, J.; Wang, R.; Cai, D.; Chen, X. Fatigue resistance of aged asphalt binders: An investigation of different analytical methods in linear amplitude sweep test. *Constr. Build. Mater.* **2020**, *241*, 118099. [CrossRef]
19. Poulikakos, L.D.; Hofko, B. A critical assessment of stiffness modulus and fatigue performance of plant produced asphalt concrete samples using various test methods. *Road Mater. Pavement Des.* **2021**, *22*, 2661–2673. [CrossRef]
20. Yu, J.; Yu, X.; Gao, Z.; Guo, F.; Wang, D.; Yu, H. Fatigue Resistance Characterization of Warm Asphalt Rubber by Multiple Approaches. *Appl. Sci.* **2018**, *8*, 1495. [CrossRef]
21. Yu, J.; Ren, Z.; Yu, H.; Wang, D.; Svetlana, S.; Korolev, E.; Gao, Z.; Guo, F. Modification of Asphalt Rubber with Nanoclay towards Enhanced Storage Stability. *Materials* **2018**, *11*, 2093. [CrossRef]

22. Benedetto, H.D.; Nguyen, Q.T.; Sauzeat, C. Nonlinearity, Heating, Fatigue and Thixotropy during Cyclic Loading of Asphalt Mixtures. *Road Mater. Pavement Des.* **2011**, *12*, 129–158. [CrossRef]
23. Cardona, D.R.; Pouget, S.; Di Benedetto, H. Influence of Air Void Content and Loading Frequency on Fatigue Test Results of Bituminous Mixtures. *J. Transp. Eng. Part B Pavements* **2021**, *147*, 04021059. [CrossRef]
24. Lamothe, S.; Perraton, D.; Di Benedetto, H. Fatigue behaviour of dry or partially saturated hot mix asphalt (HMA). *Fatigue Fract. Eng. Mater. Struct.* **2020**, *43*, 1100–1114. [CrossRef]
25. Tapsoba, N.; Sauzéat, C.; Di Benedetto, H. Analysis of Fatigue Test for Bituminous Mixtures. *J. Mater. Civ. Eng.* **2013**, *25*, 701–710. [CrossRef]
26. Nguyen, M.L.; Di Benedetto, H.; Sauzeat, C. Crack propagation characterization of bituminous mixtures using a four-point bending notched specimen test. *Road Mater. Pavements Des.* **2016**, *17*, 70–86. [CrossRef]
27. Wu, Z.; Zhang, S.; You, H. Prediction of fatigue life of asphalt mixture based on strain control. *J. South China Univ. Technol. Nat. Sci. Ed.* **2014**, *42*, 6.
28. Yang, P. Experimental Study on High Temperature and Fatigue Performance of Basalt Fiber Asphalt Mixture. Master's Thesis, Yangzhou University, Yangzhou, China, 2019.
29. *Standard Test Methods of Bitumen and Bituminous Mixtures for Highway Engineering (JTG E20-2011)*; Institute of Highway Science, Ministry of Communications, Ministry of Transport of the People's Republic of China: Beijing, China, 2011.
30. Cheng, Y.; Zhu, C.; Tan, G.; Lv, Z.; Yang, J.; Ma, J. Laboratory Study on Properties of Diatomite and Basalt Fiber Compound Modified Asphalt Mastic. *Adv. Mater. Sci. Eng.* **2017**, *2017*, 4175167. [CrossRef]
31. *Test Methods of Aggregate for Highway Engineering (JTG E42-2005)*; Institute of Highway Science, Ministry of Communications, Ministry of Transport of the People's Republic of China: Beijing, China, 2005.
32. *Technical Specifications for Construction of Highway Asphalt Pavements (JTG F40-2004)*; Institute of Highway Science, Ministry of Communications, Ministry of Transport of the People's Republic of China: Beijing, China, 2004.
33. Morova, N. Investigation of usability of basalt fibers in hot mix asphalt concrete. *Constr. Build. Mater.* **2013**, *47*, 175–180. [CrossRef]
34. Wang, W.; Cheng, Y.; Chen, H.; Tan, G.; Lv, Z.; Bai, Y. Study on the Performances of Waste Crumb Rubber Modified Asphalt Mixture with Eco-Friendly Diatomite and Basalt Fiber. *Sustainability* **2019**, *11*, 5282. [CrossRef]
35. Keleştemur, O.; Yildiz, S.; Gökçer, B.; Arici, E. Statistical analysis for freeze–thaw resistance of cement mortars containing marble dust and glass fiber. *Mater. Des.* **2014**, *60*, 548–555. [CrossRef]
36. Cong, P.; Liu, N.; Tian, Y.; Zhang, Y. Effects of long-term aging on the properties of asphalt binder containing diatoms. *Constr. Build. Mater.* **2016**, *123*, 534–540. [CrossRef]
37. Wang, W.; Cheng, Y.; Ma, G.; Tan, G.; Sun, X.; Yang, S. Further Investigation on Damage Model of Eco-Friendly Basalt Fiber Modified Asphalt Mixture under Freeze-Thaw Cycles. *Appl. Sci.* **2019**, *9*, 60. [CrossRef]
38. Liu, H.; Fu, L.; Jiao, Y.; Tao, J.; Wang, X. Short-Term Aging Effect on Properties of Sustainable Pavement Asphalts Modified by Waste Rubber and Diatomite. *Sustainability* **2017**, *9*, 996. [CrossRef]
39. Duojie, C.; Si, W.; Ma, B.; Hu, Y.; Liu, X.; Wang, X. Assessment of Freeze-Thaw Cycles Impact on Flexural Tensile Characteristics of Asphalt Mixture in Cold Regions. *Math. Probl. Eng.* **2021**, *7*, 1–10. [CrossRef]
40. Bekele, A.; Balieu, R.; Jelagin, D.; Ryden, N.; Gudmarsson, A. Micro-mechanical modelling of low temperature-induced micro-damage initiation in asphalt concrete based on cohesive zone model. *Constr. Build. Mater.* **2021**, *286*, 122971. [CrossRef]
41. Tan, G.; Wang, W.; Cheng, Y.; Wang, Y.; Zhu, Z. Establishment of Complex Modulus Master Curves Based on Generalized Sigmoidal Model for Freeze–Thaw Resistance Evaluation of Basalt Fiber-Modified Asphalt Mixtures. *Polymers* **2020**, *12*, 1698. [CrossRef]
42. Mangiafico, S.; Sauzéat, C.; Di Benedetto, H.; Pouget, S.; Olard, F.; Planque, L. Quantification of biasing effects during fatigue tests on asphalt mixes: Non-linearity, self-heating and thixotropy. *Road Mater. Pavement Des.* **2015**, *16*, 73–99. [CrossRef]
43. Tan, G.; Wang, W.; Cheng, Y.; Wang, Y.; Zhu, Z. Master Curve Establishment and Complex Modulus Evaluation of SBS-Modified Asphalt Mixture Reinforced with Basalt Fiber Based on Generalized Sigmoidal Model. *Polymers* **2020**, *12*, 1586. [CrossRef] [PubMed]
44. Chen, J.R.; Ye, J.; Wu, F.C. Study on fatigue performance of chopped basalt fiber asphalt mixture. *Highway* **2013**, *4*, 188–191.

polymers

Article

Fatigue Properties and Its Prediction of Polymer Concrete for the Repair of Asphalt Pavements

Senzhi Ren and Xin Hu *

School of Civil Engineering, Central South University of Forestry and Technology, Changsha 410004, China; rensenzhi@126.com
* Correspondence: 20201100378@csuft.edu.cn

Abstract: Polymer concrete (PC) is considered a promising repair material for asphalt pavement, since it has excellent paving performance and water stability. Although the mechanical properties of PC have been widely researched, the fatigue behavior of PC under traffic loads was still poorly understood. To predict the fatigue life and optimize the material design of PC, the semi-circular bending (SCB) tests were performed, considering different polymer content, sand ratio, aggregate features and stress condition. Two typical polymer materials were applied to prepare PC specimens, including epoxy resin (ER) and polyurethane (PU). The aggregate features were analyzed by the aggregate image measurement system. The mechanical behavior under repeated loads was investigated by the displacement, fatigue life and stiffness modulus. Results show that the flexural strength increases nonlinearly with the increasing polymer content, rapidly at first, and then slowly. The optimized polymer content and sand ratio were respectively 15% and 30%. As the loading number increases, the vertical displacement of PC shows three stages, i.e., undamaged stage, damage development stage, and fatigue failure stage. The stiffness modulus of the specimen is stress-dependent. An empirical model was developed to predict the fatigue life of PC, which can effectively capture the effects of the polymer content, sand ratio and stress level (or nominal stress ratio). It suggests that the fatigue life has a strong correlation with the mixing gradation, and the optimal sand ratio of PC can be determined by the proposed function. Moreover, the effect of aggregate shapes cannot be neglected.

Keywords: polymer concrete; fatigue property; semi-circular bending test; stress level

Citation: Ren, S.; Hu, X. Fatigue Properties and Its Prediction of Polymer Concrete for the Repair of Asphalt Pavements. *Polymers* **2022**, *14*, 2941. https://doi.org/10.3390/polym14142941

Academic Editor: Filippo Berto

Received: 30 June 2022
Accepted: 18 July 2022
Published: 20 July 2022

1. Introduction

As the main form of pavement structure in China, asphalt pavement plays an important role in the service performance of highways [1]. Under the long-term combined actions of the environment and loads, including heat, aging, rainfall, and repeated loading, asphalt pavement may develop diseases such as pit slots and cracks [2]. These diseases damage the stability of the pavement structure and reduce its operational efficiency. Statistical analysis shows that China's annual asphalt pavement maintenance mileage is over 100,000 km, and the cost of pavement repair alone is up to 2.3 billion yuan [3]. As the traffic volume and quality requirements of highways continue to increase, a fast and effective treatment method is required for accelerated repair and reopening of damaged highways. In general, pavement repair is conducted regardless of weather conditions. Hot mix asphalt is the most common repair material, yet it requires high-temperature heating and is not environmentally friendly [4]. In the meantime, the bonding between old and fresh asphalt mixtures is relatively weaker, and the penetration depth of fresh asphalt is low in small cracks, which limits its application in pavement repair [5]. As a supplement, various special polymer materials, including emulsified asphalt [6], polyurethane (PU) [7], and epoxy resin (ER) [8] have been adopted in pavement engineering. Polymer materials exhibit several excellent performances in terms of the corrosion resistance [9], inflaming retarding [10], thermal stability [11], and fracture toughness [12], etc. These new features were granted to asphalt pavement by modifying or mixing with corresponding polymers.

Polymer concrete (PC), a composite material consisting of various types of aggregates and a polymer matrix, is considered a promising material for pavement repair due to its fast curing and strong bonding with aggregates [13]. Compared with traditional asphalt mixtures, it has higher chemical resistance, better high-temperature stability, and good durability in field environments [14]. PU is a common polymeric binder that has been widely used in materials for drainage pavement structures, such as porous PU mixtures [15]. Adding fillers or changing the polymeric structure could improve the UV resistance, aging resistance, water stability, and freeze-thaw resistance of PU-based materials [16,17]. Ma, et al. [18] found through laboratory tests and field application that the high-temperature performance, low-temperature performance, deformation coordination performance, and fatigue resistance of PU–PC were better than cast asphalt concrete and epoxy asphalt concrete. Various studies have been conducted on the mechanical properties and reinforcement techniques of PC [19,20]. For example, the pavement performance of PC, such as compressive strength, fracture toughness, and flexural stiffness, has been studied through laboratory tests and numerical analysis [21]. In terms of material development, various composited and modified PC designs and blending solutions, such as fiber-reinforced PC [22] and ER/PU polymer concrete [23], have been investigated to achieve various performance enhancements. Test methods for the interface shear strength have also been developed to evaluate the bonding properties between PC and asphalt mixtures [24]. These studies have greatly improved the understanding of the mechanical properties of PC.

Fatigue cracking, a common disease in asphalt pavements, is one of the main considerations in their structural design [25]. However, the existing literature somehow neglected the fatigue properties of PC as a repair material of pavement [26]. Zaghloul, et al. [27,28] found both the stress level and the fiber content have a great influence on the tensile and fatigue behaviors of fiber-reinforced polyester. Yeon, et al. [29] and Ahn, et al. [30] investigated the fatigue performance of PC and its stiffness damage through bending beam tests and vibration measurements, but mainly focused on the effects of test methods on the fatigue characteristics of PC. The effects of component types and contents on fatigue performance were not fully revealed in previous research. Meanwhile, considering the effects of vehicle loading and temperature, the fatigue performance of PC under different conditions should be studied, since its damage mechanism is different from that under the static loading [31,32]. Due to the large number of trials and the complex processes, the construction sites lack the conditions for the required tests. Therefore, it is necessary to establish a PC fatigue life prediction equation to facilitate the performance evaluation and maintenance timing of this repair material [33]. Moreover, existing studies have not simultaneously considered the effects of polymer content, aggregate gradation, shape characteristics, and stress condition on the fatigue life of PC. The optimal design of PC lacks the sufficient data about dynamic tests.

In this study, PCs with ER and PU binders were prepared, respectively, for the rapid repair of asphalt pavements. Semi-circular bending (SCB) tests were carried out to analyze the effects of different polymer contents, aggregate characteristics, and stress levels on the mechanical properties of PC. Indexes such as displacement, fatigue life, and the stiffness modulus were used to evaluate fatigue performance. Finally, a PC fatigue life prediction equation considering the combined effect of several factors was established.

2. Materials and Methods

2.1. Materials

2.1.1. Aggregates

The fine aggregate used in this test was the natural sand produced from a quarry at Zhuzhou, China, which had a fineness modulus of 2.8 and a mud content below 1.0%. Its screen test result is shown in Figure 1. The coarse aggregate used in the tests was the continuously graded basalt gravel with a size of 5–16 mm. Table 1 shows the mechanical properties of the coarse aggregate, which conforms to the Chinese Standard for Technical

Requirements and Test Method of Sand and Crushed Stone (or Gravel) for Ordinary Concrete (JGJ 52-2006).

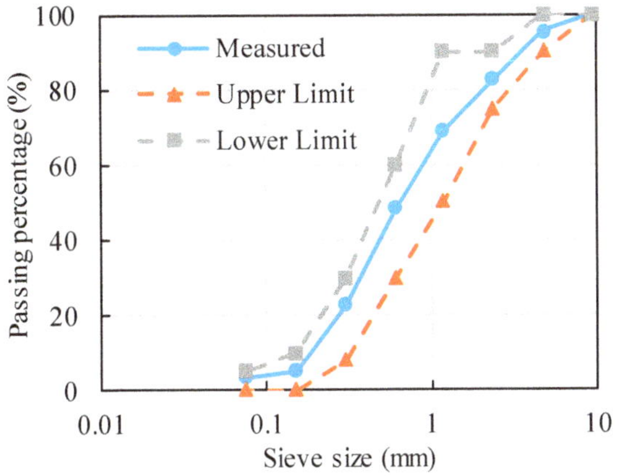

Figure 1. Result of sand screening test.

Table 1. Mechanical properties of coarse aggregates.

Items	Units	Results		Requirements
		5–10 mm	10–16 mm	
Apparent density	g/cm^3	2.718	2.735	≥2.65
Crushing value	%	16.7	17.4	≤22
Water absorption	%	0.43	0.41	≤1.5
Los Angeles attrition loss	%	16.0	16.7	≤22.0
Mud content	%	0.45	0.3	≤0.8

2.1.2. Polymer Binders

The polymer binders used in this study were ER and PU. The ER binder had a longer initial setting time and greater viscosity than typical PU materials. Meanwhile, the initial viscosity and setting time of the polymer matrix were controllable [34]. ER was prepared with the E51 ER, the triethylenetetramine curing agent, and the DMP-30 accelerator (100:9:1). The synthetic materials of PU mainly included isocyanate monomer and polyether. To ensure enough time for paving and compacting in the road repair project, the 6170F single-component PU (Wanhua Chemical Company, Yantai, China) was selected as the PU binder. The test results of polymer binders are shown in Table 2.

Table 2. Physical and chemical properties of polymer binders.

Items	Units	Epoxy Resin	Polyurethane
Density	g/cm^3	1.317	1.117
PH	—	7.5	8.1
Melting point	°C	252	175
Thermal expansion	μm/mK	54	160
Viscosity	MPa·s, 25 °C	2734	1233
Tensile strength	MPa	2.7	1.9
Elongation at break	%	200	550
Curing time	h	12	≤12

2.2. Sample Preparation

The PC specimens were prepared in the laboratory by mixing the aggregates and polymer binders according to the ASTM C192 standard. The mixing ratios of aggregates to polymer matrix were 95:5, 90:10, 85:15, and 80:20 by weight. The sand ratios were 25%, 30%, and 35%, respectively. The well-mixed PC was poured into test molds pre-applied with a release agent. Cylindrical specimens 150 mm in diameter and 180 mm in height were molded by the Superpave rotary compactor. During compaction, the compressive stress was 600 kPa, the offset angle was 1.16°, and the rotation rate was 30 r/min. Each prepared specimen was compacted 100 times to ensure uniform mass density distribution. The specimens were demolded 1 h after molding. Before use, the final specimen was conditioned at room (25 °C and a standard atmospheric pressure) for 12 h [35]. Wang, et al. [36] found that the internal stress of specimens above 50 mm in thickness tends to planar strain state, and the stress response tends to stabilize. Therefore, the specimens for SCB tests were cut into semi-circular halves with a diameter of 150 mm and a thickness of 50 mm. The preparation process of the SCB specimens is shown in Figure 2.

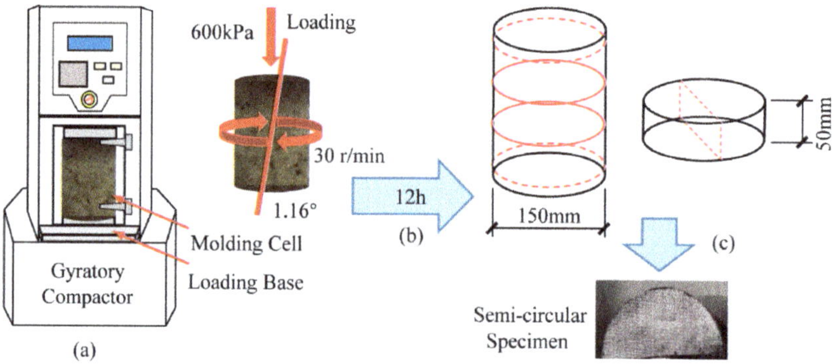

Figure 2. Preparation process of semicircular specimens: (**a**) compaction; (**b**) curing for 12 h; (**c**) cutting into semi-circular halves.

2.3. Testing Methods

2.3.1. Aggregate Feature Measurement

The morphology characteristics of the coarse aggregates were analyzed by the aggregate image measurement system (AIMS). Three different shape indicators of the aggregates were measured, namely: surface texture, gradient angularity, and sphericity. The surface texture was described through wavelet analysis based on transverse, longitudinal, and oblique images of the tested aggregates. Angularity was measured using the gradient approach to quantify the variation in size and profile of the aggregate particles. The three-dimensional shape characteristics were evaluated by sphericity, which was calculated by the long axis length, the middle axis length, and the short axis length of the aggregate bounding box. The sphericity ranges from 0 to 1, and when it approaches 1, the measured aggregate is approximately spherical in shape. The detailed descriptions can be found in the previous literature [37].

Table 3 shows the measured results of coarse aggregates with a diameter of 5–16 mm. The results show that the angularity distribution of total aggregates meets the requirements of pavement performance due to the small content of flat and elongated particles. AIMS tests revealed relatively pronounced differences in the quantitative indicators of the aggregates, such as texture and angularity, and their effects on the performance of the specimens need further investigation.

Table 3. Morphology measurement of basalt aggregates.

Particle Size (mm)	Value	Percentage of Flat-elongated Particles (%)	Texture	Gradient Angularity	Sphericity
5–10	Mean	8.34	343.6	3283	0.67
	Standard deviation	—	93.5	802.5	0.08
10–16	Mean	7.98	422.9	2961	0.69
	Standard deviation	—	124.2	671.8	0.09

2.3.2. Semi-Circular Bending Test

The SCB tests were conducted using the dynamic universal testing machine UTM-100, composed of the constant temperature incubator, loading platform, and servo controller. The selected loading mode was the three-point bending method. The specimens were set on beam support with a loading roller above and two support rollers beneath, all of which were 1 mm in diameter. The distance between the two support rollers was set to 0.8 times the specimen diameter. Before testing, a contact load of 0.2 kN was applied and held for 10 s to ensure uniform contact between the specimen and the loading roller. The test temperature was set to 15 °C, according to the annual temperature gradient variation of Hunan Province. A constant loading rate of 50 mm/min was maintained in the SCB strength tests until the cracking failure of the specimen. The tests stopped as the load dropped to 0.3 kN. The tensile strength of specimens can be determined by SCB tests with single loading. The maximum tensile stress at the bottom of the SCB specimen can be calculated with Equation (1). For the SCB fatigue tests, the loading was applied in the form of haversine waves at a frequency of 10 Hz. The stress levels were set at four different loading stress ratios (0.2, 0.3, 0.4 and 0.5), based on the SCB strength. For each stress level, three parallel tests were performed to ensure the reliability of the fatigue tests. The testing results in the figures represent an average result by parallel tests.

$$\sigma_t = \frac{4.976P}{TD} \tag{1}$$

where, σ_t stands for the maximum tensile stress at the bottom of SCB specimen; P is the vertical loading; T and D are respectively the thickness and semicircle diameter of the SCB specimen.

In the SCB fatigue tests, unrecoverable damage occurred inside the PC specimens as the number of loadings increased, which led to the mechanical property decay of the material. The effect of fatigue damage on the mechanical properties can be described by the decay of the stiffness modulus. The stiffness modulus is the stress to strain ratio at the bottom center of the SCB specimen in tension, as presented in Equation (2). Assuming that the stress distribution in the span section of the SCB specimen conforms to the flat section assumption, the tensile strain at the bottom center of the specimen can be expressed as Equation (3).

$$S_t = \frac{\sigma_t}{\varepsilon_t} \tag{2}$$

$$\varepsilon_t = \frac{6Ld}{1.14D^2\left(5.578\frac{L}{D} \quad 1.3697\right)} \tag{3}$$

where, S_t stands for the stiffness modulus; ε_t stands for the tensile strain at the center of the bottom of the SCB specimen; L is the distance between adjacent fixing supports; d is the deflection at the center of the SCB specimen.

3. Results

3.1. SCB Strength Test

Table 4 shows the SCB strength test results under the three different mixing proportions and different polymer contents. The polymer content represents the percentage of polymer binder in the PC by weight. The flexural strength of PC at different sand ratios varies significantly with polymer content. Overall, the flexural strength increases nonlinearly with the increasing polymer content, rapidly at first and then slowly. The main reason is that at low polymer content, the binder cannot completely coat the aggregate surface to develop sufficient interfacial strength. In this study, the flexural strength of PC was less affected by the sand ratio. When the sand ratio is 30%, the bending strengths of ER–PC and PU–PC peaked at 18.82 MPa and 15.24 MPa, respectively. The strength of ER–PC is higher than that of PU–PC, which is consistent with the physical test results of the polymers. As the sand ratio exceeds 30%, the flexural strength of the PC decreases. The main reason is that the excessively low sand ratio causes incompletely filled voids between coarse aggregates. At the same time, an excessive sand ratio reduces the amount of coarse aggregate and increases the total surface area of coarse and fine aggregates, leading to an increased amount of PU. As shown in Table 4, the variability of the strength test was small, with all coefficients of variation (CV) in the parallel tests below 10%. Therefore, considering the economy and design strength requirements, the sand ratio for the tests was determined to be 30%, and the initial admixture of the polymer was 15%.

Table 4. SCB strength with different sand ratio and polymer content.

Sand Ratio (%)	Polymer Content (%)	ER–PC		PU–PC	
		SCB Strength (MPa)	CV(%)	SCB Strength (MPa)	CV(%)
25	5	2.29	5.80	1.28	4.17
	10	9.88	5.40	7.34	4.29
	15	14.05	5.31	11.97	5.09
	20	16.68	5.91	13.51	4.22
30	5	2.98	4.92	1.76	4.47
	10	12.02	5.76	5.94	4.15
	15	16.27	4.30	13.19	4.43
	20	18.82	6.09	15.24	5.10
35	5	2.71	4.55	1.49	4.62
	10	10.63	5.81	4.91	4.19
	15	15.11	5.19	11.27	4.82
	20	17.83	6.44	14.45	4.83

3.2. Displacement Changes of SCB Fatigue Test

The fatigue damage process is as follows. Under repeated loading, the asphalt mixture gradually sustains plastic deformation and cumulative damage, finally leading to fracture damage. The plastic deformation trend of the PC was analyzed with SCB fatigue tests. Taking the 30% sand ratio and the 15% polymer content as an example, Figure 3 shows the displacement of the two kinds of PCs under different stress levels.

As the number of cycles increases, the vertical displacement curve shows three stages, i.e., undamaged stage, damage development stage, and fatigue failure stage. In the initial stage of loading, the specimen undergoes elastic-plastic deformation due to internal contact and void compression [38]. The permanent deformation increases significantly with the increasing number of loadings. At the second stage of loading, the vertical displacement increases linearly with the number of loadings. This time, microcracks appear and develop steadily inside the specimen. In the third stage, an inflection point appears on the loading curve, and the specimen fails. Elseifi, et al. [39] found that the microdamage concentrated mainly in the middle of the specimen and then produced penetration cracks. The criterion for fatigue damage is usually defined as the critical point between the damage development stage and the fatigue failure stage. Figure 3 shows that as the stress level increases, the fatigue life and final displacement of the specimen tend to decrease. The fatigue life of ER–PC is greater than that of PU–PC in this study.

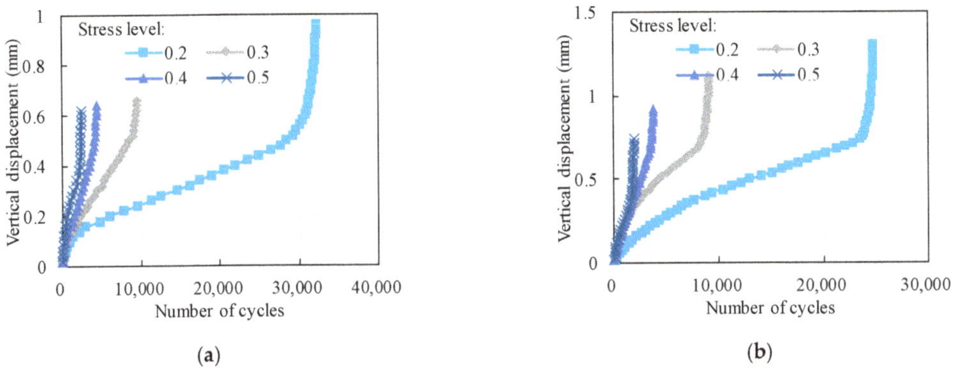

Figure 3. Displacement changes of SCB fatigue test: (**a**) ER–PC; (**b**) PU–PC.

3.3. Attenuation of Stiffness Modulus with Loading Cycles

Figure 4a shows that the stiffness modulus changes in the SCB tests under different stress levels. The sand ratio of the specimens is 30%, and the polymer content is 15%. It is obvious that the stiffness modulus of the PC decreased continuously during the tests, indicating that the reduced part of the stiffness modulus is attributed to the fatigue damage of the specimens [40]. In the damage development stage, the elastic deformation of PC can fully recover after unloading, but the damage deformation it produces is not recoverable. As a result, the stiffness modulus of PC decays, and damage deformation occurs.

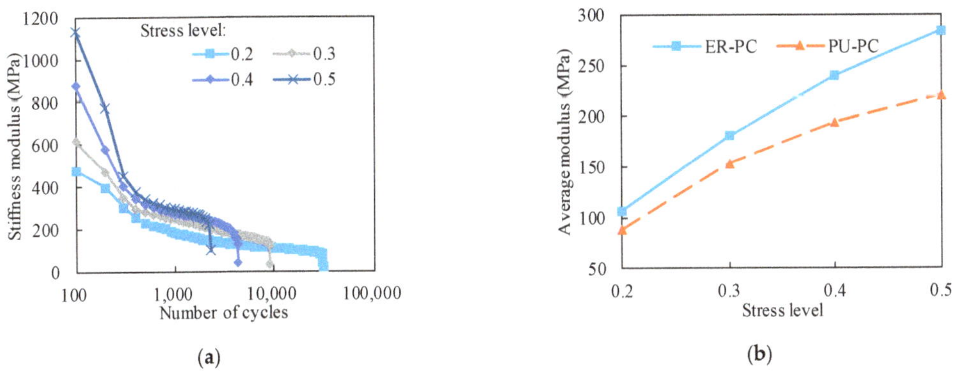

Figure 4. Stiffness modulus of SCB fatigue test: (**a**) changes of ER–PC with cycle number; (**b**) Average modulus of ER–PC and PU–PC.

The variation trends of the stiffness modulus can be divided into three stages. In the first stage, cracking occurs at the surface and internal defects of the specimen, such as cracks and voids, resulting in a sharp decrease in the stiffness modulus due to stress concentration. In the second stage, as the number of loadings increases, the high-density energy at the internal defects of the specimen is released due to the cracking. Meanwhile, the fatigue damage development is inhibited, manifested as the linear decrease in the stiffness modulus. In the third stage, the stiffness modulus of the asphalt mixture decreases sharply, manifesting as severe damage and rapidly increasing fatigue damage rate until failure.

In the damage development stage, the average stiffness modulus is displayed in Figure 4b to compare the elastic properties of the two PC. The results show that the stiffness modulus of the specimen is dependent on the stress condition. The average modulus of the specimens increases approximately linearly with the increasing stress. Due to the better adhesion between ER and aggregate, the stiffness modulus of ER–PC is larger than that of PU–PC.

3.4. Evaluation and Prediction of Fatigue Life

The above analysis reveals that the fatigue performance of the PC specimens is mainly related to three factors, i.e., sand ratio, polymer content, and stress level. Additionally, the morphological index of the aggregates was introduced in this study for fatigue life prediction. The fatigue test results are presented in Table 5. The framework of the fatigue prediction equation is established in the following steps. Firstly, the effect of stress level is considered, and the standard σ-N fatigue equation is adopted for predicting the fatigue life, which can be expressed as Equation (4). Taking 5% ER content as an example, the relationship between the fatigue life and the stress level were decided by multivariate nonlinear regression analysis, according to Equation (4). Figure 5a shows the fatigue prediction results of PC specimens with different sand ratios. Table 5 presents all predicted results of σ-N fatigue equation. The fatigue life showed a good correlation with stress levels, since the determination coefficients of Equation (4) is higher than 0.99.

$$N_f(\sigma) = K_1\left(\frac{1}{\sigma}\right)^{K_2} \tag{4}$$

where, N_f stands for the fatigue life; σ is the stress level; K_1 and K_2 is the fitting parameters.

Table 5. Fitting results of σ-N fatigue equation.

Sand Ratio (%)	Polymer Content (%)	ER–PC			PU–PC		
		K_1	K_2	R^2 (%)	K_1	K_2	R^2 (%)
25	5	110.910	3.033	99.42	81.855	3.005	99.05
	10	152.310	3.082	99.77	97.822	3.134	99.58
	15	235.350	2.824	99.95	150.340	2.883	99.77
	20	343.560	2.618	99.99	200.460	2.758	99.91
30	5	157.050	2.972	99.8	122.080	3.079	99.62
	10	176.190	3.141	99.98	170.620	3.114	99.86
	15	289.830	2.872	99.95	265.170	2.887	99.94
	20	431.950	2.643	99.85	398.970	2.657	99.93
35	5	119.320	3.109	99.4	88.262	3.083	99.6
	10	172.020	3.074	99.85	112.890	3.171	99.95
	15	296.260	2.763	99.99	189.090	2.879	99.95
	20	415.400	2.593	99.99	265.690	2.691	99.98

Secondly, the fatigue life difference of PC specimens with different ER contents has a similar power function relationship with the stress level, as shown in Figure 5b–d. Therefore, the function of fatigue life difference with the fatigue life of the 5% ER–PC as the reference can be expressed as Equation (5).

$$\Delta N_f(\sigma, \Delta c_p, S_r) = K_3(\Delta c_p, S_r)\left(\frac{1}{\sigma}\right)^{K_4(\Delta c_p, S_r)} \tag{5}$$

where, Δc_p stands for the relative content of polymer binder; S_r stands for sand ratio; K_3 and K_4 are the model parameters considering polymer content and sand ratio.

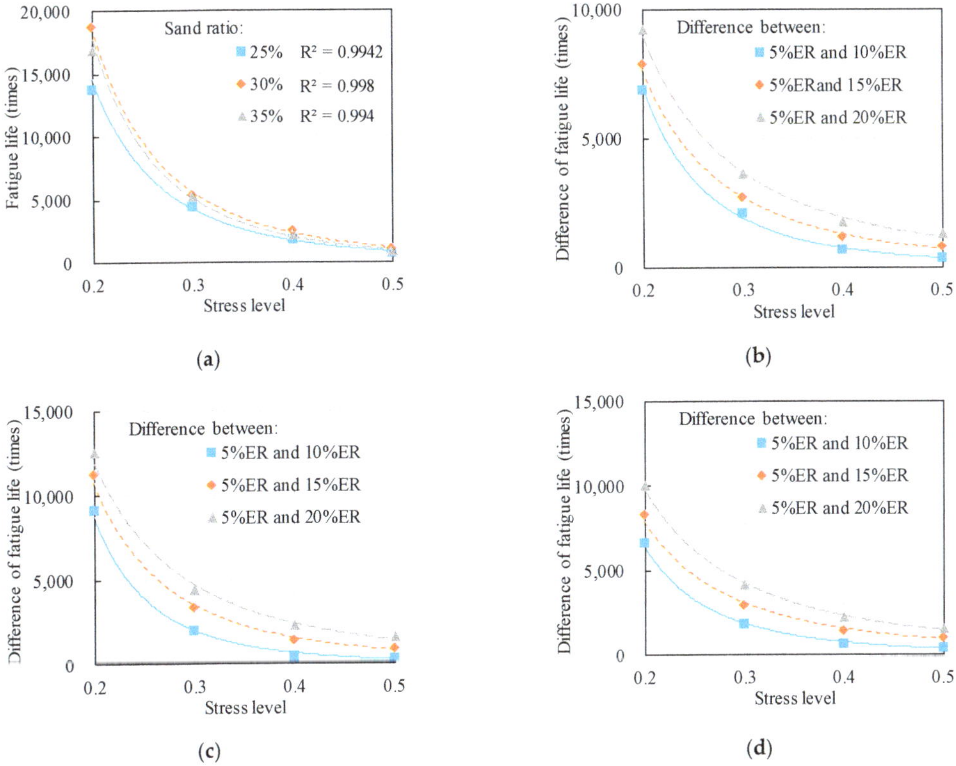

Figure 5. Effect of stress level on the fatigue life of ER–PC: (**a**) 5% ER content; (**b**) sand ratio of 25%; (**c**) sand ratio of 30%; (**d**) sand ratio of 35%.

Thus, the fatigue life of PC specimens with different ER contents can be determined by Equation (6). Moreover, the basic frame of fatigue life prediction is still a power function, since the K_3 is greater than 0. It implies that the stress level plays an essential role for the predictability and accuracy of the proposed fatigue function.

$$\begin{aligned} N_f(\sigma, \Delta c_p, S_r) &= N_f(\sigma, 0, S_r) + \Delta N_f(\sigma, \Delta c_p, S_r) \\ &= K_1\left(\frac{1}{\sigma}\right)^{K_2} + K_3(\Delta c_p, S_r)\left(\frac{1}{\sigma}\right)^{K_4(\Delta c_p, S_r)} \end{aligned} \quad (6)$$

In the case of the same sand ratio, the relationship between the coefficient K_3 and the relative polymer content was fitted in Figure 6. The fatigue life difference should be 0 at the ER content of 5% (Δc_p = 0). Thus, the y-intercept of the fatigue life difference function is set to 0. Meanwhile, to ensure the accuracy of the prediction model, the determination coefficient of the K_3 fitting results should be above 98%. Therefore, the fitted model takes the form of a quadratic function, as expressed in Equation (7). To verify the reliability of Equation (7), Table 6 shows the fitting results of ER–PC and PU–PC. The results suggest that the fitted equations are applicable to predict the fatigue life of both materials.

$$K_3(\Delta c_p) = a(\Delta c_p)^2 + b\Delta c_p \quad (7)$$

where, a and b are the fitting parameters.

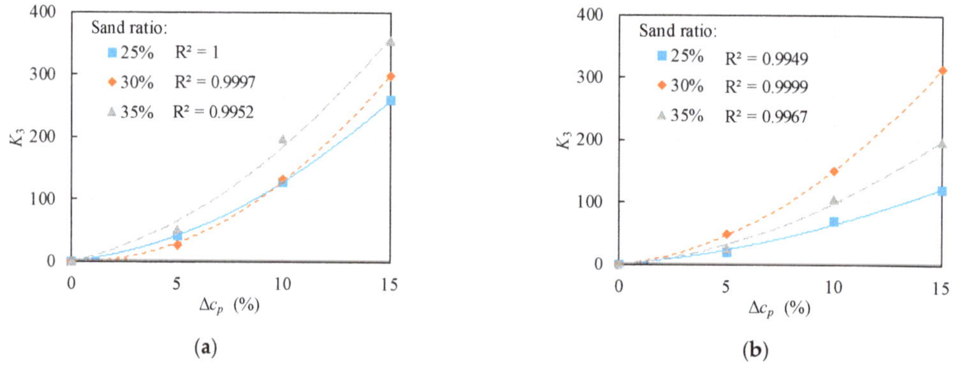

Figure 6. Fitting results of coefficient K_3 with different Δc_p: (**a**) ER–PC; (**b**) PU–PC.

Table 6. SCB strength with different sand ratio and content.

Materials	Sand Ratio (%)	a	b	R^2 (%)
ER–PC	25	0.8959	3.8454	1
	30	1.4134	−1.1837	99.97
	35	1.1071	7.3932	99.54
PU–PC	25	0.3479	2.9345	99.49
	30	1.1604	3.5212	99.99
	35	0.6810	3.0317	99.67

Thirdly, coefficients K_1 and K_2 of the fatigue equation are analyzed. The gradation (or sand ratio) of the mixture significantly contributes to the mechanical properties of PC. The optimized gradation is often determined by static mechanical tests, such as compaction tests or uniaxial compression tests. These static design methods are unreasonable since the real pavement response is under dynamic traffic loading. However, previous literature rarely mentioned the mixing design method considering the fatigue life. This study supplied a novel view about the design of PC.

Table 7 presents the details of material properties and predicted parameters, taking 5% polymer content as an example. Results shows that there is no definitive link between the sand ratio and the coefficient K_2. Nevertheless, the coefficient K_1 shows a strong correlation with the type of binder and the sand ratio. According to the results of Table 6, the empirical relationship between tensile strength, sand ratio, and coefficient K_1 was established by multivariate nonlinear regression, as expressed in Equation (8). For other polymer contents in this study, the results show similar functional relationships. It suggests that the fatigue life of the specimen is proportional to the tensile strength of the binder, and a suitable gradation could optimize the performance of the PC. Especially, in this case of 5% polymer content, the optimal sand ratio of PC should be 30.237%.

$$K_1(T_s, S_r) = T_s\left[-0.674(S_r - 30.237)^2 + 60.221\right],\ R^2 = 96.95\% \tag{8}$$

Furthermore, the fatigue performance of PC is affected by the interlocking effect of the aggregate structure and the interfacial properties [41]. Through experiments and numerical analysis [37], Yao, et al. [42] found that the shape characteristics of aggregate are related to the stress dependence of deformation behavior. The univariate predictor variables of coefficient K_2 were screened using the Bootstrap Forest model in Figure 7. It is proved that the angularity of aggregates has the strongest correlation with K_2, followed by the sand ratio and texture, whereas sphericity has the weakest correlation with K_2.

Table 7. Relationships between materials properties and predicted parameters.

Polymer Matrix	Tensile Strength (MPa)	Sand Ratio (%)	Aggregate Morphology			K_1	K_2
			Texture	Angularity	Sphericity		
ER	2.7	25	368.2	3056	0.69	110.91	3.033
	2.7	30	359.2	2991	0.69	157.05	2.972
	2.7	35	368.5	3263	0.68	119.32	3.109
PU	1.9	25	365.6	2967	0.67	81.855	3.006
	1.9	30	368.4	3197	0.68	122.08	3.079
	1.9	35	372.2	3270	0.68	88.262	3.083

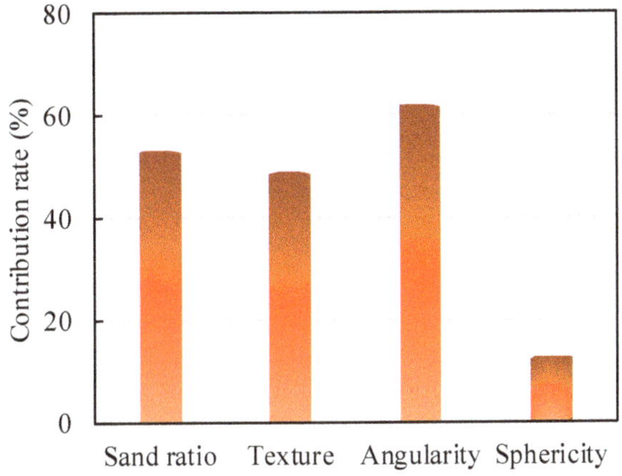

Figure 7. Contribution rate of material properties on K_2.

4. Conclusions

This study carried out the SCB tests of PC considering the polymer content, aggregate shape and stress level. The main finding can be drawn as follow:

(1) Based on the SCB strength test results, it shows that the polymer content and sand ratio have significant influence on the flexural strength. The strength increases nonlinearly with the increasing polymer content, rapidly at first and then slowly. However, as the sand ratio exceeds 30%, the flexural strength of the PC decreases.

(2) According to displacement changes of PC under repeated loadings, the testing process presents three stages, i.e., undamaged stage, damage development stage, and fatigue failure stage, as the number of cycles increases. Moreover, the stress level increases, and the fatigue life and final displacement tend to decrease.

(3) In terms of the stiffness modulus, the fatigue damage of specimens may result in the decay of the stiffness modulus. Meanwhile, the stiffness modulus is dependent on the stress level. The average modulus of the specimens increases approximately linearly with the increasing stress.

(4) A prediction model of fatigue life is established containing stress level, polymer content, tensile strength and sand ratio. The basic frame of fatigue life prediction is a power function, and the stress level plays an essential role for the predictability and accuracy.

(5) The fatigue life has a strong correlation with the type of binder and the mixing gradation. Meanwhile, the optimal sand ratio of PC can be determined by the proposed empirical function. According to aggregate shape analysis, the effects of angularity and texture on fatigue life are more significant, whereas the effect of sphericity is relatively weak.

Author Contributions: Conceptualization, S.R.; Data curation, X.H.; Formal analysis, X.H.; Funding acquisition, S.R.; Investigation, X.H.; Methodology, S.R.; Project administration, S.R.; Resources, S.R.; Software, X.H.; Supervision, S.R.; Validation, S.R.; Visualization, X.H.; Writing—original draft, S.R.; Writing—review & editing, X.H. All authors have read and agreed to the published version of the manuscript.

Funding: This research was funded by the Key Research and Development Program of Hunan Province (2021SK2050).

Data Availability Statement: Not applicable.

Conflicts of Interest: The authors declare no conflict of interest.

References

1. Liu, C.; Lv, S.; Jin, D.; Qu, F. Laboratory investigation for the road performance of asphalt mixtures modified by rock asphalt–styrene butadiene rubber. *J. Mater. Civ. Eng.* **2021**, *33*, 04020504. [CrossRef]
2. Li, J.; Wei, H.; Yao, Y.; Hu, X.; Wang, L. Contribution modeling on condition evaluation of asphalt pavement using uncertainty measurement and entropy theory. *Adv. Mater. Sci. Eng.* **2021**, *2021*, 9995926. [CrossRef]
3. Transport, E.D.o.C.J.o.H.a. Review on China's subgrade engineering research 2021. *China J. Highw. Transp.* **2021**, *34*, 1–49.
4. Byzyka, J.; Rahman, M.; Chamberlain, D.A. Thermal analysis of hot mix asphalt pothole repair by finite element method. *J. Transp. Eng. Part B Pavements* **2020**, *146*, 04020029. [CrossRef]
5. Byzyka, J.; Rahman, M.; Chamberlain, D.A. An improved interface temperature distribution in shallow hot mix asphalt patch repair using dynamic heating. *Int. J. Pavement Eng.* **2020**, *21*, 1617–1625. [CrossRef]
6. Jiang, S.; Li, J.; Zhang, Z.; Wu, H.; Liu, G. Factors influencing the performance of cement emulsified asphalt mortar—A review. *Constr. Build. Mater.* **2021**, *279*, 122479. [CrossRef]
7. Zhang, Z.; Sun, J.; Jia, M.; Ban, X.; Wang, L.; Chen, L.; Huang, T.; Liu, H. Effects of polyurethane thermoplastic elastomer on properties of asphalt binder and asphalt mixture. *J. Mater. Civ. Eng.* **2021**, *33*, 04020477. [CrossRef]
8. Li, M.; Min, Z.; Wang, Q.; Huang, W.; Shi, Z. Effect of epoxy resin content and conversion rate on the compatibility and component distribution of epoxy asphalt: A MD simulation study. *Constr. Build. Mater.* **2022**, *319*, 126050. [CrossRef]
9. Fuseini, M.; Zaghloul, M.M.Y.; Elkady, M.F.; El-Shazly, A.H. Evaluation of synthesized polyaniline nanofibres as corrosion protection film coating on copper substrate by electrophoretic deposition. *J. Mater. Sci.* **2022**, *57*, 6085–6101. [CrossRef]
10. Zaghloul, M.Y.M. Mechanical properties of linear low-density polyethylene fire-retarded with melamine polyphosphate. *J. Appl. Polym. Sci.* **2018**, *135*, 46770. [CrossRef]
11. Fuseini, M.; Zaghloul, M.M.Y. Statistical and qualitative analysis of the kinetic models using electrophoretic deposition of polyaniline. *J. Ind. Eng. Chem.* **2022**. [CrossRef]
12. Zaghloul, M.M.Y.; Zaghloul, M.Y.M.; Zaghloul, M.M.Y. Experimental and modeling analysis of mechanical-electrical behaviors of polypropylene composites filled with graphite and MWCNT fillers. *Polym. Test.* **2017**, *63*, 467–474. [CrossRef]
13. Alhazmi, H.; Shah, S.A.R.; Anwar, M.K.; Raza, A.; Ullah, M.K.; Iqbal, F. Utilization of polymer concrete composites for a circular economy: A comparative review for assessment of recycling and waste utilization. *Polymers* **2021**, *13*, 2135. [CrossRef]
14. Mayhoub, O.A.; Abadel, A.A.; Alharbi, Y.R.; Nehdi, M.L.; de Azevedo, A.R.G.; Kohail, M. Effect of polymers on behavior of ultra-high-strength concrete. *Polymers* **2022**, *14*, 2585. [CrossRef]
15. Cong, L.; Wang, T.; Tan, L.; Yuan, J.; Shi, J. Laboratory evaluation on performance of porous polyurethane mixtures and OGFC. *Constr. Build. Mater.* **2018**, *169*, 436–442. [CrossRef]
16. Chen, J.; Yin, X.; Wang, H.; Ding, Y. Evaluation of durability and functional performance of porous polyurethane mixture in porous pavement. *J. Clean. Prod.* **2018**, *188*, 12–19. [CrossRef]
17. Huang, H.; Pang, H.; Huang, J.; Zhao, H.; Liao, B. Synthesis and characterization of ground glass fiber reinforced polyurethane-based polymer concrete as a cementitious runway repair material. *Constr. Build. Mater.* **2020**, *242*, 117221. [CrossRef]
18. Ma, W.; Zhao, Z.; Guo, S.; Zhao, Y.; Wu, Z.; Yang, C. Performance evaluation of the polyurethane-based composites prepared with recycled polymer concrete aggregate. *Materials* **2020**, *13*, 616. [CrossRef]
19. Huang, H.; Pang, H.; Huang, J.; Yu, P.; Li, J.; Lu, M.; Liao, B. Influence of hard segment content and soft segment length on the microphase structure and mechanical performance of polyurethane-based polymer concrete. *Constr. Build. Mater.* **2021**, *284*, 122388. [CrossRef]
20. Zaghloul, M.Y.M.; Zaghloul, M.M.Y.; Zaghloul, M.M.Y. Developments in polyester composite materials—An in-depth review on natural fibres and nano fillers. *Compos. Struct.* **2021**, *278*, 114698. [CrossRef]
21. Hong, B.; Lu, G.; Li, T.; Lin, J.; Wang, D.; Liang, D.; Oeser, M. Gene-editable materials for future transportation infrastructure: A review for polyurethane-based pavement. *J. Infrastruct. Preserv. Resil.* **2021**, *2*, 27. [CrossRef]
22. Naser, M.; Hawileh, R.; Abdalla, J. Fiber-reinforced polymer composites in strengthening reinforced concrete structures: A critical review. *Eng. Struct.* **2019**, *198*, 109542. [CrossRef]
23. Cheng, H.T.; Lee, Y.S.; Liu, H.C.; Lee, W.J. The effect of component addition order on the properties of epoxy resin/polyurethane resin interpenetrating polymer network structure. *J. Appl. Polym. Sci.* **2021**, *138*, 49833. [CrossRef]

24. Guo, T.; Xie, Y.; Weng, X. Evaluation of the bond strength of a novel concrete for rapid patch repair of pavements. *Constr. Build. Mater.* **2018**, *186*, 790–800. [CrossRef]
25. Wei, H.; Li, J.; Hu, B.; Wang, F.; Zheng, J. Influence of temperature on deformation failure and acoustic emission characterisation of asphalt concrete under uniaxial compression. *Int. J. Pavement Eng.* **2022**, 1–10. [CrossRef]
26. Zhi, S.; Gun, W.W.; Hui, L.X.; Bo, T. Evaluation of fatigue crack behavior in asphalt concrete pavements with different polymer modifiers. *Constr. Build. Mater.* **2012**, *27*, 117–125. [CrossRef]
27. Zaghloul, M.M.Y.; Mohamed, Y.S.; El-Gamal, H. Fatigue and tensile behaviors of fiber-reinforced thermosetting composites embedded with nanoparticles. *J. Compos. Mater.* **2019**, *53*, 709–718. [CrossRef]
28. Zaghloul, M.Y.; Zaghloul, M.M.Y.; Zaghloul, M.M.Y. Influence of Stress Level and Fibre Volume Fraction on Fatigue Performance of Glass Fibre-Reinforced Polyester Composites. *Polymers* **2022**, *14*, 2662. [CrossRef]
29. Yeon, K.-S.; Choi, Y.-S.; Kim, K.-K.; Yeon, J.H. Flexural fatigue life analysis of unsaturated polyester-methyl methacrylate polymer concrete. *Constr. Build. Mater.* **2017**, *140*, 336–343. [CrossRef]
30. Ahn, S.; Jeon, E.-B.; Koh, H.-I.; Kim, H.-S.; Park, J. Identification of stiffness distribution of fatigue loaded polymer concrete through vibration measurements. *Compos. Struct.* **2016**, *136*, 11–15. [CrossRef]
31. Gullapalli, A.; Lee, J.H.; Lopez, M.M.; Bakis, C.E. Sustained loading and temperature response of fiber-reinforced polymer–concrete bond. *Transp. Res. Rec.* **2009**, *2131*, 155–162. [CrossRef]
32. Guo, Q.; Liu, Q.; Zhang, P.; Gao, Y.; Jiao, Y.; Yang, H.; Xu, A. Temperature and pressure dependent behaviors of moisture diffusion in dense asphalt mixture. *Constr. Build. Mater.* **2020**, *246*, 118500. [CrossRef]
33. Zhao, Z.; Wang, S.; Ren, J.; Wang, Y.; Wang, C. Fatigue characteristics and prediction of cement-stabilized cold recycled mixture with road-milling materials considering recycled aggregate composition. *Constr. Build. Mater.* **2021**, *301*, 124122. [CrossRef]
34. Chen, Q.; Wang, S.; Wang, C.; Wang, F.; Fu, H.; Yang, X. Modified waterborne epoxy as a cold pavement binder: Preparation and long-term working properties. *J. Mater. Civ. Eng.* **2021**, *33*, 04021079. [CrossRef]
35. Jung, K.-C.; Roh, I.-T.; Chang, S.-H. Evaluation of mechanical properties of polymer concretes for the rapid repair of runways. *Compos. Part B Eng.* **2014**, *58*, 352–360. [CrossRef]
36. Wang, H.; Zhang, C.; Li, L.; You, Z.; Diab, A. Characterization of low temperature crack resistance of crumb rubber modified asphalt mixtures using semi-circular bending tests. *J. Test. Eval.* **2016**, *44*, 847–855. [CrossRef]
37. Yao, Y.; Li, J.; Ni, J.; Liang, C.; Zhang, A. Effects of gravel content and shape on shear behaviour of soil rock mixture: Experiment and DEM modelling. *Comput. Geotech.* **2022**, *141*, 104476. [CrossRef]
38. Jiang, J.; Ni, F.; Dong, Q.; Wu, F.; Dai, Y. Research on the fatigue equation of asphalt mixtures based on actual stress ratio using semi-circular bending test. *Constr. Build. Mater.* **2018**, *158*, 996–1002. [CrossRef]
39. Elseifi, M.A.; Mohammad, L.N.; Ying, H.; Cooper III, S. Modeling and evaluation of the cracking resistance of asphalt mixtures using the semi-circular bending test at intermediate temperatures. *Road Mater. Pavement Des.* **2012**, *13*, 124–139. [CrossRef]
40. Shiri, S.; Yazdani, M.; Pourgol-Mohammad, M. A fatigue damage accumulation model based on stiffness degradation of composite materials. *Mater. Des.* **2015**, *88*, 1290–1295. [CrossRef]
41. Li, J.; Zhang, J.; Qian, G.; Zheng, J.; Zhang, Y. Three-dimensional simulation of aggregate and asphalt mixture using parameterized shape and size gradation. *J. Mater. Civ. Eng.* **2019**, *31*. [CrossRef]
42. Yao, Y.; Li, J.; Liang, C.; Hu, X. Effect of coarse recycled aggregate on failure strength for asphalt mixture using experimental and DEM method. *Coatings* **2021**, *11*, 1234. [CrossRef]

polymers

MDPI

Article

Road Performance Investigation on Fiber-Reinforced Recycled Cement Base Material

Yongcheng Ji *, Wenhao Ji, Ziyi Zhang and Rui Wang *

School of Civil Engineering, Northeast Forestry University, Harbin 150040, China
* Correspondence: yongchengji@nefu.edu.cn (Y.J.); wangrui20075039@nefu.edu.cn (R.W.); Tel.: +86-151-0457-1851 (Y.J.)

Abstract: The characteristics of the materials used in early buildings in China have led to a large proportion of discarded red bricks among the construction waste generated by demolishing abandoned buildings. The application of red brick aggregate with a particle size $\leq$5 mm and red brick powder with particle size 0.125~0.75 mm (referred to as recycled brick powder) was studied in this study after the crushing of waste red brick in road structures. The research results will provide a theoretical basis for the whole-grain recycling of waste red brick aggregate. The aggregate of red brick with a particle size smaller than 2 mm was mixed with different amounts of cement soil and fiber to prepare a cement-stable binder for the sub-base material. The recycled brick powder of 0.125~0.75 mm was used to replace the quartz sand with different substitution rates. As pavement materials, different amounts of fiber were used to prepare fiber-reinforced recycled-brick-powder cementitious composites. The optimal mixing ratio of the two materials was evaluated from the mechanical properties. The results showed that the optimal mixing ratio of the cement-stable binder was as follows: waste-red-brick-aggregate content was 50%, cement content was 4%, and fiber content was 0.2%. The optimum ratio of fiber-reinforced recycled-brick-powder cementitious composites was determined to be as follows: the replacement rate of recycled brick powder is 25%, and the content of PVA fiber is 1%. The regression analysis was used to fit the equations between the fiber content and the 7d unconfined compressive strength and the tensile strength of the cement-stabilized binder for different red-brick-aggregate admixtures at 4% cement content. A scanning electron microscope was used to observe the failure modes of the fiber. The influence of failure modes, such as pulling out, fracture, and plastic deformation, on the mechanical properties was expounded.

Keywords: fiber cementitious materials; mechanical property; pavement structure; waste red brick; fiber microstructure

Citation: Ji, Y.; Ji, W.; Zhang, Z.; Wang, R. Road Performance Investigation on Fiber-Reinforced Recycled Cement Base Material. *Polymers* **2022**, *14*, 4102. https://doi.org/10.3390/polym14194102

Academic Editors: Wensheng Wang, Yongchun Cheng, Heping (Fred) Chen and Guojin Tan

Received: 31 August 2022
Accepted: 24 September 2022
Published: 30 September 2022

1. Introduction

The annual output of construction waste reached 3 billion tons and continued to show an increasing trend in 2020 [1]. In the past, influenced by economic development and traditional construction habits, many sintered clay red bricks were used in old buildings. However, in recent years, with the rapid development of urban–rural integration, a large amount of construction waste, mainly composed of red bricks, will be generated during the demolition and renovation of old rural houses, accounting for about 35–45% of the total construction waste [2]. The specific proportion is shown in Figure 1, below. At the same time, with the progress of road infrastructure projects, many fillers are needed, and plain-soil backfilling destroys cultivated land and occupies many land resources [3]. Therefore, the waste red brick is applied to the road backfilling material. It not only can effectively alleviate a large amount of construction waste produced yearly but also reduce a large amount of earth and rock required by the road infrastructure project.

Scholars at home and abroad have conducted extensive studies on recycling and using recycled red brick. Ren et al. [4] used orthogonal experiments to explore the preparation of recycled aggregate permeable concrete with recycled red brick aggregate of different particle sizes (5–10 mm and 10–20 mm) and different substitution rates (5%, 10%, 15%, and 20%). The results show that the mechanical properties and water permeability of recycled aggregate pervious concrete are better when the red brick replacement rate is 15% and the sand rate is 6%. The strength of recycled aggregate pervious concrete with 10–20 mm is lower than that with 5–10 mm, but the water permeability is better than that with 5–10 mm. Li [5] prepared cement mortar by replacing recycled concrete fine aggregates with recycled red bricks of a particle size less than 5 mm in different proportions. The results showed that each performance index of recycled fine aggregate decreased with the increase of red brick content. The water requirement of recycled cement mortar was linearly and positively correlated with the red brick content. The compressive and flexural strengths were reduced to different degrees, and the compressive strength of recycled cement mortar was highly linearly and negatively correlated with the red brick content. Sun [6] used red waste bricks as coarse aggregate for the preparation of concrete, which was analyzed in a compressive-strength experiment and abrasion-resistance experiment, and determined that the recycled concrete prepared from the coarse aggregate of recycled red bricks could meet the requirements. Kim [7] et al. used the broken red brick as coarse aggregate and prepared concrete with 0%, 30%, and 60% content, respectively. The results showed that the concrete with 30% coarse red brick aggregate performed similarly to ordinary concrete. Ji [8] et al. used discarded red bricks instead of natural aggregates. They combined them with steel fiber to establish a finite element model of steel-fiber/recycled-brick-aggregate concrete under uniaxial compression.

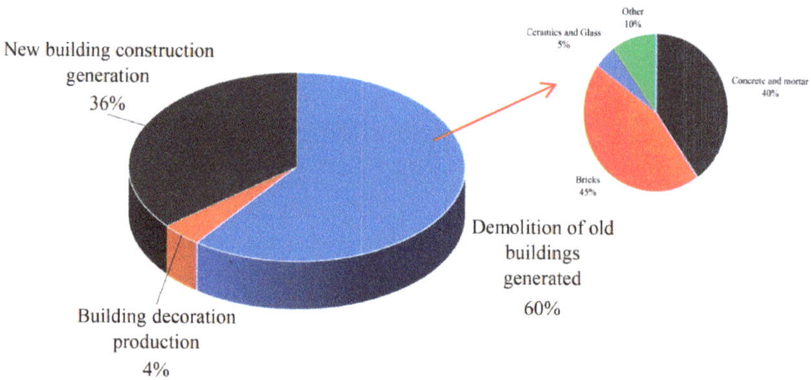

Figure 1. Proportion of various construction wastes.

Scholars have also paid close attention to improved road structural materials. Zhou [9] conducted an indoor experiment by mixing construction waste of three granular combinations into expanded soil at 0%, 20%, 30%, 40%, and 50%. The results showed that, when the compaction degree meets the specification requirements, choosing a suitable combination of aggregate and mixing ratio and construction waste improved expansive soil satisfies the requirement of highway roadbed material strength and expansion rate of total demand. Xu [10] studied the characteristics of the unconfined compressive strength of soil–cement mixed with glass fiber under the freeze–thaw cycle. The study showed that the unconfined compressive strength of soil–cement could be improved when the mass fraction of glass fiber was 0.5%, and the strength was the highest when the length of glass fiber was 3 mm. Khiem [11] studied the influence of corn fiber content on the mechanical properties of soil–cement, found that the incorporation of corn fiber enhanced the compressive and tensile strength of soil–cement, and determined that the optimal incorporation range of corn fiber in soil–cement was 0.25~0.5%. Yu [12] et al. blended waste concrete:waste red bricks = 100%:0%, 50%:50%, and 10%:90% into cement-stabilized aggregates and conducted experimental studies at cement doses of 3%, 5%, and 7%, respectively. The results showed that using recycled aggregate as a road base was feasible. Li [13] et al., using construction waste as aggregate, verified the feasibility of applying recycled aggregate made of construction waste in subgrade through experiments such as unconfined compressive strength and California bearing ratio.

Cement concrete pavement is widely used because of its high strength, good integrity, and strong carrying capacity. However, it also has certain defects. For example, a certain degree of cracking and plate breaking will appear during use. In order to overcome the defects of the cement-concrete pavement, the researchers have made various improvements to the cement-based pavement materials. Li [14] studied the natural brucite mineral fiber and industrial-solid-waste fly ash in pavement concrete. The results show that the early performance of the pavement concrete is not apparent, but 28d-age-and-above strength, impact resistance, and toughness are significantly higher than those of the base group. Simultaneously, the incorporation of brucite fiber can significantly improve the durability of concrete. Ma [15] studied the influence rule of steel slag powder content on the road performance of a cement composite. The results showed that the incorporation of steel slag powder reduces the water consumption of standard consistency and prolongs the setting time of cement. When the mixing capacity of steel slag powder is less than 40%, the stability of the steel slag cement composite is qualified. The 28d compressive and flexural strengths showed a rising and decreasing trend with the increase of steel-slag-powder dosing, with a peak at 20% of steel-slag-powder dosing. Qi [16] used waterborne epoxy resin as a modifying agent and added an appropriate amount of fly ash to prepare the modified mortar for pavement materials. This researcher studied composite materials' optimal mix ratio and road performance through a compressive test, anti-folding test, bond strength test, and anti-skid performance test. The results showed that fly ash mixed at 10% and waterborne epoxy resin mixed at 10% have the best mechanical properties, bonding strength, and anti-skid performance to meet the standard requirements. Qian [17] et al. showed that fiber-cement-based composite paving could reduce the thickness of pavement to a great extent compared with ordinary concrete pavement. It also effectively curbs the upward reflection crack expansion of the subgrade, and its fatigue life is more than double that of ordinary concrete pavement.

Red waste bricks are mainly applied as recycled aggregates for preparing cementitious materials and filling roadbeds. The red brick aggregates' particle size is concentrated above 5 mm. However, waste red bricks will inevitably produce particles or dust with particle size $\leq$5 mm during the crushing process, which will cause secondary damage to the environment if not utilized. Therefore, in order to make a recycled red brick aggregate of all sizes that can be used, this study adopted the regeneration of the particle size of 5 mm or less red brick particles with different dosages mixed with soil, with 9 mm long polyvinyl alcohol (PVA) fiber [18]. A prepared cement-stabilizing binder was applied to the road sub-base, with an unconfined-compressive-strength test and split test as the evaluation index, to explore the influence of fiber, cement, and recycled red brick content on the mechanical properties of the stable inorganic binder. The recycled brick powder with a particle size of 0.075~0.125 mm was used to replace quartz sand with different substitution rates. Combined with 6 mm–long PVA fiber, the fiber-reinforced recycled-brick-powder cementitious composite was prepared with low cost, low carbon, and environmental protection. A mechanical-property test determined the optimal mixing amount of recycled brick powder and PVA fiber. The experiment results will provide a theoretical basis for fully recycling red waste bricks at all levels in the future and help achieve a "carbon peak" in 2030.

2. Experimental Program

2.1. Material Properties

The raw materials used in the experiment include cement, fly ash, quartz sand, PVA fiber, waste red brick aggregate, plain soil, a water-reducing agent, and a thickening agent. The cement used in this experiment is P.O. 42.5 ordinary silicate cement, and its performance index is in line with GB175-2007 "Common Portland Cement" specification standard research, as shown in Table 1. The fly ash is the grade I fly ash produced by Shanxi Longhui Building Materials Co., LTD. The specific parameters of the material are shown in Table 2. The particle size of quartz sand is 120–200 mesh produced by Zhengzhou Hanhai Environmental Protection Technology Co., LTD. PVA fiber is produced by Shanghai Chenqi Chemical Technology Co., LTD. The specific parameters are shown in Table 3. The waste red brick aggregate is the waste red brick obtained from the demolition of a specific building in Harbin; it is obtained by crushing and screening. The plain soil is clay soil distributed from 1.5–3 m underground in Harbin city, and the specific indexes are shown in Table 4. The water-reduction rate of the polycarboxylic acid superplasticizer was 28.5%; hydroxypropyl methylcellulose thickening agent was used, at viscosity class 200000. The water used in the experiment is Harbin ordinary tap water. The red brick aggregate with particle size $\leq$5 mm, recycled brick powder with a particle size of 0.075~0.125 mm, plain soil, quartz sand, and PVA fiber are shown in Figure 2.

Table 1. Main technical indicators of cement.

Setting Time/Min		Compressive Strength/MPa		Flexural Strength/MPa		Water Requirement of Normal Consistency/%	Stability	Fineness Modulus
Initial setting	Final setting	3d	27d	3d	27d	25.4	qualified	3.2
160	210	25.8	45.2	5.6	9.4			

Table 2. Parameter index of fly ash.

Fineness/%	Water Requirement/%	Loss on Ignition/%	SO_3 Content/%	Chloride Ion/%	Water Content/%	CaO Content/%	Free CaO Content/%	Stability
9.0	91	2.0	1.6	0.007	0.1	1.6	0.1	0.2

Table 3. Performance index of PVA fiber.

Length/mm	Density/g/cm^3	Tensile Strength/MPa	Elongation/%	Modulus of Elasticity/GPa
6, 9	1.3	1900	8	35

Table 4. Technical indexes of plain soil.

Water Content/%	Particle Size > 0.075 mm/%	Liquid Limit/%	Plastic Limit/%	Plasticity Index	Type of Soil
5.5	3.7	32.89	20.96	11.93	Low liquid limit clay

(a) (b) (c)

Figure 2. *Cont.*

(**d**)

(**e**) Recycled red brick aggregate.

Figure 2. Test raw materials. (**a**) Recycled brick powder. (**b**) Quartz sand. (**c**) PVA fiber. (**d**) Plain soil. (**e**) Recycled red brick aggregate.

2.2. *Proportioning Design and Test Method*

2.2.1. Compaction Test

Proportioning Design

The compaction test explored the samples' optimal moisture content and maximum dry density under different mixing ratios. With the content of recycled red brick aggregate (0%, 50%, and 100%) and cement content (3%, 4%, and 5%) as the research variables, five groups of samples with a moisture content of 10%, 12%, 14%, 16%, and 18% were prepared for each group, so 45 compaction tests were needed. Table 5 shows the specific mix ratio of the compaction test.

Table 5. Mix ratio of compaction test.

Group Number	Plain Soil/g	Red Brick Aggregate/g	Cement/g	Water/g
1	2300	0	3%	10%, 12%, 14%, 16%, 18%
2	1150	1150	3%	10%, 12%, 14%, 16%, 18%
3	0	2300	3%	10%, 12%, 14%, 16%, 18%
4	2300	0	4%	10%, 12%, 14%, 16%, 18%
5	1150	1150	4%	10%, 12%, 14%, 16%, 18%
6	0	2300	4%	10%, 12%, 14%, 16%, 18%
7	2300	0	5%	10%, 12%, 14%, 16%, 18%
8	1150	1150	5%	10%, 12%, 14%, 16%, 18%
9	0	2300	5%	10%, 12%, 14%, 16%, 18%

Experimental Method

The compaction test can analyze the compaction characteristics of the soil, and the variation of the moisture content of the soil with the dry density under certain fixed conditions can be obtained to determine the optimum moisture content and the maximum dry density value of the plain soil used. This test adopts a heavy compaction test, the instrument used for the laboratory multifunctional electric compaction instrument. The height of the compaction cylinder is 12.7 cm, the volume is 997 cm^3, the hammer is 4.5 kg, the falling height is 45 cm, and the compaction times of each layer are 27 times.

Soil samples and waste brick aggregates were sieved. Soil samples and red brick aggregate below 5 mm were taken, dried, and added with different moisture (in 2% water content increments); they were then mixed well and set for 12 h. Apply a layer of petroleum jelly on the compaction cylinder, and tightly fix the compaction cylinder and the compaction instrument. Pour the preprepared compacted material into the compaction cylinder in five times. After 27 times of compaction, the surface of the soil sample should be "brushed" to make the five layers of the soil sample closely connected. After the compaction is completed, the soil height on the top surface of the compaction cylinder is measured. If the height is less than 5 mm, the compaction is qualified. The top and bottom surfaces of the compacted soil were cut flat to make the volume of the soil the same as the volume of the compaction cylinder. The soil samples were removed from the compaction cylinder with a multifunctional electric demold machine, and two soil samples were taken from the center to measure the moisture content. The above steps were repeated for all ratios and water-content specimens, and the specific operation flowchart is shown in Figure 3.

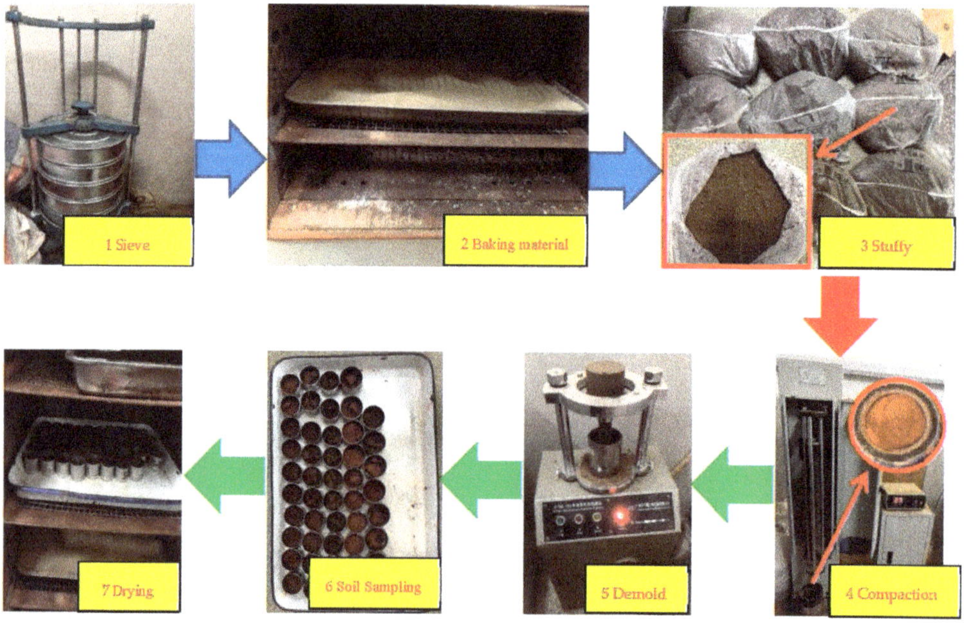

Figure 3. Compaction test process.

2.2.2. Tests for 7d Unconfined Compressive Strength and Splitting Strength Proportioning Design

The 7d unconfined compressive strength and splitting strength are essential indicators to verify the performance of inorganic bonded stabilized materials as sub-base materials. Therefore, based on the best moisture content and maximum dry density of each ratio obtained from the compaction test, each ratio was mixed with PVA fiber (0%, 0.1%, 0.2%, and 0.3%) to prepare the PVA-fiber bond-test specimens. The amount of each material was based on Formulas (1)–(5), and the specific ratio is shown in Table 6.

The standard mass of each specimen is calculated as follows:

$$m_0 = v \times \rho_{max} \times \left(1 + \omega_{opt}\right) \times \gamma \quad (1)$$

The total mass of dry material (dry soil + cement) for each specimen is calculated as follows:

$$m_1 = \frac{m_0}{1 + \omega_{opt}} \quad (2)$$

The mass of inorganic binding material (cement) in each specimen is calculated as follows:

$$m_2 = m_1 \times \alpha \quad (3)$$

The dry soil mass in each specimen is calculated as follows:

$$m_3 = m_1 - m_2 \quad (4)$$

The water added to each specimen is calculated as follows:

$$m_\omega = (m_2 + m_3) \times \omega_{opt} \quad (5)$$

where v is the volume of the specimen (cm^3), taken as 98.125; ρ_{max} and ω_{opt} are the optimum moisture content (%) and the maximum dry density (g/cm^3), respectively; γ is used for the mix compaction standard (%). According to the "Technical Guidelines for Construction of Highway Roadbases" (JTG/T F20-2015), the requirements for the construction of highway sub-base compaction should not be less than 95%, and 95% was used in this study. Moreover, α is the content of inorganic binder (%).

Table 6. Mix ratio of unconfined-compressive- and splitting-tensile-test specimens.

Group Number		Mass of Single Specimen /g	Dry Material (Soil + Cement)/g	Inorganic Binding Material/g	Soil or Red Brick Quality/g	Water/g	Fiber Quality/g
Plain soil:red brick aggregate = 100%:0%	C3P0	199.33	176.18	5.29	170.90	23.15	0
	C4P0	204.90	178.98	7.16	171.82	25.92	0
	C5P0	208.01	180.84	9.04	171.80	27.16	0
	C3P0.1	199.33	176.18	5.29	170.90	23.15	0.17
	C4P0.1	204.90	178.98	7.16	171.82	25.92	0.17
	C5P0.1	208.01	180.84	9.04	171.80	27.16	0.17
	C3P0.2	199.33	176.18	5.29	170.90	23.15	0.34
	C4P0.2	204.90	178.98	7.16	171.82	25.92	0.34
	C5P0.2	208.01	180.84	9.04	171.80	27.16	0.34
	C3P0.3	199.33	176.18	5.29	170.90	23.15	0.51
	C4P0.3	204.90	178.98	7.16	171.82	25.92	0.52
	C5P0.3	208.01	180.84	9.04	171.80	27.16	0.52
Plain soil:red brick aggregate = 50%:50%	C3P0	188.81	165.00	4.95	160.05	23.81	0
	C4P0	190.59	165.93	6.64	159.29	24.66	0
	C5P0	191.61	166.68	8.33	158.34	24.93	0
	C3P0.1	188.81	165.00	4.95	160.05	23.81	0.16
	C4P0.1	190.59	165.93	6.64	159.29	24.66	0.16
	C5P0.1	191.61	166.68	8.33	158.34	24.93	0.16
	C3P0.2	188.81	165.00	4.95	160.05	23.81	0.32
	C4P0.2	190.59	165.93	6.64	159.29	24.66	0.32
	C5P0.2	191.61	166.68	8.33	158.34	24.93	0.32
	C3P0.3	188.81	165.00	4.95	160.05	23.81	0.48
	C4P0.3	190.59	165.93	6.64	159.29	24.66	0.48
	C5P0.3	191.61	166.68	8.33	158.34	24.93	0.48
Plain soil:red brick aggregate = 0%:100%	C3P0	154.87	139.83	4.19	135.63	15.05	0
	C4P0	157.41	140.48	5.62	134.86	16.93	0
	C5P0	162.67	143.84	7.19	136.64	18.84	0
	C3P0.1	154.87	139.83	4.19	135.63	15.05	0.11
	C4P0.1	157.41	140.48	5.62	134.86	16.93	0.13
	C5P0.1	162.67	143.84	7.19	136.64	18.84	0.14
	C3P0.2	154.87	139.83	4.19	135.63	15.05	0.27
	C4P0.2	157.41	140.48	5.62	134.86	16.93	0.27
	C5P0.2	162.67	143.84	7.19	136.64	18.84	0.27
	C3P0.3	154.87	139.83	4.19	135.63	15.05	0.41
	C4P0.3	157.41	140.48	5.62	134.86	16.93	0.40
	C5P0.3	162.67	143.84	7.19	136.64	18.84	0.41

Note: C stands for cement, and P for PVA fiber. For example, C3P0.1 represents a sample with cement content of 3% and PVA-fiber content of 0.1%.

Experimental Method

The specimen preparation of the 7d-unconfined-compressive-strength test and splitting-strength test was carried out by taking waste red brick aggregate and plain soil particles below 2 mm. The test followed the "Test Method of Materials Stabilized with Inorganic Binders for Highway Engineering" (JTG E51-2009), from the preparation of specimens to the health preservation. The sample size was a 50 × 50 mm cylinder, and the prepared raw materials were mixed evenly, and then the material was stuffed for 12 h. After the stuffy material was finished, the corresponding cement and PVA-fiber quality were added. Next, a planetary cement mortar mixer fully and evenly mixed the material. The specimen was hydrostatically formed by the pressure/demolding tester, with a range of 300 KN. After demolding, the specimens were wrapped with plastic and placed in a standard constant temperature and humidity curing box for seven days before the test. The day before the end of the regimen, the specimens should be immersed in water for 24 h.

Next, a universal testing machine performed the unconfined-compressive-strength and unsplit-tensile-strength tests. The loading rate of the press was controlled to be 1 mm/min, and the maximum pressure at the failure of the specimen was recorded. The seven-day unconfined compressive strength and splitting tensile strength of the specimen were calculated according to Equations (6) and (7). A total of eight steps are required, and the specific test operation process is shown in Figure 4.

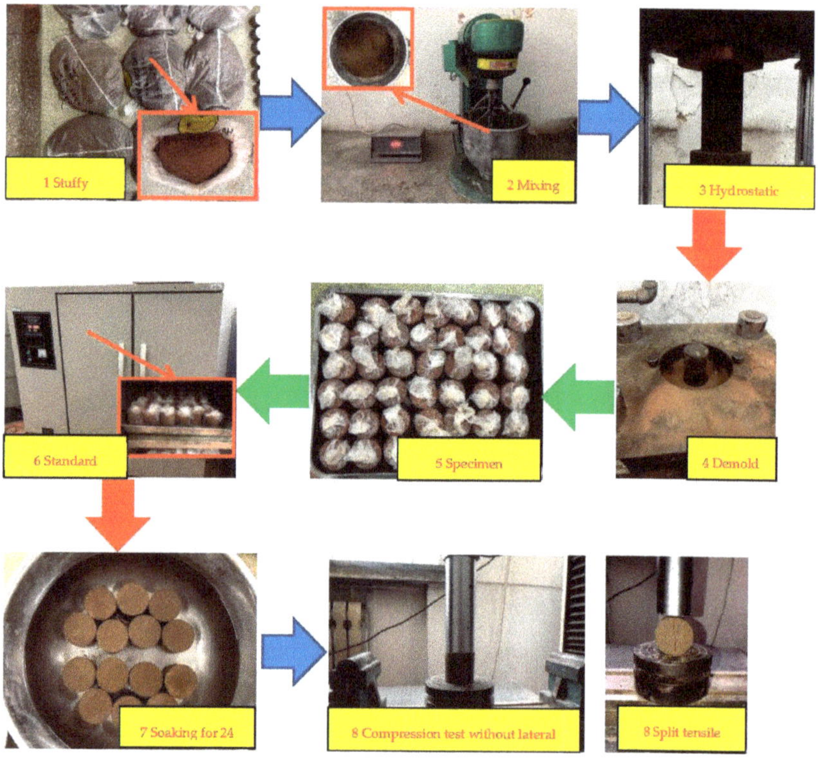

Figure 4. Flowchart for 7d unconfined compressive test and splitting tensile test.

The 7d unconfined compressive strength is calculated according to the following equation:

$$R_c = \frac{P}{A} = \frac{4P}{\pi D^2} \tag{6}$$

where R_c is the compressive strength of the sample under different mix ratios, MPa; P is the maximum pressure at the time of specimen failure, N; A is the cross-sectional area of the specimen, mm^2; and D is the diameter of the specimen, mm.

The splitting tensile strength is calculated according to the following equation:

$$R_i = \frac{2P}{\pi dh} = 0.01273\frac{P}{h} \tag{7}$$

where R_i is the splitting tensile strength of the sample at different mix ratios, MPa; P is the maximum pressure at the time of specimen failure, N; d is the diameter of the specimen, mm; and h is the height of the specimen soaked in water, mm.

2.2.3. Mechanical Properties of Fiber-Reinforced Cementitious Materials

Proportioning Design

This experiment studied the effect of the substitution rate of recycled brick powder replacing quartz sand and the volume of PVA-fiber content on the fiber-reinforced recycled-brick-powder cementitious composites' performance. Based on the water–binder ratio of 0.32 and the rubber–sand ratio of 0.36, the substitution rates of recycled brick powder replacing quartz sand are 0%, 25%, 50%, 75%, and 100%, respectively. The fiber volume content was 0%, 1.0%, 1.5%, and 2.0%, respectively. A total of 20 groups of mix ratios were designed by using the control variable method, and the specific mix ratio is shown in Table 7.

Table 7. Fiber-reinforced recycled-brick-powder cementitious composites' fitting ratio.

Group Number	Cement Quality/g	Fly Ash Quality/g	Recycled Brick Powder Quality/g	Quartz sand Quality/g	Quality of Water/g	PVA Fiber Quality/g	Water Reducer/%	Thickening Agent/%
Z0P0	439.84	527.81	0	348.36	309.66	0	0.2	0.1
Z25P0	439.84	527.81	87.09	261.27	309.66	0	0.2	0.1
Z50P0	439.84	527.81	174.18	174.18	309.66	0	0.2	0.1
Z75P0	439.84	527.81	261.27	87.09	309.66	0	0.2	0.1
Z100P0	439.84	527.81	348.36	0	309.66	0	0.2	0.1
Z0P1	439.84	527.81	0	348.36	309.66	3.33	0.2	0.1
Z25P1	439.84	527.81	87.09	261.27	309.66	3.33	0.2	0.1
Z50P1	439.84	527.81	174.18	174.18	309.66	3.33	0.2	0.1
Z75P1	439.84	527.81	261.27	87.09	309.66	3.33	0.2	0.1
Z100P1	439.84	527.81	348.36	0	309.66	3.33	0.2	0.1
Z0P1.5	439.84	527.81	0	348.36	309.66	4.99	0.2	0.1
Z25P1.5	439.84	527.81	87.09	261.27	309.66	4.99	0.2	0.1
Z50P1.5	439.84	527.81	174.18	174.18	309.66	4.99	0.2	0.1
Z75P1.5	439.84	527.81	261.27	87.09	309.66	4.99	0.2	0.1
Z100P1.5	439.84	527.81	348.36	0	309.66	4.99	0.2	0.1
Z0P2	439.84	527.81	0	348.36	309.66	6.66	0.2	0.1
Z25P2	439.84	527.81	87.09	261.27	309.66	6.66	0.2	0.1
Z50P2	439.84	527.81	174.18	174.18	309.66	6.66	0.2	0.1
Z75P2	439.84	527.81	261.27	87.09	309.66	6.66	0.2	0.1
Z100P2	439.84	527.81	348.36	0	309.66	6.66	0.2	0.1

Note: Z stands for recycled brick powder, and P for PVA fiber. For example, Z25P0 represents a recycled brick powder content of 25% and a PVA-fiber content of 0%.

Experimental Method

Cement, fly ash, quartz sand, and recycled brick powder were added to the mixing tank. Stir for 120 s so that it is fully mixed. Then add some water mixed with the admixture and stir for 120 s, and add the remaining water and stir for 120 s. After stirring evenly, add PVA fiber, dispersing manually and stirring for 180 s until no fiber agglomerates. Next, the mold is filled in layers, formed by vibration, covered with plastic film, cured for 24 h, and then demolded, as shown in Figure 5. Follow up with mechanical test conditions after 28 days of standard curing.

The compressive and flexural tests were conducted by using the YAW-300H testing machine produced by the Jinan Hengruijin Company, as shown in Figure 6. The test procedure was carried out according to the "Test Method of Cement Mortar Strength (ISO method)" (GB/T 17671-2020), and the loading speed of the flexural test was 50 N/s ± 10 N/s. The two half sections after breaking were taken out for a compressive test, and the loading rate was 2400 N/s ± 200 N/s. Therefore, the specimen size for the flexural test was 40 mm × 40 mm × 160 mm, and the size was 40 mm × 40 mm × 40 mm for the compressive test, as shown in Figure 7.

Figure 5. Specimens of fiber-reinforced recycled-brick-powder cementitious composite.

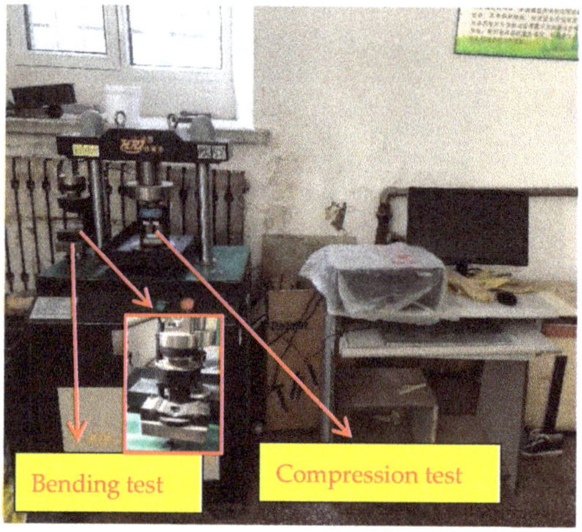

Figure 6. Compression- and bending-test machine.

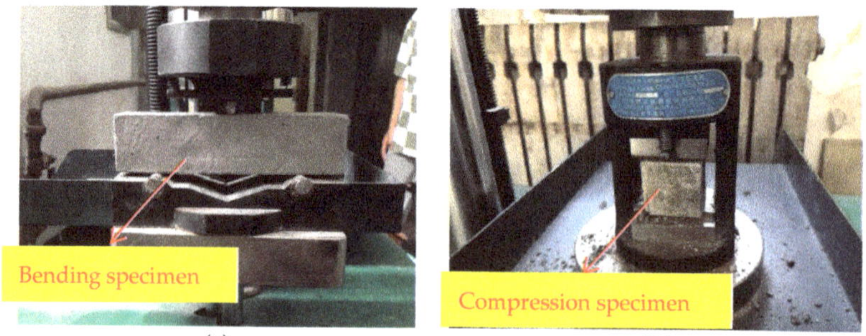

(a) (b)

Figure 7. Mechanical test. (**a**) Bending test. (**b**) Compression test.

2.3. Microscopic Experiments

The SEM test was carried out in the Analysis and Testing Center Laboratory of Northeast Forestry University, using JMF-7500F (Cold Field Emission Scanning Electron Microscope), as shown in Figure 8. The surface of the flat sheet sample was first gilded to improve its electrical conductivity. After the gold injection, the charge plate was fixed under the electron microscope for scanning.

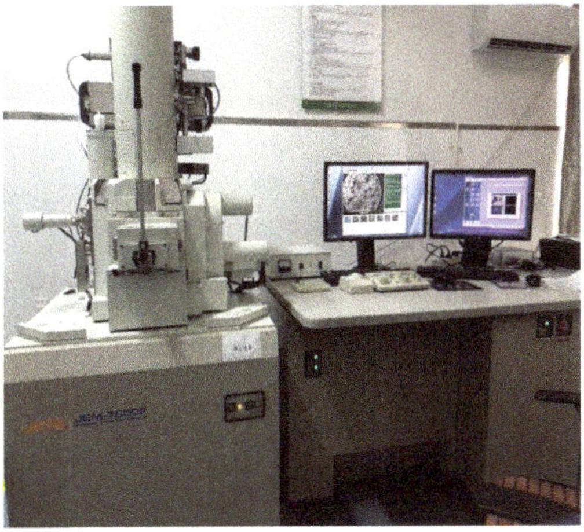

Figure 8. Scanning electron microscope.

3. Experimental Results and Analysis

3.1. Compaction Test

The compaction test is performed to obtain the maximum dry density and optimal moisture content of the test sample, provide the basis for the design of the raw-material mix ratio of the pavement structure layer, and guide and control the construction. The samples with different proportions were subjected to a compaction test according to the method in "Test Method of Materials Stabilized with Inorganic Binders for Highway Engineering" (JTG E51-2009). The maximum dry density and optimal moisture content of each mixture ratio are shown in Table 8.

Table 8. Compaction test results.

Group Number	Content of Plain Soil/%	Red-Brick-Aggregate Content/%	Cement Content/%	Maximum Dry Density/(g/cm^3)	Optimum Moisture Content /%
1	100	0		1.89	13.14
2	50	50	3	1.77	14.43
3	0	100		1.50	10.76
4	100	0		1.92	14.48
5	50	50	4	1.78	14.86
6	0	100		1.51	12.05
7	100	0		1.94	15.02
8	50	50	5	1.79	14.96
9	0	100		1.54	13.10

It can be seen from Figure 9a that, when the cement content is constant, the maximum dry density of the specimen decreases with the increase of the red-brick-aggregate content. This is because the appearance of the red brick particles is sharp, and the gap between the particles is more significant than the plain soil. In addition, the hardness is higher than the plain soil, so the density of the red brick aggregate is less than that under the same compaction action of the plain soil. As seen in Figure 9b, the optimal moisture content of the sample increases first. It then decreases with the incorporation of red brick aggregate, reaching the maximum value when the incorporation of red brick aggregate is 50%. When the red-brick-aggregate content is at 50%, soil particles can be filled with red brick particles, which will lock more water, but when the red-brick-aggregate content is greater than 50%, the red brick aggregate dominates in the sample. Due to the nature of the red brick aggregate, there is less water absorption, and the ability to lock water is weak, so the optimal moisture content will gradually decline after the red brick aggregate exceeds 50%. As can be seen from Figure 9c,d, when the red brick aggregate's mix ratio is certain, the maximum dry density and the optimal moisture content both increase with the increasing cement content. The increase of the optimum water content may be due to the chemical reaction between the cement and plain soil particles, which consume water in the sample during the reaction process, leading to the increase of the optimum water content. The increase in the maximum dry density occurs because the cement is mixed into the sample at different doses, equivalent to replacing some plain soil or red brick particles with cement. The relative density of cement is higher than that of plain soil and red brick, so the incorporation of cement increases the maximum dry density of the sample [19].

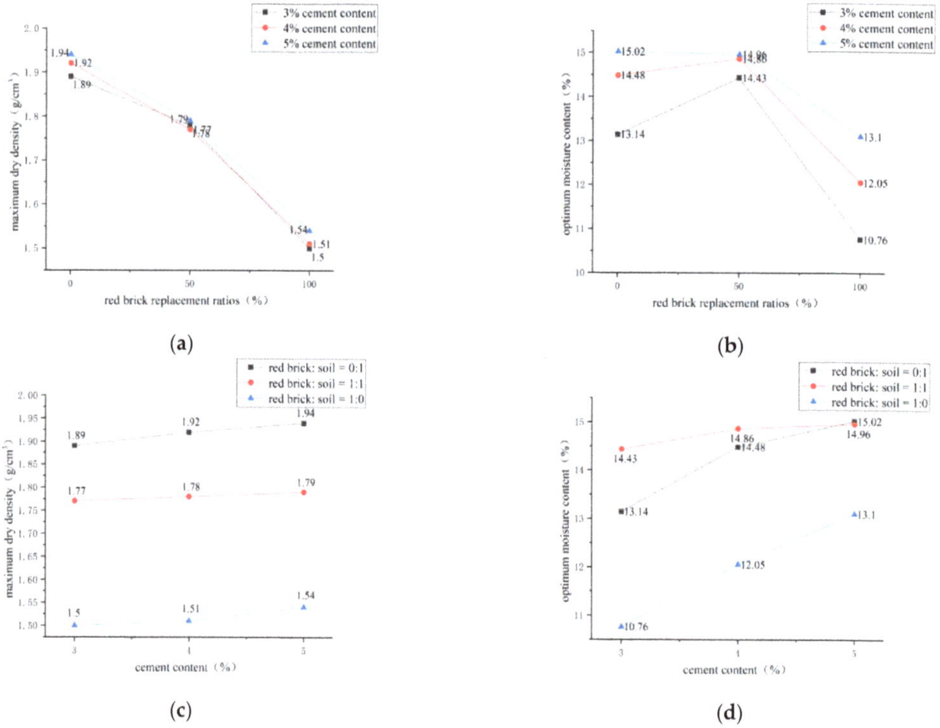

Figure 9. Compaction test analysis. (**a**) Maximum dry density and red-brick-aggregate replacement ratio. (**b**) Optimal moisture content and red-brick-aggregate replacement ratio. (**c**) Cement content and maximum dry density under different ratios. (**d**) Cement content and optimal moisture content under different ratios.

3.2. Tests for 7d Unconfined Compressive Strength and Splitting Strength

Cement-stabilized soil is generally used as the sub-base of road construction. In the design code of asphalt pavement or concrete pavement structure, the strength of the road's sub-base has specific strength requirements. For example, the size of the 7d unconfined compressive strength can reflect the limited strength of the cement-stabilized soil resistance to the axial pressure. In contrast, the splitting strength will reflect the indirect tensile properties of the pavement base. The compressive strength and split tensile strength of each mixing ratio test are listed in Table 9.

Table 9. Results of the 7d-unconfined-compressive-strength and splitting-strength test.

Group Number		7d Unconfined Compressive Strength/MPa	Splitting Strength /MPa
Plain soil:red brick aggregate = 1:0	C3P0	0.62	0.06
	C4P0	1.18	0.10
	C5P0	1.39	0.14
	C3P0.1	0.73	0.09
	C4P0.1	1.52	0.15
	C5P0.1	1.62	0.19
	C3P0.2	0.79	0.10
	C4P0.2	2.04	0.17
	C5P0.2	2.23	0.21
	C3P0.3	0.96	0.11
	C4P0.3	2.18	0.21
	C5P0.3	2.38	0.23
Plain soil:red brick aggregate = 1:1	C3P0	0.85	0.08
	C4P0	1.16	0.13
	C5P0	1.35	0.17
	C3P0.1	1.29	0.13
	C4P0.1	1.79	0.20
	C5P0.1	2.51	0.26
	C3P0.2	1.36	0.15
	C4P0.2	2.16	0.23
	C5P0.2	2.72	0.32
	C3P0.3	1.47	0.18
	C4P0.3	2.76	0.29
	C5P0.3	2.93	0.38
Plain soil:red brick aggregate = 0:1	C3P0	0.32	0.04
	C4P0	0.87	0.08
	C5P0	1.13	0.12
	C3P0.1	0.50	0.05
	C4P0.1	1.20	0.12
	C5P0.1	1.31	0.13
	C3P0.2	0.66	0.07
	C4P0.2	1.28	0.13
	C5P0.2	1.47	0.15
	C3P0.3	0.84	0.08
	C4P0.3	1.33	0.15
	C5P0.3	1.55	0.16

3.2.1. Effect of Cement Content on 7d Unconfined Compression Strength and Splitting Tensile Strength

Figures 10–12 show the influence of cement content on 7d unconfined compressive strength and splitting tensile strength under three conditions: plain soil:red brick aggregate = 1:0, 1:1, and 0:1, respectively. It can be seen from the figure that, under a certain amount of fiber and red brick aggregate, the unconfined compressive strength and splitting tensile strength of 7d show an upward trend with the increase of cement content. However, cement content from 3% to 4% and from 4% to 5% caused by the strong growth trend is different. Cement content increased from 3% to 4%, 7d unconfined compression strength and splitting tensile strength increased by more than 60%, and the mix ratio increased by more than 100%. For example, for a plain soil:red brick aggregate ratio of 0:1, with a PVA-fiber incorporation of 0%, the compressive strength and tensile strength increased by 120% and 100%, respectively. When the cement content increased from 4% to 5%, the 7d unconfined compression strength and splitting tensile strength increased by less than 30%. For example, when the plain soil:red brick aggregate = 1:0, after a PVA-fiber incorporation of 0.3%, its compressive strength and tensile strength only increased by 9.2% and 9.5%, respectively. This is because the reaction of cement and water produces many cement hydration products and soil particles and brick aggregate to form a three-dimensional mesh of cement slurry gradually. The condensation between the reactants can make the particles close and make the specimen more compact, thus improving the 7d unconfined compression strength and splitting tensile strength of the specimen [20]. However, the strength improvement is not apparent as the cement content increases from 4% to 5%. For the consideration of economy, the cement content is 4%, and the strength meets the requirements of the code.

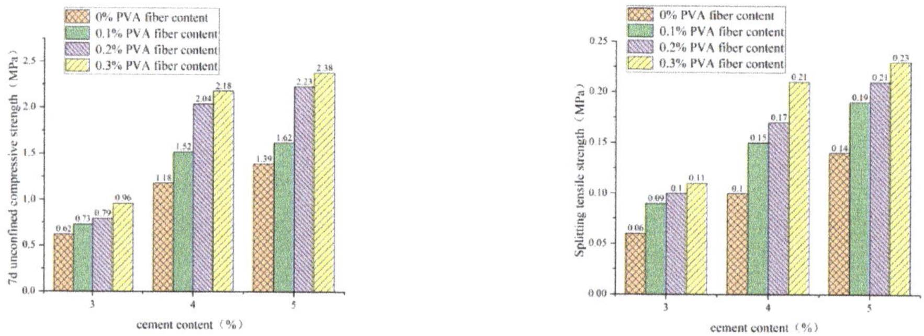

(a) (b)

Figure 10. Cement admixture with 7d unconfined compressive strength and splitting tensile strength for plain soil:red brick aggregate = 1:0. (**a**) Cement content and 7-day unconfined compressive strength. (**b**) Cement content and splitting tensile strength.

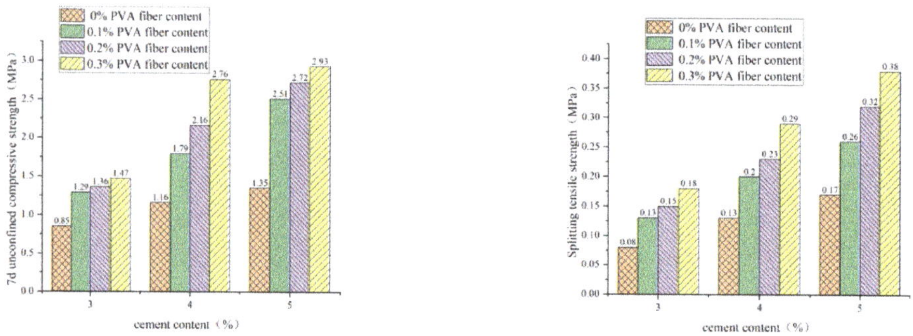

(a) (b)

Figure 11. Cement admixture with 7d unconfined compressive strength and splitting tensile strength for plain soil:red brick aggregate = 1:1. (**a**) Cement content and 7-day unconfined compressive strength. (**b**) Cement content and splitting tensile strength.

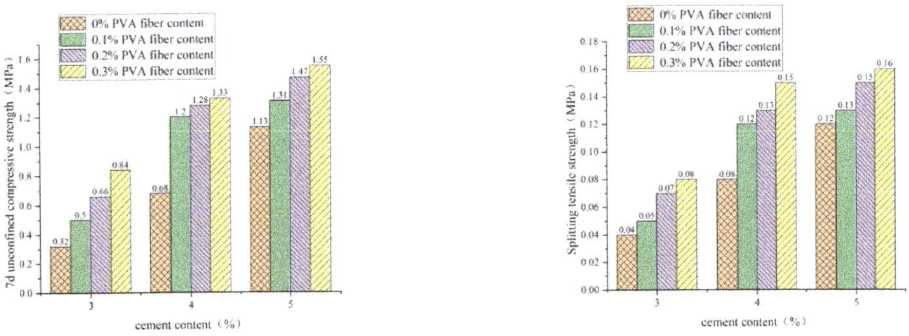

(a) (b)

Figure 12. Cement admixture with 7d unconfined compressive strength and splitting tensile strength for plain soil:red brick aggregate = 0:1. (**a**) Cement content and 7-day unconfined compressive strength. (**b**) Cement content and splitting tensile strength.

3.2.2. Influence of PVA-Fiber Content on 7d Unconfined Compressive Strength and Splitting Tensile Strength

It is clear from Figures 13–15 that the incorporation of PVA fibers positively affects the 7d unconfined compressive strength and splitting strength of the specimens at any red-brick-aggregate admixture and cement content. When the amount of PVA-fiber incorporation is not greater than 0.3%, the 7d of unlimited compressive strength and split tensile strength gradually increase with the increase of PVA-fiber incorporation. PVA fiber improves the 7d unconfined compressive strength because the fibers are distributed in a three-dimensional disorderly direction in the cement-stabilized soil, so there will be fibers with the transverse distribution. Considering the deformation coordination relationship, when subjected to axial pressure, the specimen will be subjected to transverse tensile force, and the fibers with transverse distribution will also be subjected to transverse tensile force. Since fibers will play the constraint role of bearing tensile force, the fiber-and-cement-stabilized soil will jointly bear the horizontal tension. Thus, the ultimate axial pressure that the test block can bear will also increase accordingly, thus improving the unconfined compressive strength of the specimen [21]. Fiber-enhanced splitting tensile strength: ① PVA fiber itself has an elastic modulus and tensile strength that are higher than those of cement-stabilized soil. The fiber can share a large part of the load when the cement stabilizes the soil load. After the cracking load, the fiber between the cracks plays a bridge role, making the specimen continue to load [22]. ② The surface extrusion plastic deformation of PVA fiber was influenced by soil particles and red brick aggregate from the micro perspective. When the test is under external load and deformation, the surface roughness of PVA fiber increases to overcome greater friction, thus improving the splitting tensile strength.

Based on the analysis results in Section 3.2.1, the optimal cement content selected is 4%. The regression analysis is carried out according to the test results of unconfined compressive strength and splitting tensile strength in 7 days, as shown in Figures 16–18. The relationships between PVA-fiber content and 7d unconfined compressive strength and splitting tensile strength were obtained, respectively, under the conditions of cement content of 4%; plain soil:red brick aggregate = 1:0, 1:1, and 0:1; and PVA-fiber content being within 0.3%, as shown in Equations (8a,b)–(10a,b).

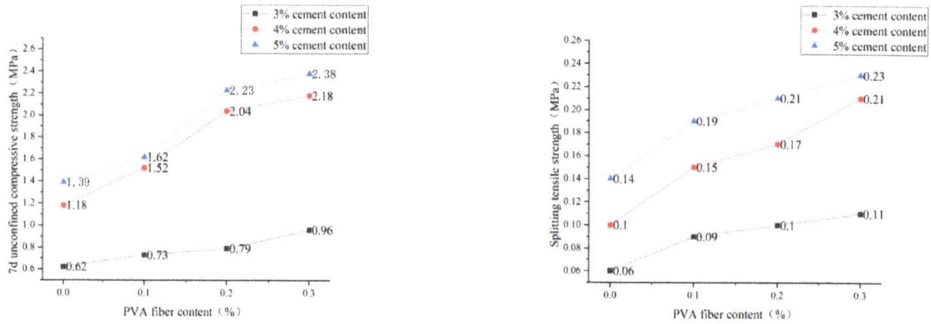

(**a**) (**b**)

Figure 13. PVA-fiber content with 7d unconfined compressive strength and splitting tensile strength for plain soil:red brick aggregate = 1:0. (**a**) PVA-fiber content and 7d unconfined compressive strength. (**b**) PVA-fiber content and splitting tensile strength.

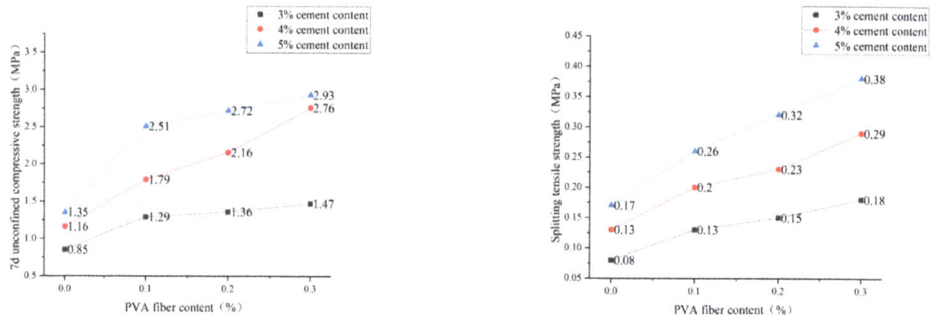

(**a**) (**b**)

Figure 14. PVA-fiber content with 7d unconfined compressive strength and splitting tensile strength for plain soil:red brick aggregate = 1:1. (**a**) PVA-fiber content and 7d unconfined compressive strength. (**b**) PVA-fiber content and splitting tensile strength.

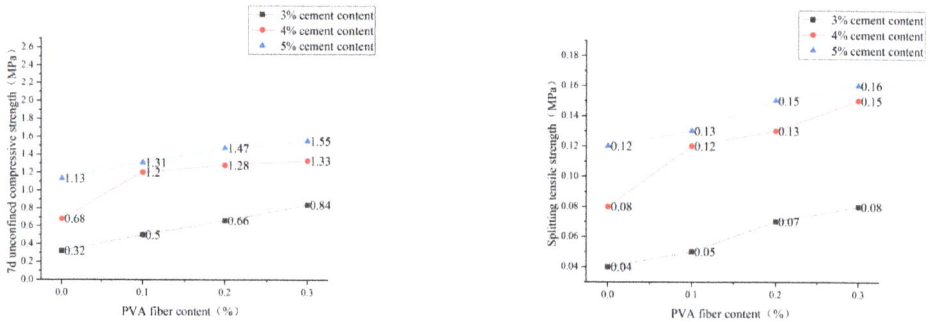

(**a**) (**b**)

Figure 15. PVA-fiber content with 7d unconfined compressive strength and splitting tensile strength for plain soil:red brick aggregate = 0:1. (**a**) PVA-fiber content and 7d unconfined compressive strength. (**b**) PVA-fiber content and splitting tensile strength.

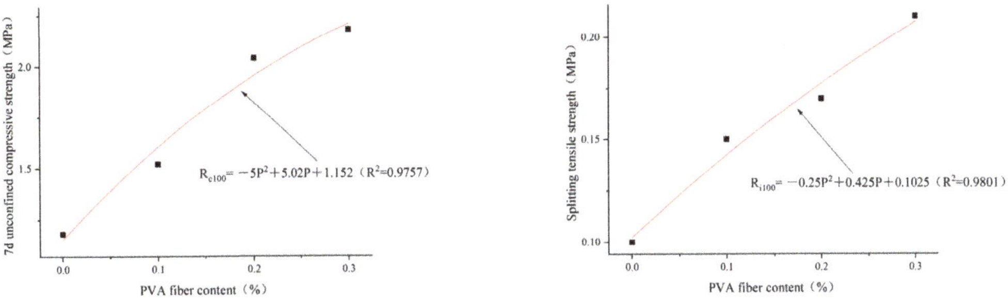

(a) (b)

Figure 16. Regression curves of PVA content and 7d unconfined compressive strength and splitting tensile strength for plain soil:red brick aggregate = 1:0, cement content 4%, and PVA-fiber content within 0.3%. (**a**) Regression curve of PVA-fiber content and 7d unconfined compressive strength. (**b**) Regression curve of PVA-fiber content and splitting tensile strength.

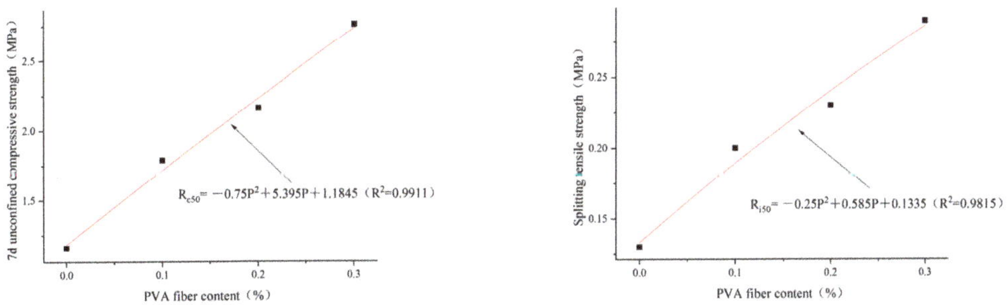

(a) (b)

Figure 17. Regression curves of PVA content and 7d unconfined compressive strength and splitting tensile strength for plain soil:red brick aggregate =1:1, cement content 4%, and PVA-fiber content within 0.3%. (**a**) Regression curve of PVA-fiber content and 7d unconfined compressive strength. (**b**) Regression curve of PVA-fiber content and splitting tensile strength.

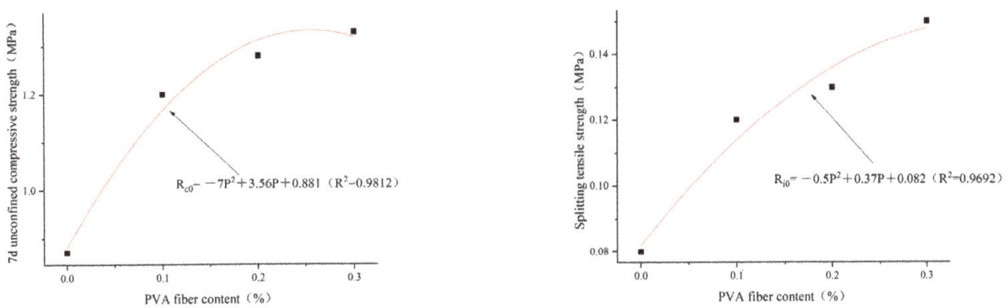

(a) (b)

Figure 18. Regression curves of PVA content and 7d unconfined compressive strength and splitting tensile strength for plain soil:red brick aggregate = 0:1, cement content 4%, and PVA-fiber content within 0.3%. (**a**) Regression curve of PVA-fiber content and 7d unconfined compressive strength. (**b**) Regression curve of PVA-fiber content and splitting tensile strength.

The cement content is 4%. The ratio of plain soil to red brick aggregate is 1:0. The content of PVA fiber is 0~0.3%. The relationship among PVA-fiber content, 7d unconfined compressive strength, and splitting tensile strength can be expressed as follows:

$$R_{c100} = -5P^2 + 5.02P + 1.152 \left(R^2 = 0.9757\right) \quad (8a)$$

$$R_{i100} = -0.25P^2 + 0.425P + 0.1025 \left(R^2 = 0.9801\right) \quad (8b)$$

The cement content is 4%. The ratio of plain soil to red brick aggregate is 1:1. The content of PVA fiber is 0~0.3%. The relationship among PVA-fiber content, 7d unconfined compressive strength, and splitting tensile strength can be expressed as follows:

$$R_{c50} = -0.75P^2 + 5.395P + 1.1845 \left(R^2 = 0.9911\right) \quad (9a)$$

$$R_{i50} = -0.25P^2 + 0.585P + 0.1335 \left(R^2 = 0.9815\right) \quad (9b)$$

The cement content is 4%. The ratio of plain soil to red brick aggregate is 0:1. The content of PVA fiber is 0~0.3%. The relationship among PVA-fiber content, 7d unconfined compressive strength, and splitting tensile strength can be expressed as follows:

$$R_{c0} = -7P^2 + 3.56P + 0.881 \left(R^2 = 0.9812\right) \quad (10a)$$

$$R_{i0} = -0.5P^2 + 0.37P + 0.082 \left(R^2 = 0.9692\right) \quad (10b)$$

where R_c and R_i represent the 7d compressive strength and splitting tensile strength, respectively, MPa; and P represents the content of PVA fiber, %.

According to Formulas (8)–(10), the ratios of plain soil to red brick aggregate are 1:0, 1:1, and 0:1 when the cement content is 4%. PVA-fiber content is 0~0.3% and has a quadratic function relationship with the 7d unconfined compressive strength and splitting tensile strength. Within the PVA-fiber-content test range, the ratios of plain soil to red brick aggregate are 1:0 and 1:1 when the cement content is 4%. The strength is positively related to the PVA-fiber content, and the optimal PVA-fiber content is 0.3%. However, the ratio of plain soil to red brick aggregate is 0:1, and the PVA-fiber content is 0~0.3%. The highest 7d unconfined compressive strength is obtained when the PVA-fiber content is 0.25%, and the highest splitting tensile strength occurs when the PVA-fiber content is 0.3%. Combined with 7d unlimited compressive strength and tensile strength of split, the ratios of plain soil to red brick aggregate are 1:0, 1:1, and 0:1. When the content of PVA fiber is within 0.3%, the optimal PVA-fiber content is 0.3%, 0.3%, and 0.25%, respectively. Therefore, a positive correlation is found between the intensity and PVA-fiber incorporation with PVA-fiber incorporation $\leq$0.25%, which is not affected by the amount of waste red brick aggregate.

3.2.3. Influence of Red-Brick-Aggregate Content on 7d Unconfined Compressive Strength and Splitting Tensile Strength

As seen from Figure 19, when the cement mixture is 4%, the 7d unconfined compressive strength and splitting tensile strength show a trend of first increasing and then decreasing with the increase of red-brick-aggregate incorporation, which is not affected by the fiber incorporation. When the amount of red brick aggregate is 50% and the fiber mixture is 0.3%, the 7d unconfined compressive strength and splitting tensile strength reach the maximum value, and the amount of red brick continues to increase, leading to a rapid decline in strength. The effect of fiber incorporation on strength was mentioned in the previous section. The more the incorporation is within 0.3%, the higher the intensity. When the red brick aggregate is 50% mixed, both the red brick particles and the plain soil particles have previously filled each other, increasing the compactness of the sample and, thus, increasing the strength. When the red brick aggregate increases, the red brick particles will dominate the specimen. Because the cohesion of the red brick particles is less than that of the plain soil particles, the strength drops sharply. Thus, the best amount of red brick aggregate is 50%.

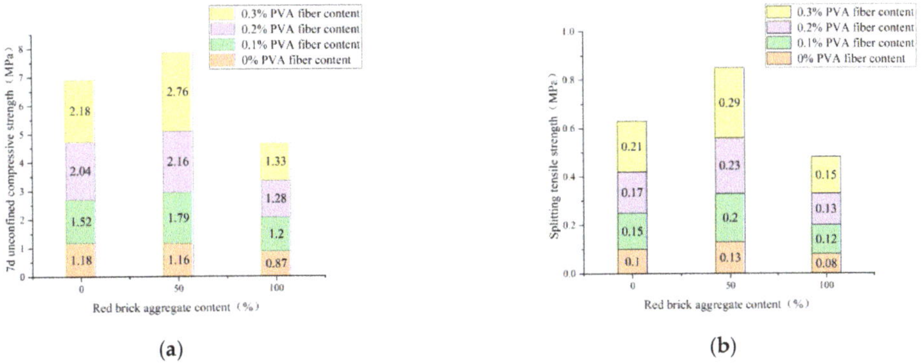

Figure 19. Red-brick-aggregate admixture with 7d unconfined compressive strength and splitting tensile strength at 4% cement admixture and different PVA-fiber content levels. (**a**) Red-brick-aggregate content and 7d unconfined compressive strength. (**b**) Red-brick-aggregate content and splitting tensile strength.

In summary, combined with the "Technical Guidelines for Construction of Highway Roadbases" (JTG/T F20-2015) for highways and primary highways, light traffic road subgrade 7d unconfined compressive strength requirements for 2~4 MPa. Within the scope of the experimental study, the strength of the test groups with red-brick-aggregate content ≤50%, cement content ≥4%, and fiber content ≥0.2% were all in line with the formal study and had a specific splitting tensile strength. Therefore, the optimal mixing ratio is 50% of the waste red brick aggregate, 4% cement, and 0.2% fiber considering the strength and material cost.

3.3. Mechanical Properties of Fiber-Reinforced Recycled Brick Powder Cementitious Composites

The most basic evaluation index of engineering materials is strength. The fiber-reinforced recycled-brick-powder cementitious composites must meet the strength design requirements to be applied to the road pavement. For cement concrete pavement, the pavement is bent when damaged, and flexural strength is a crucial factor to be considered when designing the pavement. Therefore, the compressive and flexural strength are two indispensable indexes to evaluate the road performance of fiber reinforced recycled brick powder cementitious composites. Therefore, a strength test was carried out on the specimens of each mixing ratio, and the test results are shown in Table 10.

Figure 20 establishes the three-dimensional surface plots of the two variables of PVA content and recycled-brick-powder content with the mechanical properties of fiber-reinforced recycled-brick-powder cementitious composites. Figure 19a,b show the relationship between PVA fiber and recycled-brick-powder content with compressive and flexural strength, respectively. According to the color depth of the projection area, it can be seen that when the mass content of recycled brick powder is 0%, and the volume content of PVA is 1%, the fiber-reinforced recycled brick powder cementitious composites compressive strength reaches the highest, at 38.12 MPa, recorded as (0,1,38.12). The highest flexural strength of fiber-reinforced recycled-brick-powder cementitious composites reached 17.56 MPa when the dose of recycled brick powder was 0% and the volume dose of PVA was 1.5%, which was recorded as (0,1.5,17.56).

Table 10. Compressive and flexural strength of fiber-reinforced recycled-brick-powder cementitious composites.

Group Number	Compressive Strength/MPa	Flexural Strength/MPa
Z0P0	37.86	14.56
Z25P0	35.63	13.12
Z50P0	31.76	12.34
Z75P0	28.32	11.74
Z100P0	25.66	10.56
Z0P1	38.12	16.47
Z25P1	36.03	15.32
Z50P1	32.31	14.47
Z75P1	28.47	13.23
Z100P1	26.34	11.96
Z0P1.5	36.47	17.56
Z25P1.5	34.32	16.14
Z50P1.5	29.61	15.37
Z75P1.5	27.54	14.43
Z100P1.5	25.21	12.54
Z0P2	34.78	16.15
Z25P2	34.97	14.23
Z50P2	28.73	13.15
Z75P2	25.43	12.65
Z100P2	25.01	11.32

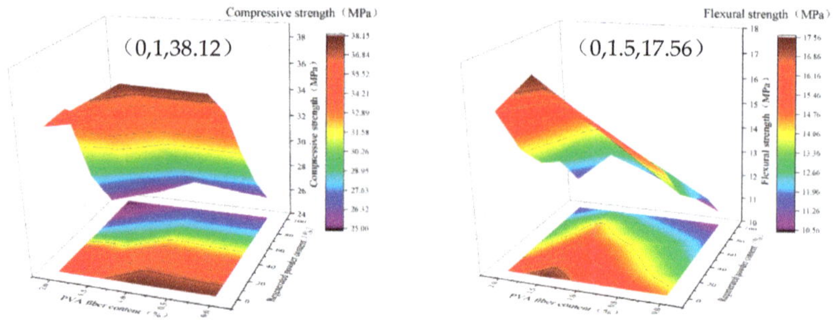

(a) (b)

Figure 20. Three-dimensional surface with projection. (**a**) PVA-fiber volume admixture and recycled-brick-powder mass admixture with compressive strength. (**b**) PVA-fiber volume admixture and recycled-brick-powder mass admixture with flexural strength.

As seen from the color of the projection area in Figure 20a, the general trend of the whole surface graph shows a downward trend from the lower right corner to the upper left corner. It indicates that, with the increase of the mass content of recycled brick powder and the volume content of PVA fiber, the fiber-reinforced recycled-brick-powder cementitious composites' compressive strength shows an overall downward trend. However, within a specific range in the lower right corner of the projection area, the color shades are relatively close, and the point (0,1) shows the deepest color of the entire projection area. It shows that, under a certain amount of PVA, recycled brick powder has little influence on the fiber-reinforced recycled-brick-powder cementitious composites compressive strength in a small range of mixing amounts. When the mixing amount of recycled brick powder is 25%, the compressive strength decreases by about 5–8%, and even part of the mixing ratio shows a strength increase. The recycled red brick aggregate has large porosity and low hardness. It often produces more cracks internally during the crushing process of preparing recycled brick powder. When the mass admixture of recycled brick powder is less than

25%, the recycled brick powder has a specific volcanic ash activity [23]. It can react with the hydration product CH to produce gelling products, which can play a good filling role and alleviate the material defects of the recycled brick powder itself. However, when the content of recycled brick powder exceeds 25%, the pozzolanic reaction of recycled brick powder is not enough to make up for the defects of the material itself. Thus, the strength will decrease. In the case of a certain amount of recycled brick powder, the fiber-reinforced recycled-brick-powder cementitious composites compressive strength showed a trend of increasing and then decreasing with the increase of PVA-fiber volume content and reached the maximum value when the volume dosing of PVA fiber was 1%. When the specimen is subjected to vertical load, the compression specimen is complete, and the bearing surface drops slightly. There is no breakage, and the specimen has a slight transverse displacement. A small amount of fiber will limit the transverse displacement of the specimen to share part of the axial pressure, thus improving the specimen's compressive strength. However, when the fiber content exceeds the optimal amount, the fiber will appear to clump, resulting in uneven fiber distribution. The high fiber content will lead to the introduction of increased bubbles, and the increased air content in the specimen will harm the compactness, thus reducing the compressive strength [24].

As shown in Figure 20b, the projected color gradually becomes lighter from the lower part to the upper part of the projected area, indicating that the flexural strength shows a decreasing trend with the increase of the mass content of the recycled brick powder. From the analysis of the projection area's left and right directions, the line's color at the point where the fiber volume content is 1.5% is darker than that of the left and right sides. Therefore, the three-dimensional surface graph formed by the volume content of PVA fiber, the mass content of recycled brick powder, and the flexural strength presents a "ridge" shape. It indicates that the fiber-reinforced recycled-brick-powder cementitious composites flexural strength increases first and decreases with the increase of PVA—fiber volume content. It can be seen from the color change in the figure that the increase of flexural strength caused by PVA—fiber volume content from 0% to 1% is more evident than that caused by 1% to 1.5%. The micro-cracks of different sizes are generated inside the cementitious material, and a large stress concentration is generated near the micro-cracks during tension. It can be explained by the brittleness of the cementitious material itself and the self-shrinkage in the process of setting and hardening [25]. When the volume content of PVA fiber is less than 1.5%, the PVA—fiber incorporation reduces the stress intensity factor at the crack tip. It can fully play the connection role in the early stage of fracture emergence, ease the stress concentration intensity at the fracture tip, and limit the width of the fracture, thus improving the flexural strength [26]. When the volume content of PVA fiber exceeds 1.5%, the excessive fiber will increase the fiber interface in cement-based composites. However, an excessive interface between fiber and slurry will occupy a large amount of cement slurry, which will greatly slow down the mobility of the matrix and increase the material's porosity, thus leading to the increase of initial defects inside the specimen. For example, cavitation caused by low vibration, initial cracks between fiber and interface, and the interface between fibers will adversely affect the matrix's compactness, reducing the component's flexural strength.

According to the "Specification for Design of Highway Cement Concrete Pavement" (JTG-D40-2011), the standard value of flexural tensile strength of cement concrete pavements for extremely heavy, extra heavy, and heavy traffic loads is $\geq$ 5.0 MPa, and the corresponding compressive strength is C35. Therefore, considering compressive strength, flexural strength, and economic and environmental protection, the best substitution rate of recycled brick powder for quartz sand in fiber-reinforced recycled-brick-powder cementitious composites is 25%, and the best mixing amount of PVA fiber is 1%.

3.4. Micro-Analysis

Figure 21 shows PVA fibers' overall appearance and morphology and three types of destructive patterns of samples under a scanning electron microscope. Firstly, it can be seen in Figure 20b–d that a large number of irregular particles are wrapped on the surface of the fiber, and this improves the roughness of the surface of the PVA fiber, increases the friction between the fiber and the matrix, and thus improves the strength of the specimen. Figure 20a shows the appearance of the PVA fiber pulled out from matrix morphology; part of the hydration products adhered to the fiber's surface and had serious scraping marks. The closer to the end, the smaller the diameter of the fiber, and the more scraping needed. This suggests that fiber is pulled up after a long distance between sliding [27], and cumulative scraping is severe, leading to a decrease in the diameter. The friction force generated by the fiber during sliding helps to share part of the axial pressure, thus improving the strength of the specimen. Figure 20b is a microscopic photo of a PVA-fiber fracture. The section is uneven. Compared to Figure 20a, the side of the fiber was scraped to a lesser extent, indicating that the fiber did not go through a long distance of sliding. However, under the action of external forces, fiber is constantly stretched by the role of external tensile stress. When the external stress exceeds the ultimate tensile strength that the fiber can withstand, it will be directly pulled off. As the ultimate tensile strength of PVA fibers is much greater than that of cementitious materials, it will require more axial pressure to pull off the fibers, so a certain amount of fiber incorporation will enhance the strength of the specimen. Figure 20c is enlarged, and a fiber surface can be found on the fiber surface. There are a lot of uneven areas, and these areas are sliding failure patterns. Fiber produced by scraping marks is different, more like a fiber in the brick particles or quartz sand particles under the extrusion of plastic deformation, increasing the fiber surface roughness. The friction between the fiber and the matrix is indirectly increased to improve the strength of the specimen.

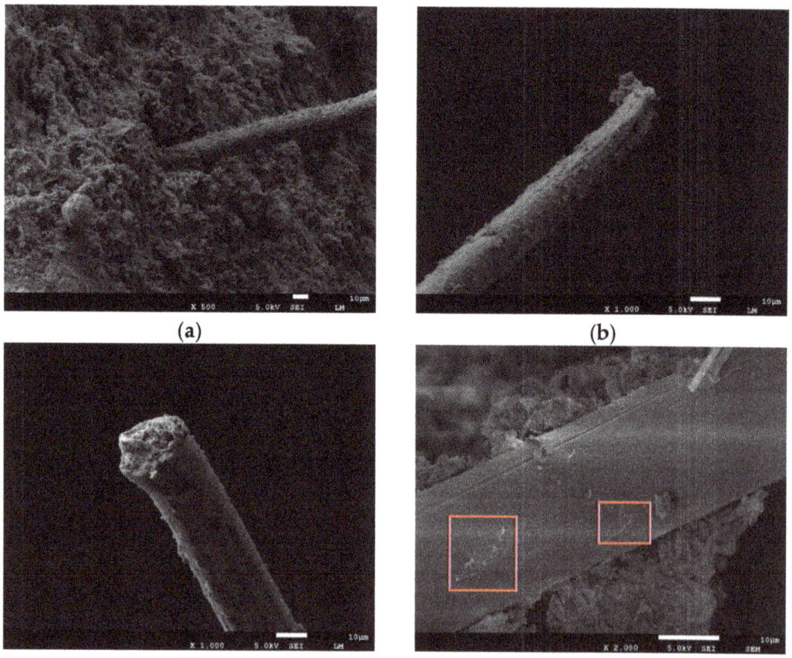

Figure 21. SEM image of PVA fiber. (**a**) SEM chart. (**b**) Pull-out. (**c**) Fracture. (**d**) Plastic damage.

4. Conclusions

The recycling and utilization of waste brick aggregate with a particle size less than 5 mm was mainly studied. The waste brick aggregate with a diameter of less than 2 mm was mixed into the soil, with different amounts, and combined with fibers to prepare the highway sub-base material meeting the specification requirements. The 7d unconfined compressive strength and splitting tensile strength were taken as the evaluation criteria. Recycled brick powder of 0.075~0.125 mm milled from waste brick aggregate was used to replace quartz sand with different replacement rates, together with different doses of PVA fiber to prepare environmentally friendly fiber-reinforced recycled-brick-powder cement-based composite pavement material. The relationship between the mass content of recycled brick powder, the volume content of PVA fiber, and the fiber-reinforced recycled-brick-powder cementitious composites' strength was investigated. The possibility of application of waste red brick aggregate with different particle sizes in pavement structure is verified, and the following conclusions are drawn:

(1) The red brick aggregate was mixed into the plain soil with different content levels (0%, 50%, and 100%) and different content levels of cement (3%, 4%, and 5%). After 45 comprehensive compaction tests, the optimal moisture content and maximum dry density under nine different mixing ratios were obtained.
(2) Within the scope of the experimental study, according to the obtained 7d unconfined compressive strength and splitting tensile strength, the strength of the test group when the red-brick-aggregate admixture was $\leq$50%, cement admixture was $\geq$4%, and fiber admixture was $\geq$0.2% followed the specification study. Therefore, combined with the careful consideration of strength and material cost, an ideal ratio for the test is 50% of the waste red brick aggregate, 4% of cement, and 0.2% of fiber.
(3) Based on the test results of the 7d unconfined compressive strength and splitting tensile strength, a regression analysis was used to fit the mathematical relationship between the PVA-fiber content and 7d compressive and splitting tensile strength. It was determined that, when the cement content is 4%, then the PVA-fiber content is within 0.3%; and the ration of plain soil to red brick aggregate = 1:0, 1:1, and 0:1, the PVA-fiber content is 0.3%, 0.3%, and 0.25%, and the 7d unconfined compressive strength is the best value. When cement content is 4%; and the PVA-fiber content is within 0.3%; and the plain soil:red brick aggregate = 1:0, 1:1, and 0:1, then the PVA-fiber content is 0.3%, 0.3%, and 0.3%, and the maximum splitting tensile strength is the best value. In the scope of the test, comprehensive consideration, when cement content is 4%, plain soil:red brick aggregate = 1:0, 1:1, and 0:1, the optimal PVA content is 0.3%, 0.3%, and 0.25%, respectively.
(4) The flexural and compressive strength of the fiber-reinforced recycled-brick-powder cement-based composite pavement material decreases with increased quartz sand replaced by recycled brick powder. However, when the replacement rate is less than 25%, the strength does not decrease significantly and is controlled within 10%, which meets the strength requirements. Therefore, the flexural and compressive strength of fiber-reinforced recycled-brick-powder cementitious composites increased first and then decreased with the content of PVA fiber. However, when the flexural and compressive strength reach the best value, the content of PVA fiber is different. For example, when the content of PVA fiber is 1% and 1.5%, the material's compressive strength and flexural strength reach the maximum value. Therefore, the best mixing ratio of fiber-reinforced recycled-brick-powder cementitious composites as pavement material is as follows: the replacement rate of recycled brick powder is 25%, and the content of PVA fiber is 1%.
(5) The influence of fibers on the material's mechanical properties was analyzed from the microstructure. The mechanism of fiber pull-out, fracture, and plastic damage in three different forms of damage to the material's mechanical properties was analyzed.

Author Contributions: Conceptualization, Y.J.; methodology, Y.J. and W.J.; validation, R.W.; formal analysis, Y.J.; data curation, Y.J., W.J. and Z.Z.; writing—original draft preparation, Y.J., W.J. and Z.Z.; writing—review and editing, R.W.; funding acquisition, Y.J. All authors have read and agreed to the published version of the manuscript.

Funding: The authors would like to acknowledge support from the Fundamental Research Funds for the Central Universities (No. 2572022BJ03), the 67th Postdoctoral Foundation in China (No. 2020M670872), and the Heilongjiang Province Studying Abroad Student (Startup Class) Scholarships.

Institutional Review Board Statement: Not applicable.

Informed Consent Statement: Not applicable.

Data Availability Statement: All data generated or analyzed during this study are included in this published article.

Conflicts of Interest: The authors declare no conflict of interest.

References

1. Zhang, Z.-H. Practice of Resource Utilization of Construction Waste in Waster Conservancy Projects. *Water Conserv. Constr. Manag.* **2022**, *42*, 64–67. [CrossRef]
2. Xiao, X.-W.; Feng, D.-K.; Tian, W. State and Suggestion on Construction Waste Recycling in China. *Constr. Technol.* **2015**, *44*, 6–8. [CrossRef]
3. Mao, Z.-Q.; Tong, T.; Liu, Q.; Wang, S.-T.; Wang, Y.-M.; Yin, S.-Q. Effect of Particle Size and Gradation on Shear Strength of Saturated Construction Waste. *Yangtze River* **2021**, *52*, 190–195. [CrossRef]
4. Ren, Y.-N.; Zhu, Y.-G.; Zhang, Q.-Y. Experimental Study on Mechanical Properties and Water Permeability of Recycled Aggregate Permeable Concrete. *Low Temp. Archit. Technol.* **2019**, *41*, 16–18. [CrossRef]
5. Li, Z.-G.; Ji, W.-J.; Zhang, H.; Li, Y.; Li, Y.-L. Study on the Properties of Recycled Fine Aggregate Containing Bricks and the Properties of Mortar. *Concrete* **2022**, *4*, 110–115.
6. Sun, Y. Comparative Study of Recycled Concrete Prepared with Construction Waste. *Bull. Chin. Ceram. Soc.* **2013**, *32*, 2637–2641. [CrossRef]
7. Kim, J.-S.; Yong, S.-S.; Cho, C.-H.; NO, S.-Y. Effect of the Broken Red Bricks on the Mechanical Properties of Reinforced Concrete Beams. *J. Korea Inst. Struct. Maint. Insp.* **2008**, *12*, 83–90. Available online: https://www.koreascience.or.kr/article/JAKO200810103476880.page (accessed on 30 August 2022).
8. Ji, Y.-C.; Zhang, H.-R.; Li, W. Investigation on Steel Fiber Strengthening of Waste Brick Aggregate Cementitious Composites. *Case Stud. Constr. Mater.* **2022**, *17*, e01240. [CrossRef]
9. Zhou, Z.-Q.; Wang, D.-P. Feasibility Study on Filling Roadbed with Expensive Soil Improved by Construction Waste. *J. Xi'an Univ. Archit. Technol.* **2021**, *53*, 716–722. [CrossRef]
10. Xu, L.-N.; Niu, L.; Zhang, T. Experimental Study on Frost Resistance of Glass Fiber Reinforced Cemented Soil. *Build. Struct.* **2020**, *50*, 822–825. [CrossRef]
11. Tran, K.-Q.; Satomi, T.; Takahashi, H.; Hiroshi, T. Improvement of Mechanical Behavior of Cemented Soil Reinforced with Waste Cornsilk Fibers. *Constr. Buildong Mater.* **2018**, *178*, 204–210. [CrossRef]
12. Yu, L.; Liu, W.-D.; Qian, Y.-L.; Yang, L.-X. Research on the Performance of Construction Waste Recycled Aggregates as Road Base Materials. *China Water Transp.* **2020**, *20*, 251–253.
13. Li, Y.; Zhou, H.; Su, L.-J.; Hou, H.; Li, D. Investigation into the Application of Construction and Demolition Waste in Urban Roads. *Adv. Mater. Sci. Eng.* **2017**, *2017*, 9510212. [CrossRef]
14. Li, P. Experimental Research on the Mechanical Properties and Durabilities of the Fiber Brucite Pavement Concrete Mixed with Fly Ash. Chongqing Jiaotong University, Chongqing, China. 2016. Available online: https://cdmd.cnki.com.cn/Article/CDMD-10618-1016165342.htm (accessed on 30 August 2022).
15. Ma, W.-B. Research on the Influence of Steel Slag Powder on Cement Road Performance. *Higways Transp. Inn. Mong.* **2021**, *47*, 7–9+26. [CrossRef]
16. Qi, S.-X.; Wang, X.-S.; Chen, K.-Y.; Wang, M.-J.; Xu, J.-Y. Application of Waterborne Epoxy Resin Modified Cement-based Material to Colored Pavement. *Sci. Technol. Eng.* **2022**, *22*, 4977–4983.
17. Qian, S.-Z.; Li, V.-C.; Zhang, H.; Keoleian, G.-A. Life Cycle Analysis of Pavement Overlays Made with Engineered Cementitious Composites. *Cem. Concr. Compos.* **2013**, *35*, 78–88. [CrossRef]
18. Liu, Y.-T. Experimental Study on Strength and Stability of Fiber Reinforced Soil with TG Curing Agent. Master's Thesis, Northeast Forestry University, Harbin, China, 2018. Available online: https://wap.cnki.net/touch/web/Dissertation/Article/10225-1018253648.nh.html (accessed on 30 August 2022).
19. Wang, Z. Research on Application Technology of Cement Soil Subgrade Engineering in Seasonal Frozen Area. Master's Thesis, Jilin University, Jilin, China, 2022. [CrossRef]
20. You, S.-J.; Yue, P.-C.; Luo, F.; Wang, H.; Xu, A. Study on Road Performance of Cement Stabilized Construction Waste. *Higheays Transp. Inn. Mong.* **2019**, *44*, 18–21. [CrossRef]
21. Zhou, S.-Z. Experimental Study on Strength Properties of the Polypropylene Fiber Reinforced Cement Soil[D]. Guangdong University Of Technology, Guangzhou, China. 2017. Available online: http://cdmd.cnki.com.cn/Article/CDMD-11845-1017841289.htm (accessed on 30 August 2022).

22. Li, F.H.; Gao, H.; Tang, H.Q.; Jiang, Y.L.; Zhan, Y.L.; Shen, D. Basic Properties and Shrinkage Model of Chopped Basalt Fiber Concrete. *J. Railw. Sci. Eng.* **2022**, *19*, 419–427. [CrossRef]
23. Tian, Y. Study on Basic Mechanical Properties of the Activated Recycled Brick ECC. Master's Thesis, Zhengzhou University, Zhengzhou, China, 2021. [CrossRef]
24. Yuan, C.-F.; Wang, D.; Li, H.-F.; Zhang, Z. Effect of Fiber Content on Fluidity and Mechanical Properties of Recycled Brick Powder ECC. *J. Archit. Civ. Eng.* **2021**, *38*, 74–82. [CrossRef]
25. Wang, Q. Experimental Research on Self-Consolidating PVA Fiber Reinforced Cementitious Composites. Master's Thesis, Dalian Maritime University, Dalian, China, 2016.
26. Sun, J.-Y. Effect of Polypropylene Fiber on Flexural Strength and Impact Resistance of High Performance Concrete. *Concrete* **1999**, *21*, 19–21.
27. Niu, H.-M.; Wu, W.-H.; Xing, Y.-M.; Zhao, Y.-R. Effect of the Process of Polyvinyl Alcohol(PVA) Fiber Bridging Cracks and Fiber Distribution on the Bending Properties of Cementitious Composites. *J. Build. Mater.* **2016**, *19*, 352–358. [CrossRef]

polymers

MDPI

Article

Structural Design Calculation of Basalt Fiber Polymer-Modified RPC Beams Subjected to Four-Point Bending

Yafeng Gong [1], Jianxing Yang [1], Xin He [1,*], Xiang Lyu [1,2] and Hanbing Liu [1]

1 Department of Road and Bridge Engineering, Nanling Campus, College of Transportation, Jilin University, Changchun 130025, China; gongyf@jlu.edu.cn (Y.G.); jxyang20@mails.jlu.edu.cn (J.Y.); lvxiang18@mails.jlu.edu.cn (X.L.); lhb@jlu.edu.cn (H.L.)

2 China Construction Second Engineering Bureau Ltd., Beijing 100160, China

* Correspondence: hexin@jlu.edu.cn; Tel.: +86-0431-8509-5446

Abstract: In this paper, a basalt fiber surface was treated with coupling agent KH-550 and hydrochloric acid, and the basalt fiber polymer-modified active powder concrete (RPC) material was prepared. There are significant differences in material composition and properties between basalt fiber polymer-modified RPC and ordinary concrete, and the structural design calculation (cracking moment and normal section bending bearing capacity) of an ordinary reinforced concrete beam is no longer applicable. Thus, mechanical parameters such as displacement and strain of reinforcement basalt fiber polymer-modified RPC beams subjected to four-point bending were tested. The excellent compressive and tensile strengths of basalt fiber polymer-modified RPC were fully utilized. The tensile strength of basalt fiber polymer-modified RPC in the tensile zone of the beam was considered in the calculation of normal section bending bearing capacity of reinforcement basalt fiber polymer-modified RPC beams. The results showed that the measured values of the cracking moment and ultimate failure bending moment of reinforcement basalt fiber polymer-modified RPC beams were in good agreement with the calculated values. The established formulas for cracking moment and normal section bending bearing capacity can provide references for the design of reinforcement basalt fiber polymer-modified RPC simply supported beam and promote the wide application of basalt fiber polymer-modified RPC materials in practical engineering.

Keywords: construction materials; basalt fiber polymer-modified reactive powder concrete; cracking moment; normal section bending bearing capacity

Citation: Gong, Y.; Yang, J.; He, X.; Lyu, X.; Liu, H. Structural Design Calculation of Basalt Fiber Polymer-Modified RPC Beams Subjected to Four-Point Bending. *Polymers* **2021**, *13*, 3261. https://doi.org/10.3390/polym13193261

Academic Editor: Pablo Marcelo Stefani

Received: 23 August 2021
Accepted: 21 September 2021
Published: 24 September 2021

1. Introduction

Basalt fiber polymer-modified reactive powder concrete (RPC) is considered to be an innovative high property cement-based concrete material prepared by the theory of densified particle packing [1–4]. Basalt fiber polymer-modified RPC replaces the coarse aggregate in ordinary concrete with millimeter-level aggregate (quartz sand) and fills the gaps between quartz sand aggregates with micron-sized cementitious materials (cement) and submicron-sized cementitious materials (silica fume) [5,6]. In this way, the internal defect of basalt fiber polymer-modified RPC are controlled, and both the density and homogeneity are improved. Besides, basalt fiber polymer-modified RPC composites are cured by heat treatment and blended with high-performance, acid and alkali-resistant basalt fibers [7]. Owing to this, the microstructure of basalt fiber polymer-modified RPC materials is more compact and the mechanical properties, durability and toughness of basalt fiber polymer-modified RPC composites are improved. Therefore, basalt fiber polymer-modified RPC has the characteristics of ultra high strength, high durability and high temperature adaptability [8–10]. Basalt fiber polymer-modified RPC can effectively reduce the self-weight of structures and increase the spanning capacity, which has a broad application prospect in various infrastructure construction. At present, the engineering application of basalt fiber polymer-modified RPC materials is relatively few, mainly due to the lack of corresponding structural design calculation methods, while the calculation of cracking moment and normal section bending bearing capacity is an important part of structural design.

From the present research, it can be seen that researchers have proposed different methods for calculating cracking moments and bending bearing capacity for RPC structures and composite structures of RPC with other materials [11,12]. Zingaila et al. [13] experimentally investigated the mechanical properties of a combined beam made of ordinary concrete and ultra-high performance concrete and calculated the cracking moment of the combined beam using the layered method. Zhang et al. [14] conducted flexural tests on damaged bridge decks reinforced by ultra-high performance concrete and established analytical equations for the cracking and ultimate flexural bearing capacity of the composite structure. Yang et al. [15] proposed an analytical method for predicting the bending response of ultra-high-performance fiber-reinforced concrete structures, which can accurately predict the bending strength of ultra-high-performance fiber-reinforced concrete beams. Ujike et al. [16] used RPC to strengthen a part of the tensile zone of reinforced concrete beams and used elasticity theory to estimate the cracking moment of this composite material. Turker et al. [17] also researched the bending performance of ultra-high-performance fiber-reinforced concrete beams by four-point bending tests and proposed two numerical methods for predicting the nominal moment bearing capacity of these materials. Sim et al. [18] proposed flexural design guidelines for precast prestressed concrete members and found that the traditional equivalent rectangular stress block in compression can still be used to produce satisfactory results in prestressed concrete members. Prem et al. [19] developed an integrated nonlinear fracture mechanics model to predict the moment carrying capacity of ultra-high performance concrete reinforced damaged reinforcement concrete composite beams. Based on the planar section assumption and specific damage criterion, Guo et al. [20] used the numerical integration method to simulate the full process of flexural behavior of RPC and ordinary concrete composite beams and obtained the full section moment-curvature curves for different design scenarios. Chi et al. [21] investigated the effect of different water-cement ratios and different fiber compositions on the flexural performance of RPC beams using finite element analysis and deduced the formulae for calculating the flexural bearing capacity of RPC beams with different fiber compositions. Similarly, Chen et al. [22] derived an equation for calculating the ultimate flexural bearing capacity of ultra-high performance concrete beams utilizing ANSYS finite element simulation analysis. Qi et al. [23] carried out an experimental and analytical study of steel-ultra-high performance fiber concrete composite beams and calculated the flexural strength of the specimens using a simplified analytical method. Hasgul et al. [24] proved through an experimental study that the simplified numerical method for the flexural design of fiber-reinforced concrete was also applied to ultra-high-performance fiber-reinforced concrete beams. Cao et al. [25] investigated the flexural behavior of prestressed RPC in-situ panels suffering from four-point bending and established a modified formula for the flexural bearing capacity of prestressed RPC in-situ panels, considering the tensile strength of the RPC and prestressing stress. In addition, Cao et al. [26] also analyzed the effect of longitudinal reinforcement ratio and reinforcement diameter on the cracking moment of high-strength reinforcement RPC beams, and established formulas for calculating the resistance coefficient of plasticity in section and the cracking moment of high-strength reinforcement RPC beams.

Existing research shows that there is a trend to develop concrete composite materials such as fiber-reinforced cementitious matrix and steel reinforced grout [27,28]. The basalt fiber polymer-modified RPC composites studied in this paper belong to the fiber-reinforced cement matrix. The basic mix proportion design, mechanical properties and durability of basalt fiber polymer-modified RPC composites have been discussed in previous studies of this research group. This paper focuses on the structural design calculation equations of basalt fiber polymer-modified RPC simply supported beams.

Basalt fiber polymer-modified RPC is a new material and differs significantly from other concrete materials in terms of material composition and properties. The formulae for calculating cracking moment and normal section bending bearing capacity of other concrete materials are not fully applicable to basalt fiber polymer-modified RPC. This paper thus investigated the mechanical properties of reinforcement basalt fiber polymer-modified RPC simply supported beams subjected to four-point bending. Based on the test data, the calculation methods of cracking moment and normal section bending bearing capacity were proposed for reinforcement basalt fiber polymer-modified RPC beam to provide a reference for the structural design and specification of reinforcement basalt fiber polymer-modified RPC beams.

2. Experimental Details

2.1. Materials and Mixture

The material used to prepare Basalt fiber polymer-modified RPC beams consists of quartz sand, quartz powder, cement, silica fume, water reducer, basalt fiber, coupling agent, hydrochloric acid, water, and reinforcement. Three kinds of quartz sand with the sizes of 20~40 mesh, 40~80 mesh, and 80~120 mesh produced by the Zhenxing quartz sand factory (Luoyang, China) were used. These quartz sands were mixed in the ratio of 2:2:1 and filtered with square hole sieves of 0.6 mm, 0.3 mm, and 0.15 mm in order. The specification of quartz powder was 400 mesh. The silica fume SF92 produced by Jilin Changchun Si-Ao Technology Co., Ltd. (Jilin, China) and silicate cement P.II 52.5 produced by Jilin Yatai Cement Co., Ltd. (Jilin, China) were adopted. The chemical composition of silica fume and cement was determined by X-ray fluorescence spectrometer (manufactured by HORIBA Jobin Yvon, Paris, France) according to the "Method of Chemical Analysis of Silicate Rocks, Part 28: Determination of 16 Major and Minor Element Quantities" (GB/T 14506.28-2010) [29], and the results are shown in Figure 1. The coupling agent KH-550 is produced by Nanjing Chuangshi Chemical Additives Co., Ltd. (Jilin, China). The mix proportion of basalt fiber polymer-modified RPC is shown in Table 1. This mix proportion is the best mix proportion obtained from the previous research results of the research group [30]. According to the specification "Method of Testing Cement-Determination of Strength" (GB/T 17671-1999) [31], the compressive strength and flexural strength of basalt fiber polymer-modified RPC with this mix proportion were tested by computerized full-automatic compressive and flexural one piece testing machine (manufactured by Meters Testing Machine Factory, Tianjin, China). The loading rates of compressive strength and flexural strength are 2.4 kN/s and 0.05 kN/s, respectively. The compressive strength and flexural strength of basalt fiber polymer-modified RPC prepared by this mix proportion were 115 MPa and 17 MPa, respectively.

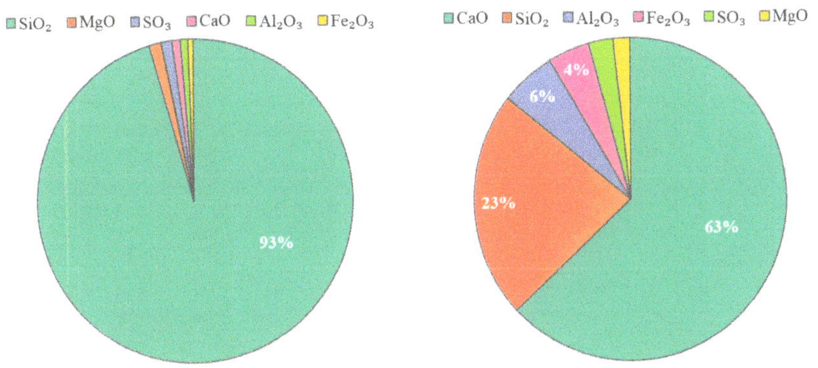

Figure 1. Chemical composition of cementing materials: (**a**) Silica fume; (**b**) Cement.

Table 1. Mix proportion of basalt fiber polymer-modified RPC (kg/m^3).

Water	Cement	Silica Fume	Quartz Sand		Quartz Powder	Basalt Fiber	Water Reducer
			0.15 mm~0.3 mm	0.3 mm~0.6 mm			
151.5	841.8	210.4	364.2	582.8	311.4	12	52.6

2.2. Specimen Preparation and Experimental Procedure

In this paper, basalt fiber was first etched in hydrochloric acid, and then treated according to the ratio of KH-550 coupling agent to basalt fiber mass ratio of 3:100. Finally, three reinforcement basalt fiber polymer-modified RPC beams with different reinforcement ratios were prepared. The designed size of test beams was 1500 mm × 150 mm × 300 mm. Figure 2 shows the dimensions of three test beams and the relative positions of the rebar.

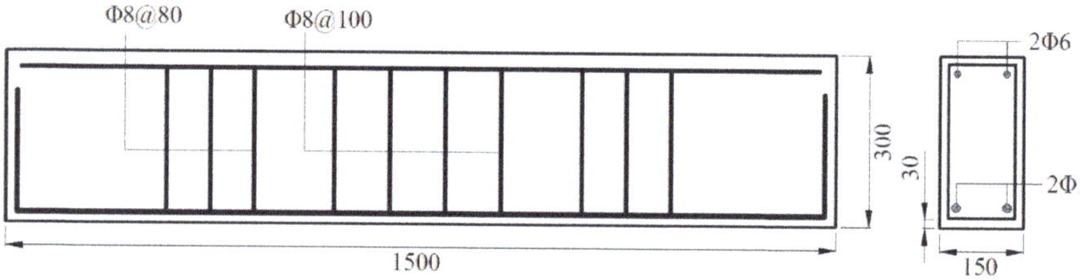

Figure 2. Dimension of the reinforcement basalt fiber polymer-modified RPC beam (mm).

The types and diameters of erective steel bars, stirrups, and main rebar of each test beam are shown in Table 2. In each test beam, two HPB300 steel bars with a diameter of 6 mm were used as erecting steel bars; ten HPB300 steel bars with a diameter of 8 mm were used as stirrups; two HRB335 rebar with a diameter of 12 mm, 14 mm, and 16 mm respectively were used as the main rebar.

Table 2. Types and diameters of steel bars in test beams.

Teat Beam Number	Erecting Steel Bar		Stirrup		Main Rebar	
	Type	Diameter (mm)	Type	Diameter (mm)	Type	Diameter (mm)
Beam-1	HPB 300	6	HPB 300	8	HRB 335	12
Beam-2	HPB 300	6	HPB 300	8	HRB 335	14
Beam-3	HPB 300	6	HPB 300	8	HRB 335	16

The prepared reinforcement basalt fiber polymer-modified RPC beams firstly were cured in 90 °C steam for 48 h. After curing, all test beams were performed the four-point bending test. The static response values (displacements, strains, etc.) of a reinforcement basalt fiber polymer-modified RPC beams subjected to four-point bending were recorded. Finally, considering the tensile strength of basalt fiber polymer-modified RPC in the tensile zone of the beam, the formulae for calculating the cracking moment and the normal section bending bearing capacity were established from the calculation principle applicable to the reinforcement basalt fiber polymer-modified RPC beam.

2.3. Experimental Setups

The reinforcement basalt fiber polymer-modified RPC beam were tested with reference to the specification "Standard for Test Methods for Concrete Structures" (GBJ50152-2012) [32]. Loading method adopts four-point bending loading, loading device selection the WAW-1000 kN type electro-hydraulic servo universal testing machine produced by Jilin Jinli test technology Co., Ltd. (Jilin, China). The experimental setup is schematically shown in Figure 3. Displacement and strain were measured by dynamic and static strain testing machine DH3817 and static strain testing machine DH3818Y produced by Jiangsu donghua Test Technology Co. (Jiangsu, China).

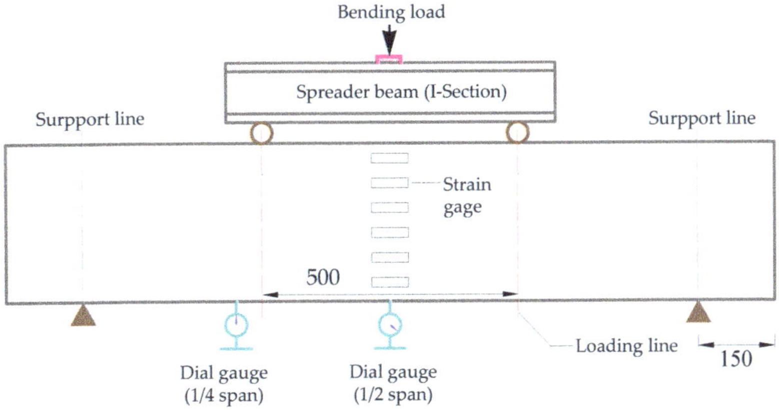

Figure 3. Sketch of the experimental setup for the four-point bending test (unit: mm).

As shown in Figure 3, the dial gauges were arranged at the bottom of the 1/2 and 1/4 span of the beams for recording the deflection values during the loading process. Simultaneously, Six strain gauges were arranged along the beam height of 1/2 span to record the strain change during the test. Additionally, a multi-step loading process (20 kN for each step) was adopted, and the loading rate was controlled to be 0.037 kN/s.

3. Results

3.1. Test Results

Based on loads of the three reinforcement basalt fiber polymer-modified RPC beams at cracking and failure, the measured values of the cracking moment and ultimate failure bending moment of all test beams can be calculated as shown in Table 3.

Table 3. Measured values of the cracking moment and ultimate failure bending moment.

Teat Beam	Cracking Moment (kN· m)	Ultimate Failure Bending Moment (kN· m)
Beam-1	15.05	39.40
Beam-2	15.75	47.00
Beam-3	17.50	52.20

3.2. Cracking Moment Calculation

3.2.1. Calculation Principle of Cracking Moment

The cracking moment is the main index of cracking resistance and the foundation for the analysis of the mechanical properties of reinforcement basalt fiber polymer-modified RPC beams. The cracking moment is usually defined as the bending moment at the appearance of the first visible crack in the test beam. When the test beam is cracked, the tensile zone of the test beam shows a certain plastic deformation. The stress distribution graph is curved. The maximum stress in the tensile zone of the test beam reaches the ultimate tensile strength f_t, and the strain at the tensile edge reaches the ultimate tensile strain ε_{tu}. The basalt fiber polymer-modified RPC in the compression zone of the test beam is in the elastic phase, and the stress distribution graph is linear. The strain of basalt fiber polymer-modified RPC at the edge of the compressed zone is ε_c. In view of the above analysis, the calculation model of the cracking moment of the reinforcement basalt fiber polymer-modified RPC beam is shown in Figure 4.

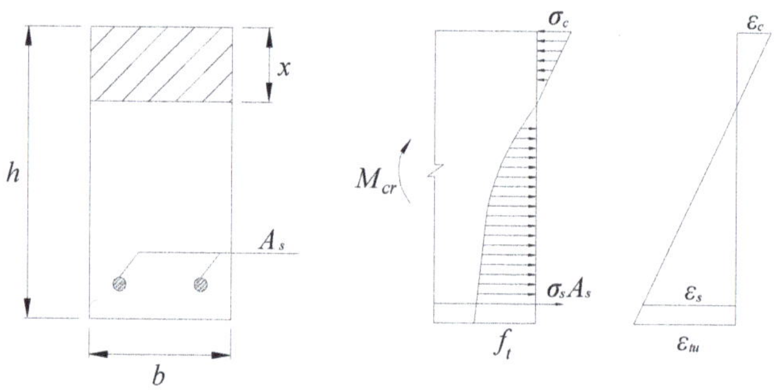

Figure 4. Calculation model of the cracking moment.

According to the calculation model in Figure 4, Equations (1) and (2) can be derived. The cracking moment can be divided into two parts: the bending moment borne by the basalt fiber polymer-modified RPC matrix and the bending moment borne by the main rebars.

$$M_{cr} = f_t \cdot W_s, \tag{1}$$

$$M_{cr} = M_s + M_m, \tag{2}$$

where: M_{cr} is the cracking moment of basalt fiber polymer-modified RPC; f_t is the ultimate tensile strength of basalt fiber polymer-modified RPC; W_s is the elastic–plastic resistance moment of the basalt fiber polymer-modified RPC beam section to the tensile edge; M_s is the bending moment borne by the reinforcement; M_m is the bending moment borne by the basalt fiber polymer-modified RPC matrix.

The section resistance moment plasticity influence coefficient γ was used to reflect the elasticplastic development degree in the tensile zone of the beam. The cracking moment formula is established by the principle of material mechanics. The formula for calculating the cracking moment is shown in Equation (3)

$$M_{cr} = \gamma \cdot f_t \cdot W_0, \tag{3}$$

where: γ is the section resistance moment plasticity influence coefficient; W_0 is the elastic-plastic resistance moment of the test beam's converted section to the tensile edge.

Combining Equations (1) and (3), the formula of section resistance moment plasticity influence coefficient γ is shown in Equation (4).

$$\gamma = \frac{W_S}{W_0}, \tag{4}$$

3.2.2. Calculation of W_s

The stress distribution in the tensile zone was divided into the elastic and plastic tensile zones. The simplified calculation model of the normal section is shown in Figure 5.

Based on test data, the flexural initial cracking tensile strains of Beam-1, Beam-2 and Beam-3 are 495.0 με, 491.8 με and 478.4 με, respectively. The ratios of the flexural initial cracking tensile strain to the peak tensile strain (213 με) of the basalt fiber polymer-modified RPC are 2.32, 2.31 and 2.25, respectively. The average value was taken as 2.29. Thus, $\varepsilon_{tu} = 2.29\varepsilon_{t0}$ can be obtained. The combined force and bending moment of each part of the reinforcement basalt fiber polymer-modified RPC beam are shown in Equation (5) and Equation (6), respectively.

$$\begin{cases} F_c = 1.15\frac{x_c^2}{(h-x_c)}f_t b \\ F_{te} = 0.22 f_t b(h - x_c) \\ F_{tp} = 0.56 f_t b(h - x_c) \\ F_s = 2.29\alpha_E f_t \rho b h_0 \frac{h_0 - x_c}{h - x_c} \end{cases}, \tag{5}$$

$$\begin{cases} M_c = 0.76 f_t b\frac{x_c^3}{(h-x_c)} \\ M_{te} = 0.07 f_t b(h - x_c)^2 \\ M_{tp} = 0.4 f_t b(h - x_c)^2 \\ M_s = 2.29\alpha_E f_t \rho b h_0 \frac{(h_0 - x_c)^2}{h - x_c} \end{cases}, \tag{6}$$

where: F_c and M_c are the combined force and bending moment of the beam in the compression zone; F_{te} and M_{te} are the combined force and bending moment of the beam in the elastic tension zone; F_{tp} and M_{tp} are the combined force and bending moment of the beam in the plastic tension zone; F_s and M_s are the combined force and bending moment of the main rebars; x_c is the relative height of compression zone; b is the width of the beam section; h is the height of the beam section; h_0 is the effective height of the beam section; α_E is the section conversion factor; ρ is the reinforcement rate of the main rebars.

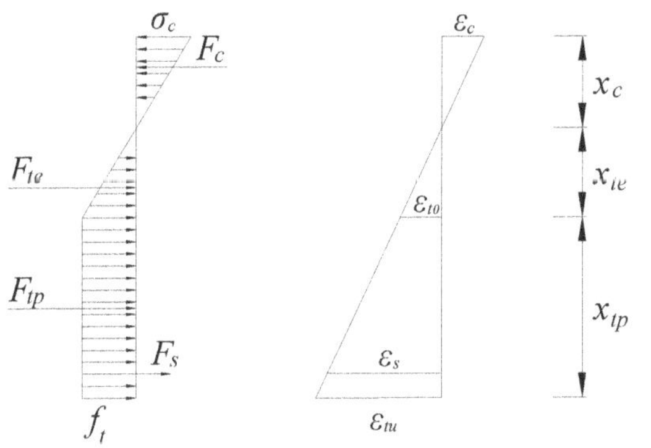

Figure 5. Simplified calculation model of normal section.

Equation (7) can be derived from the equilibrium condition of the force.

$$F_c = F_{te} + F_{tp} + F_s, \tag{7}$$

Bringing Equation (5) into Equation (7) to obtain Equation (8).

$$x_c = \frac{-B + \sqrt{B^2 - 4AC}}{2A} \tag{8}$$

The parameters in Equation (8) are shown in Equation (9).

$$\begin{cases} A = (a-1)^2 \\ B = (4ah - 2h + 2a^2\alpha_E \rho h_0) \\ C = -(2ah^2 - h^2 + 2a^2\alpha_E \rho h_0^2) \end{cases} \tag{9}$$

Equation (10) can be obtained from the equilibrium conditions of moments.

$$M_{cr} = bf_t\left[0.76\frac{x_c^3}{(h-x_c)} + 0.47(h-x_c)^2 + 2.29\alpha_E \rho h_0 \frac{(h_0 - x_c)^2}{(h-x_c)}\right] \tag{10}$$

Combining Equations (1) and (10), the formula of W_s is shown in Equation (11).

$$W_s = b\left[0.76\frac{x_c^3}{(h-x_c)} + 0.47(h-x_c)^2 + 2.29\alpha_E\rho h_0\frac{(h_0-x_c)^2}{(h-x_c)}\right] \tag{11}$$

3.2.3. Calculation of W_0

Using the equivalent conversion method, the main rebar area is equated to the basalt fiber polymer-modified RPC area with the same elasticity modulus. The converted section and its stress distribution are shown in Figure 6.

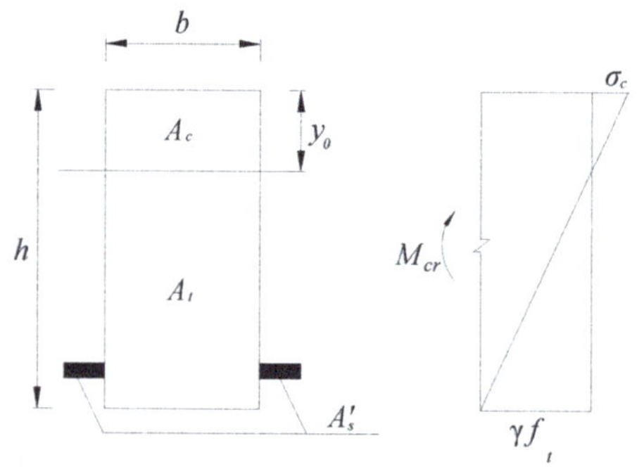

Figure 6. Conversion section and its stress distribution.

The conversion coefficient α_E for converting the main rebar area to the basalt fiber polymer-modified RPC area and the converted section area A'_s is shown in Equation (12) and Equation (13), respectively

$$\alpha_E = \frac{E_s}{E_c} = \frac{1.95\times10^5}{42.9\times10^3} = 4.55, \tag{12}$$

$$A'_s = A_s(\alpha_E - 1) = 3.55A_s, \tag{13}$$

According to the formula of material mechanics, the calculation of W_0 is shown in Equation (16).

$$I_0 = \frac{by_0^3}{3} + \frac{b(h-y_0)^3}{3} + 3.55A_s(h_0-y_0)^2, \tag{14}$$

$$y_0 = \frac{0.5bh^2 + 3.55A_sh_0}{bh + 3.55A_s}, \tag{15}$$

$$W_0 = \frac{I_0}{h-y_0} = \frac{by_0^3 + b(h-y_0)^3 + 10.65A_s(h_0-y_0)^2}{3(h-y_0)}, \tag{16}$$

3.2.4. Calculations of γ and Cracking Moment

Taking Equations (11) and (16) into Equation (4), the section resistance moment plasticity influence coefficient γ of Beam-1, Beam-2 and Beam-3 can be calculated as 1.67, 1.71 and 1.75, respectively. It can be noticed that with the increase of reinforcement ratio, the section resistance moment plasticity influence coefficient γ increases gradually, which indicates that the increase of reinforcement ratio makes its plasticizing effect on the basalt fiber polymer-modified RPC around the reinforcement increase.

The calculated value of cracking moment for all test beams can be obtained by bringing γ and Equation (16) into Equation (3) and comparing it with the test value of cracking moment, as shown in Table 4.

Table 4. Calculated and test values of the cracking moment.

Teat Beam	Calculated Value M^c_{cr}(kN· m)	Test Value M^t_{cr}(kN· m)	M^c_{cr}/M^t_{cr}
Beam-1	15.78	15.05	1.05
Beam-2	16.23	15.75	1.03
Beam-3	16.73	17.50	0.96

From Table 4, it can be observed that the ratio between the calculated and test values of the cracking moment has a small variation, and the mean value of this ratio is 1.01, the standard deviation is 0.05, and the variation coefficient is 0.05. The data error satisfies the requirements, which shows that the established formula for calculating the cracking moment of the reinforcement basalt fiber polymer-modified RPC beam is accurate.

3.3. *Calculation of Normal Section Bending Bearing Capacity*

3.3.1. Basic Assumptions

Similar to the calculation of the normal section bending bearing capacity of the ordinary reinforced concrete beam, that of reinforcement basalt fiber polymer-modified RPC beam needs to adopt the plane section assumption in the process of reaching the limit state of bending bearing capacity. This can meet the error requirement of engineering calculation, and also can clarify the logic of the calculation process and the physical meaning of the calculation formula. For basalt fiber polymer-modified RPC materials, the standard value of tensile strength is significantly higher than that of ordinary concrete. Therefore, in this paper, the tensile strength of basalt fiber polymer-modified RPC in the tensile zone was considered in the calculation of the normal section bending bearing capacity of the reinforcement basalt fiber polymer modified RPC beam.

3.3.2. Equivalent Rectangular Stress Pattern

A rectangular stress diagram is used to equate the graphs of compressive and tensile stress curves in the cross-section of reinforcement basalt fiber polymer-modified RPC beam. The stress distribution at the time of damage of the reinforcement basalt fiber polymer-modified RPC beam normal section is shown in Figure 7.

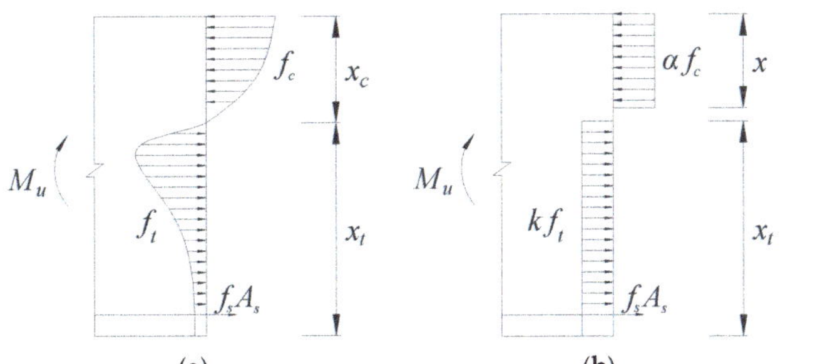

(a) (b)

Figure 7. Stress distribution of normal section of properly reinforced beams under failure: (**a**) Actual stress distribution; (**b**) Equivalent stress distribution.

According to the equivalent rectangular stress pattern of concrete in the compressed area, the combined compressive stress C and its distance y_c from the neutral axis of concrete in the compressed area are shown in Equation (17) and Equation (18), respectively.

$$C = \int_0^{x_c} \sigma_c(\varepsilon) \cdot b \cdot dy, \tag{17}$$

$$y_c = \frac{\int_0^{x_c} \sigma_c(\varepsilon) \cdot b \cdot y \cdot dy}{\int_0^{x_c} \sigma_c(\varepsilon) \cdot b \cdot dy}, \tag{18}$$

In accordance with the plane section assumption, the distance y from the neutral axis is related to the compressive strain of concrete at this location as in Equation (19).

$$\frac{\varepsilon}{\varepsilon_{cu}} = \frac{y}{x_c}, \tag{19}$$

Substitute Equation (19) into Equations (17) and (18) to obtain the resultant force C and its distance y_c from the neutral axis is shown in Equation (20) and Equation (21) respectively.

$$C = \frac{x_c \cdot b}{\varepsilon_{cu}} \cdot \int_0^{\varepsilon_{cu}} \sigma_c(\varepsilon_c) \cdot d\varepsilon, \tag{20}$$

$$y_c = \frac{x_c}{\varepsilon_{cu}} \cdot \frac{\int_0^{\varepsilon_{cu}} \sigma_c(\varepsilon) \cdot b \cdot \varepsilon \cdot d\varepsilon}{\int_0^{\varepsilon_{cu}} \sigma_c(\varepsilon) \cdot b \cdot d\varepsilon}, \tag{21}$$

Let the maximum value of the combined compressive stress and its distance from the neutral axis be Equation (22) and Equation (23), respectively.

$$\sigma_{cu} = \int_0^{\varepsilon_{cu}} \sigma_c(\varepsilon) d\varepsilon, \tag{22}$$

$$y_{cu} = \frac{\int_0^{\varepsilon_{cu}} \sigma_c(\varepsilon) \cdot \varepsilon \cdot d\varepsilon}{\int_0^{\varepsilon_{cu}} \sigma_c(\varepsilon) d\varepsilon}, \tag{23}$$

Then the combined force C and its distance y_c from the neutral axis are Equation (24) and Equation (25), respectively.

$$C = x_c \cdot b \cdot \frac{\sigma_{cu}}{\varepsilon_{cu}}, \tag{24}$$

$$y_c = \frac{x_c}{\varepsilon_{cu}} \cdot y_{cu}, \tag{25}$$

Let $k_1 = \frac{\sigma_{cu}}{f_c \varepsilon_{cu}}, k_2 = \frac{y_{cu}}{\varepsilon_{cu}}$, then the bending moment assumed by the basalt fiber polymer-modified RPC in the pressure zone is shown in Equation (26).

$$M_c = C \cdot (h_0 - x_c + y_c) = k_1 f_c x_c b[h_0 - (1 - k_2)x_c], \tag{26}$$

Given the parameter α, the bending moment assumed by the basalt fiber polymer-modified RPC in the compression zone is given in Equation (27).

$$M_c = \alpha f_c b x \cdot \left(h_0 - \frac{\beta}{2} x\right), \tag{27}$$

Combining Equations (26) and (27) to obtain Equation (28).

$$\begin{cases} \alpha = \frac{k_1}{2(1-k_2)} \\ \beta = 2(1 - k_2) \end{cases}, \tag{28}$$

According to the basalt fiber polymer-modified RPC compressive stress–strain relationship showed in Equation (29), k_1 and k_2 were calculated to be 0.693 and 0.615, respectively. Then bring the k_1 and k_2 into Equation (28), α and β were calculated to be 0.92 and 0.76, respectively.

$$\begin{cases} y = 1.12x + 0.11x^2 - 0.65x^6 & 0 \le x < 1 \\ y = 1 & x \ge 1 \end{cases}, \tag{29}$$

Similarly, the curved tensile stress diagram of the tensile zone is equated to a rectangular tensile stress diagram, and the stress height kf_t of the equivalent graph can be deduced backward from the equilibrium conditions and the test data. Equations (30) and (31) can be derived from the equilibrium conditions of forces and moments in Figure 7

$$0.92 f_c b x = f_y A_s + k f_t b \left(h - \frac{x}{0.76}\right), \tag{30}$$

$$M_u = 0.92 f_c b x \left(h_0 - \frac{x}{2}\right) - k f_t b \left(h - \frac{x}{0.76}\right) \left[0.5\left(h - \frac{x}{0.76}\right) - a_s\right], \tag{31}$$

The parameters b, h, h_0, α, β, f_c, f_y, A_s, a_s, and f_t of Equations (30) and (31) are known quantities, and the relative compressive zone height x can be obtained by bringing the test data into Equations (30) and (31). The equivalent coefficient k can be calculated by bringing the obtained relative compressive zone height x into Equation (30). The calculated values of k for Beam-1, Beam-2, and Beam-3 are 0.32, 0.33, and 0.29, respectively. In order to simplify the calculation, the value of k for the reinforcement basalt fiber polymer-modified RPC beam is safely taken as 0.2.

3.3.3. Calculation of Normal Section Bending Bearing Capacity

Combining Equations (30) and (31), and bringing in the value of k, the ultimate failure bending moment of the reinforcement basalt fiber polymer-modified RPC beam is obtained from the equilibrium condition in Equation (32).

$$\begin{cases} 0.92 f_c b x = f_y A_s + 0.2 f_t b \left(h - \frac{x}{0.76}\right) \\ M_u = 0.92 f_c b x \left(h_0 - \frac{x}{2}\right) - 0.2 f_t b \left(h - \frac{x}{0.76}\right)\left[0.5\left(h - \frac{x}{0.76}\right) - a_s\right] \end{cases}, \tag{32}$$

The calculated values of the ultimate failure bending moment of each test beam can be calculated by Equation (32) and compared it with the test value of the ultimate failure bending moment, as shown in Table 5

Table 5. Calculated and test values of ultimate failure bending moment.

Teat Beam	Calculated Value M_u^c(kN· m)	Test Value M_u^t(kN· m)	M_u^c/M_u^t
Beam-1	43.00	39.40	1.09
Beam-2	49.02	47.00	1.04
Beam-3	55.88	52.20	1.07

From Table 5, it can be seen that the ratio between the calculated and tested values of the ultimate failure bending moment of the reinforcement basalt fiber polymer-modified RPC beams have a small variation, and the mean value of this ratio is 1.07, the standard deviation is 0.02, and the coefficient of variation is 0.02. The data error meets the requirements, which shows that the established formula for calculating the normal section bending bearing capacity of the reinforcement basalt fiber polymer-modified RPC beam is accurate.

3.3.4. Relative Pressure Zone Height and Reinforcement Ratio Range

Using the data of the test beams in this paper and the parameters of the main rebar used into Equation (33), the relative height of compressive area ξ_b is 0.48.

$$\xi_b = \frac{\beta}{1 + \frac{f_y}{\varepsilon_{cu} E_s}}, \tag{33}$$

In order to ensure that the beam structure meets the moderate reinforcement ratio damage, it is also necessary to calculate the maximum and minimum reinforcement ratio of basalt fiber polymer-modified RPC to HRB335 main rebar.

The reinforcement rate of basalt fiber polymer-modified RPC simply supported beam is the maximum reinforcement rate when meets $\xi = \xi_b$. Bringing the test beam parameters and results into Equation (34) leads to a maximum reinforcement ratio of about 18% for HRB335 main rebar for basalt fiber polymer-modified RPC.

$$\rho_{\max} = \frac{A_s}{bh_0} = \frac{\xi_b bh_0(0.92f_c + 0.26f_t) - 0.2f_t bh}{f_y bh_0}, \tag{34}$$

The HRB335 reinforcement rate of the basalt fiber polymer-modified RPC beam is the minimum reinforcement rate when the bending moment limit value of the test beam is equal to the bending moment of the plain basalt fiber polymer-modified RPC beam with the same section without reinforcement when it is about to crack. Bringing the parameters and results of the test beam into Equation (35), the minimum reinforcement rate of HRB335 reinforced basalt fiber polymer-modified RPC is about 0.71%.

$$\rho_{\min} = \frac{A_s}{bh_0} = \frac{\gamma f_t W_0 - 0.2f_t b\left(h - \frac{x}{0.76}\right)\left[0.5\left(h - \frac{x}{0.76}\right) + \frac{x}{2}\right]}{f_y\left(h_0 - \frac{x}{2}\right)bh_0}, \tag{35}$$

4. Discussion

In this paper, basalt fiber treated with a coupling agent was used to modify the RPC material. Visualized scanning electron microscopy (SEM) was adopted to analyze the microstructure of basalt fiber modified polymer RPC, and the results are shown in Figure 8. Figure 8a shows that the thickness of the interfacial transition zone between the aggregate quartz sand and the cement matrix is negligible. Basalt fiber polymer-modified RPC replaces coarse aggregates in ordinary concrete with fine quartz sand, which eliminates mechanical, physical and chemical differences between normal concrete aggregates and mortar. As a result, the internal defects of basalt fiber polymer-modified RPC were reduced and the compactness was improved. From Figure 8b, it can be seen that the microstructure of basalt fiber polymer-modified RPC is denser and the hydration products are mainly dense C–S–H matrix. This is due to the fact that the basalt fiber polymer-modified RPC is cared for by heat treatment, which makes the hydration process more complete, changing the properties of the hydration products and reducing the porosity. It can be observed from Figure 8c that the basalt fibers are distributed in a random direction in the RPC material, and there is no agglomeration. Figure 8d reveals that the basalt fibers are tightly bonded to the cement matrix. These characteristics illustrate that basalt fibers form a tight monolith with the cement matrix, which can jointly withstand stress changes caused by external loads and temperature. The above-mentioned microstructural characteristics of basalt fiber polymer-modified RPC are important reasons for its excellent mechanical and durability properties.

It is owing to the excellent mechanical properties and durability of basalt fiber polymer-modified RPC composites that this study applies basalt fiber polymer-modified RPC to simply supported beams and investigates its structural design calculation equations. In contrast to ordinary concrete, the compressive and flexural strengths of basalt fiber polymer-modified RPC are substantially increased. Therefore, the flexural strength of concrete in the tensile zone needs to be considered in the design calculation of basalt fiber polymer-modified RPC simply supported beams. In this paper, from the calculation principle, the contribution coefficient of concrete in the tensile zone of basalt fiber polymer-modified RPC to the bending bearing capacity is 0.2 from the experimental data.

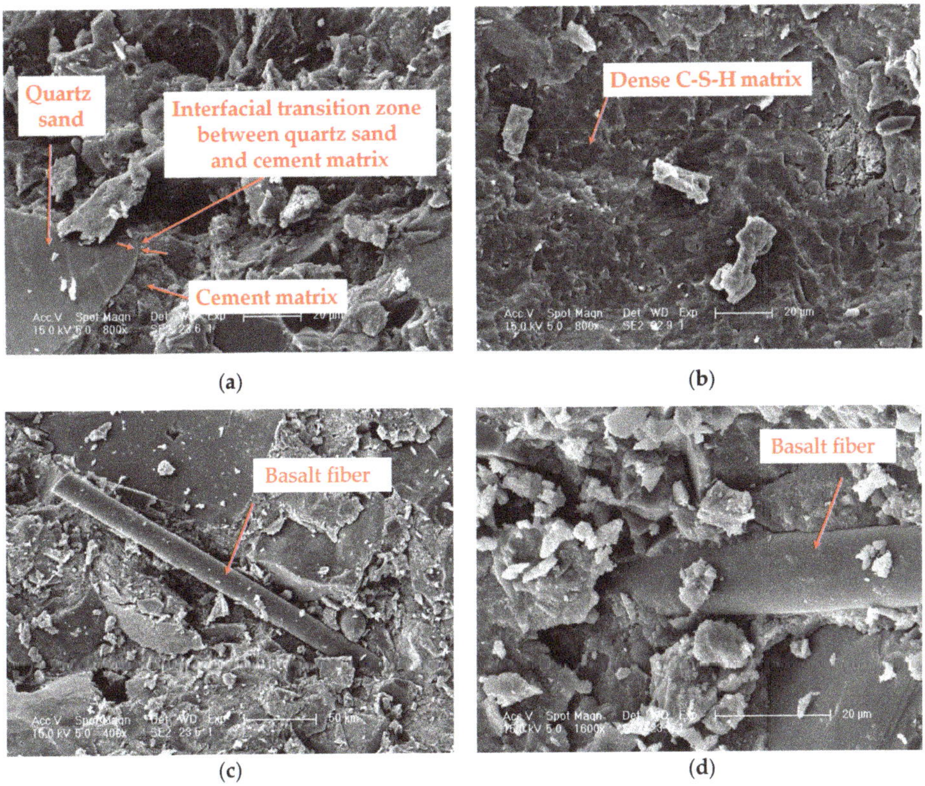

Figure 8. Microstructure of basalt fiber polymer-modified RPC: (**a**) Interfacial transition zone between quartz sand and cement matrix; (**b**) Dense C–S–H matrix; (**c**) Basalt fiber 400×; (**d**) Basalt fiber 1600×.

5. Conclusions

In this study, design calculations of basalt fiber polymer-modified RPC simply supported beams subjected to four-point bending were studied. Several conclusions derived based on the experimental results may be summarized as follows:

(1) The hydration products of basalt fiber polymer-modified RPC are mainly dense C–S–H matrix and the thickness of the interfacial transition zone between the aggregate quartz sand and the cement matrix is negligible.
(2) Basalt fibers treated with coupling agents in RPC materials are distributed in a random direction, and the basalt fibers are tightly bonded to the RPC matrix.
(3) The section resistance moment plasticity influence coefficient of the reinforcement basalt fiber polymer-modified RPC simply supported beam is 1.7; the relative height of the compressive area is 0.48; the minimum and maximum reinforcement ratios of HRB335 rebar are 0.71% and 18%, respectively.
(4) The established formulas for cracking moment and normal section bending bearing capacity is reasonably accurate. The research results of this paper can provide references for the design of reinforcement basalt fiber polymer-modified RPC simply supported beam and promote the wide application of basalt fiber polymer-modified RPC materials in practical engineering.

Author Contributions: Conceptualization, Y.G. and J.Y.; methodology, Y.G. and X.L.; software, X.L.; validation, Y.G.; formal analysis, J.Y.; investigation, X.H.; resources, X.L.; data curation, X.H.; writing—original draft preparation, Y.G. and J.Y.; writing—review and editing, Y.G.; visualization,

X.L.; supervision, H.L.; project administration, H.L.; funding acquisition, X.H. All authors have read and agreed to the published version of the manuscript.

Funding: This research was funded by the Industrial Technology Research & Development Special Project of Jilin Province (2018C042-1), the Transportation Science and Technology Program of Jilin Province (2018-1-9), the Science and Technology Development Program of Jilin Province (20200403157SF), the Transportation Science and Technology Project of Jilin Province (2021-1-1), the Special Funding for Basic Scientific Research Operation Fees of Central Universities, and the Project funded by China Postdoctoral Science Foundation (2021T140262).

Institutional Review Board Statement: Not applicable.

Informed Consent Statement: Not applicable.

Data Availability Statement: Not applicable.

Acknowledgments: The authors would like to express their appreciation for the anonymous reviewers for their constructive suggestions and comments to improve the quality of the paper.

Conflicts of Interest: The authors declare no conflict of interest.

References

1. Edwin, R.; Gruyaert, E.; de Belie, N. Influence of intensive vacuum mixing and heat treatment on compressive strength and microstructure of reactive powder concrete incorporating secondary copper slag as supplementary cementitious material. *Constr. Build. Mater.* **2017**, *155*, 400–412. [CrossRef]
2. Chen, X.; Wan, D.; Jin, L.; Qian, K.; Fu, F. Experimental studies and microstructure analysis for ultra high-performance reactive powder concrete. *Constr. Build. Mater.* **2019**, *229*, 116924. [CrossRef]
3. Zdeb, T. Effect of vacuum mixing and curing conditions on mechanical properties and porosity of reactive powder concretes. *Constr. Build. Mater.* **2019**, *209*, 326–339. [CrossRef]
4. Sultan, H.; Alyaseri, I. Effects of elevated temperatures on mechanical properties of reactive powder concrete elements. *Constr. Build. Mater.* **2020**, *261*, 120555. [CrossRef]
5. Chan, Y.; Chu, S. Effect of silica fume on steel fiber bond characteristics in reactive powder concrete. *Cement. Concr. Res.* **2004**, *34*, 1167–1172. [CrossRef]
6. Hassan, A.; Kawakami, M. Steel-free composite slabs made of reactive powder materials and fiber-reinforced concrete. *Aci. Struct. J.* **2005**, *102*, 709–718.
7. Dhand, V.; Mittal, G.; Rhee, K.; Park, S.-J.; Hui, D. A short review on basalt fiber reinforced polymer composites. *Compos. Part. B-Eng.* **2015**, *73*, 166–180. [CrossRef]
8. Li, W.; Liu, H.; Zhu, B.; Lyu, X.; Gao, X.; Liang, C. Mechanical Properties and Freeze–Thaw Durability of Basalt Fiber Reactive Powder Concrete. *Appl. Sci.* **2020**, *10*, 5682. [CrossRef]
9. Liu, H.; Lyu, X.; Zhang, Y.; Luo, G.; Li, W. Bending Resistance and Failure Type Evaluation of Basalt Fiber RPC Beam Affected by Notch and Interfacial Damage Using Acoustic Emission. *Appl. Sci.* **2020**, *10*, 1138. [CrossRef]
10. Kannan, R.; Mathangi, D.; Sudha, C.; Neelamgam, M. Experimental investigation of reactive powder concrete exposed to elevated temperatures. *Constr. Build. Mater.* **2020**, *261*, 119593.
11. Sakr, M.A.; Osama, B.; El Korany, T.M. Modeling of ultra-high performance fiber reinforced concrete columns under eccentric loading. *Structures* **2021**, *32*, 2195–2210. [CrossRef]
12. Chand, U. Prediction of Flexural Behavior of Fiber-Reinforced High-Performance Concrete. In *Recycled Waste Materials—Proceedings of EGRWSE 2019*; Agnihotri, A.K., Reddy, K.R., Bansal, A., Eds.; Springer: Berlin/Heidelberg, Germany, 2019; pp. 193–198.
13. Zingaila, T.; Augonis, M. Influence of Partial Use of UHPC on the Cracking Moment of Flexural Composite Beams. In Proceedings of the 20th International Scientific Conference (Mechanika 2015), Kaunas, Lithuania, 23–24 April 2015; p. 281.
14. Zhang, Y.; Zhu, Y.; Yeseta, M.; Meng, D.; Shao, X.; Dang, Q.; Chen, G. Flexural behaviors and capacity prediction on damaged reinforcement concrete (RC) bridge deck strengthened by ultra-high performance concrete (UHPC) layer. *Constr. Build. Mater.* **2019**, *215*, 347–359. [CrossRef]
15. Yang, I.-H.; Joh, C.; Kim, B.-S. Flexural response predictions for ultra-high-performance fibre-reinforced concrete beams. *Mag. Concr. Res.* **2012**, *64*, 113–127. [CrossRef]
16. Ujike, I.; Sogo, H.; Takasuga, D.; Konishi, Y.; Numata, M. Cracking and deformation of RC beams strengthened with reactive powder composite. In Proceedings of the 6th International Conference on Fracture Mechanics of Concrete and Concrete Structures, Catania, Italy, 17–22 June 2007; Volume 2, p. 903.
17. Turker, K.; Hasgul, U.; Birol, T.; Yavas, A.; Yazici, H. Hybrid fiber use on flexural behavior of ultra high performance fiber reinforced concrete beams. *Compos. Struct.* **2019**, *229*, 111400. [CrossRef]
18. Sim, C.; Tadros, M.; Gee, D.; Asaad, M. Flexural design of precast, prestressed ultra-high-performance concrete members. *PCI J.* **2020**, *65*, 35–61. [CrossRef]

19. Prem, P.R.; Murthy, A.R.; Verma, M. Theoretical modelling and acoustic emission monitoring of RC beams strengthened with UHPC. *Constr. Build. Mater.* **2018**, *158*, 670–682. [CrossRef]
20. Guo, M.; Ji, W. Study on nonlinear flexural behavior of RPC-NC composite beams. In Proceedings of the 3rd International Conference on Railway Engineering: Construction and Maintenance of Railway Infrastructure in Complex Environment, Beijing, China, 1 July 2014; pp. 531–537.
21. Chi, C.; Pei, C. Finite element analysis of Reactive Powder Concrete beam flexural properties. In Proceedings of the 2017 6th International Conference on Energy, Environment and Sustainable Development, Zhuhai, China, 11–12 March 2017; pp. 231–236.
22. Chen, M.; Pei, C. Influence of different kinds of steel fibers on flexural behavior of ultra-high performance concrete beam by ANSYS. In Proceedings of the 2017 5th International Conference on Machinery, Materials and Computing Technology, Beijing, China, 25 March 2017; Hou, H., Han, Z., Eds.; Atlantis Press: Paris, France, 2017; pp. 162–167.
23. Qi, J.; Cheng, Z.; Wang, J.; Tang, Y. Flexural behavior of steel-UHPFRC composite beams under negative moment. *Structures* **2020**, *24*, 640–649. [CrossRef]
24. Hasgul, U.; Turker, K.; Birol, T.; Yavas, A. Flexural behavior of ultra-high-performance fiber reinforced concrete beams with low and high reinforcement ratios. *Struct. Concr.* **2018**, *19*, 1577–1590. [CrossRef]
25. Cao, X.; Yang, Z.-X.; Zhu, W.-X.; Peng, H.Z.; Fu, F.; Wang, L.; Qian, K. Experimental study on flexural behavior of new type of prestressed reactive powder concrete sound barrier for high-speed rail. *Struct. Concr.* **2021**, *22*, 623–636. [CrossRef]
26. Cao, X.; Wang, H.; Chang, J.; Lingzhi, J. Computational Analysis of Cracking Moment of High Strength Reinforced Reactive Powder Concrete Beam. In *Electrical and Control Engineering & Materials Science and Manufacturing, Proceedings of Joint Conferences of the 6th (Icece 2015) and the 4th (Icmsm 2015), Shanghai, China, 14–15 August 2015*; WSPC: Singapore, 2016; pp. 629–638.
27. Funari, M.F.; Verre, S. The Effectiveness of the DIC as a Measurement System in SRG Shear Strengthened Reinforced Concrete Beams. *Crystals* **2021**, *11*, 265. [CrossRef]
28. Di Ludovico, M.; Cascardi, A.; Balsamo, A.; Aiello, M.A. Uniaxial Experimental Tests on Full-Scale Limestone Masonry Columns Confined with Glass and Basalt FRCM Systems. *J. Compos. Constr.* **2020**, *24*, 04020050. [CrossRef]
29. General Administration of Quality Supervision, Inspection and Quarantine of the People's Republic of China. *Methods for Chemical Analysis of Silicate Rocks—Part 28: Determination of 16 Major and Minor Elements Content (GB/T 14506.28-2010)*; China Standards Press: Beijing, China, 2010. (In Chinese)
30. Liu, H.; Liu, S.; Wang, S.; Gao, X.; Gong, Y. Effect of mix proportion parameters on behaviors of basalt fiber RPC based on box-behnken model. *Appl. Sci.* **2019**, *9*, 2031. [CrossRef]
31. State Bureau of Quality Technical Supervision. *Method of Testing Cement-Determination of Strength (GB/T 17671-1999)*; China Construction Industry Press: Beijing, China, 1999. (In Chinese)
32. Ministry of Housing and Urban-Rural Development of the People's Republic of China. *Test Standard for Test Methods for Concrete Structures (GBJ50152-2012)*; China Construction Industry Press: Beijing, China, 2012; pp. 11–25. (In Chinese)

polymers

MDPI

Article

Value Coefficient of Polyethylene Fiber Soil Embankment Slope Based on Response Surface Analysis

Yafeng Gong [1], Jiaxiang Song [1], Yulong He [2,*] and Guirong Ma [1]

[1] College of Transportation, Jilin University, Changchun 130025, China
[2] College of Civil Engineering, Hunan University, Changsha 410082, China
* Correspondence: heyl@hnu.edu.cn; Tel.: +86-177-7311-8985

Abstract: The utilization of polymers can strengthen soil, but at a high price. In this study, value coefficients were proposed to evaluate the cost-effectiveness of fiber-reinforced roadbeds, and the effects of embankment-slope-influencing factors on the value coefficients were analyzed by response surface methodology. Ultrahigh-molecular-weight polyethylene fiber (UPEF) was used as the reinforcement material for soil. First, the shear strength parameters of fiber soil with different fiber diameters were obtained from the direct shear tests to set the parameters of the finite element models. Second, three factors, namely filling height, slope angle, and fiber diameter, were selected as input parameters based on the Box–Behnken Design (BBD) experimental design method, and their effects on the value coefficient of the fiber soil embankment slope were investigated. Finally, the design parameters at the maximum value coefficient of the fiber soil embankment slope were determined based on the results of the response surface analysis. The results indicated that the addition of UPEF could effectively improve the cohesion of the soil; the interaction between the filling height and fiber diameter is most obvious. The optimization of design parameters based on the value coefficient of the fiber soil slope is a slope-engineering design method considering comprehensive benefits.

Keywords: polyethylene; fiber soil; embankment slope; response surface

Citation: Gong, Y.; Song, J.; He, Y.; Ma, G. Value Coefficient of Polyethylene Fiber Soil Embankment Slope Based on Response Surface Analysis. *Polymers* **2022**, *14*, 4295. https://doi.org/10.3390/polym14204295

Academic Editors: Victor Tcherdyntsev and Francesco Galiano

Received: 31 August 2022
Accepted: 8 October 2022
Published: 13 October 2022

1. Introduction

High-grade roads require a high load-bearing capacity for embankment slopes, and nearby available soils may not always meet the requirements of high-grade roadbed soils. Therefore, soil often has to be reinforced to meet the relevant design criteria. Reinforcement methods for embankment slope soils are mainly divided into chemical stabilization and physical reinforcement [1]. Chemical stabilization is achieved by adding additives, such as cement and lime, to soil to improve its strength and stability [2]. However, there are some disadvantages to chemical stability. For example, the pH of the soil changes after the additives are mixed with the soil, and this will cause environmental pollution [3]. Therefore, methods of physical reinforcement are increasingly being chosen. Research related to fiber-reinforced soils as a physical-reinforcement method has been carried out extensively [4].

Fibers for soil reinforcement include natural fibers and synthetic fibers. Recently, many studies have been conducted on the use of natural fibers as soil reinforcement materials to improve the physical and mechanical properties of soils due to their advantages such as wide distribution of resources, good economic benefits, and environmental protection [5,6]. The incorporation of coir fiber into soil not only enhances its strength and stiffness [7] but also reduces the seepage velocity of the soil, thereby increasing the piping resistance of the soil [8]. Fine-grained soil mixed with jute fibers can improve its unconfined compressive strength under freeze–thaw cycles [9]. In addition, jute fibers with a content of 0.6% and a length of 6 mm have been added to expansive soil to effectively improve its shear strength [10]. Palm-fiber-reinforced silty sand with a fiber length of 30 mm and fiber content of 0.5% has the best shear-strength reinforcement performance [11]. The shear strength and deformation of silty clay can be improved with the addition of sisal fibers [12]. The addition of corn silk and corn starch to soft soils can improve their compaction properties, and the optimal contents of the additive are 0.5% corn silk and 4% corn starch [13]. A series of free pressure tests was conducted to examine the effects of human hair fibers on the reinforcement of clay soils under freeze–thaw cycles, and the results showed a significant increase in unconfined compressive strength due to the addition of 1.5% human hair fibers [14]. Similarly, when adding wool to clay, the fiber content of 1.5% remains the optimal content for compressive-strength enhancement after freeze–thaw action [15].

Despite their impressive performance, natural fibers generally have little moisture and few biodegradable characteristics, and this can affect the life of the fiber and increase the cost of the project [16]. Therefore, polymer fiber has gained momentum as one of the high potential reinforcement materials in road engineering due to its stable and easy construction [17]. Currently, polypropylene fibers and polyethylene fibers are commonly used in engineering. Diambra [18] conducted an experimental study on the addition of polypropylene fibers to Hostun RF sand with different densities and found that the triaxial compressive strength of fiber-reinforced sand increases significantly with the increase in polypropylene fiber admixture. A finite element model of the polypropylene fiber soil embankment slope was established, and the effect of freeze–thaw cycling on it was analyzed by Gong and He [19]. A series of consolidated drained triaxial tests were carried out by Med Bouteben to obtain the mechanical parameters of the polypropylene fiber-reinforced cement–sand soils, which were applied to the numerical analysis of the embankment finite element model [20]. Li conducted a model roadbed test to investigate the effect of polypropylene fiber reinforcement on the settlement of the model roadbed, and the experimental results illustrated that the settlement of the fiber-reinforced soil roadbed is substantially reduced under high pressure [21]. Akbulut used polyethylene fibers as a reinforcing material for clay and found that polyethylene fibers can effectively increase the strength of clay [22]. Notably, from the perspective of environmental protection, some scholars stirred waste bags or plastics with polyethylene as the main component into soil, and this method was found to increase the shear strength and deformation resistance of the soil to a certain extent [23,24]. However, the research results showed that the reinforcement effect of waste polyethylene fiber is poor, and the quality is difficult to be unified. Therefore, waste plastic as soil-reinforcement material is not recommended [25].

In embankment-slope filling, the strength of the fiber-reinforced soil and the resistance of soil to deformation are critical [26]. The increase in strength can improve the stability of the slope, and the high modulus fiber can reduce the settlement of the slope. The optimal content is sought to achieve maximum strength. However, from the perspective of comprehensive benefits, higher fiber content means higher cost, which leads to poorer economic benefits, even though the structural load-bearing capacity is increased. Therefore, a comprehensive benefit evaluation method of fiber soil slope based on the value engineering method is proposed in this paper. The research process is shown in Figure 1. On the basis of the material parameters obtained from the direct shear test, a finite element model of the fiber soil embankment slope was established to analyze the stability of the slope model. From the perspective of cost-effectiveness, the response surface model was established based on the value coefficient of the fiber soil slope, and the best design parameters were fitted by the model.

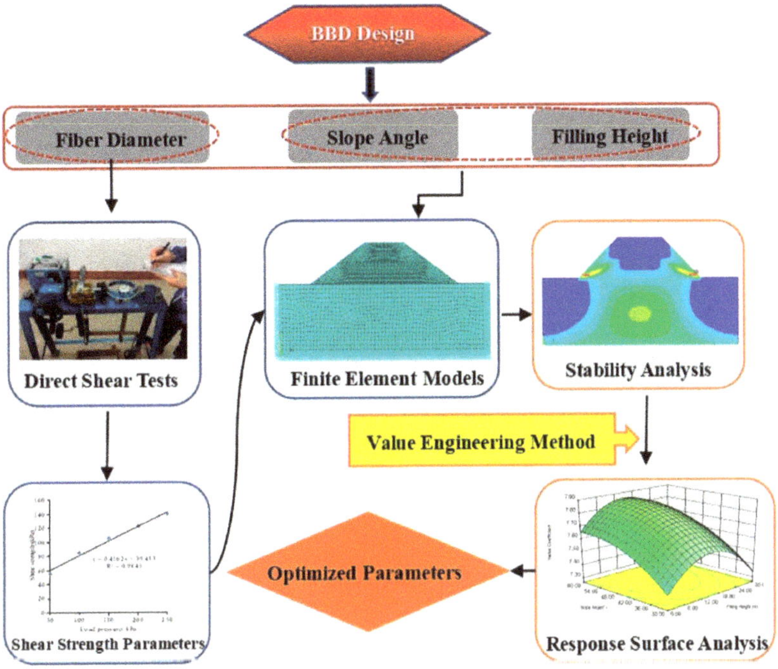

Figure 1. Technology roadmap.

2. Direct Shear Test

2.1. Materials and Sample Preparation

Soil sampling was performed in a subgrade fill in the northeast seasonal freezing area, China, which is yellowish in color and has a certain cohesiveness. Prior to the direct shear test, a series of tests were performed on the soil samples to obtain the basic physical properties of the soil, and this included specific gravity tests, compaction tests, and limit water-content tests. The basic physical properties of the soil samples are summarized in Table 1.

Table 1. Basic physical properties of soil samples.

Properties	Specific Gravity	Maximum Dry Density (g/cm^3)	Optimum Water Content (%)	Liquid Limit (%)	Plastic Limit (%)	Plasticity Index (%)
Value	2.42	1.725	12.2	34.3	25.0	9.5

Ultrahigh-molecular-weight polyethylene fiber (UPEF) is a high-performance fiber with high specific strength, high modulus, and corrosion resistance, which can effectively improve the strength and deformation resistance of fiber soil as a reinforcing fiber. The UPEF used in the test was made by Shandong Fiber Building Materials Technology Co., Ltd, Qindao, China. It is slightly white and has good flexibility. The molecular weight of UPEF is 1.5 million. UPEF has a density of 0.98, very high tensile strength (3300 MPa) and elasticity modulus (95 GPa), and good high-temperature stability (melting point of 160 °C). The diameter of the fiber is an important factor for the shear strength, and the UPEF with different diameters can affect the project cost. The commonly used UPEF diameter specifications in the market are 0.02, 0.12, and 0.2 mm. In the direct shear test, the fiber content with the mixing ratio of 6‰ and the fiber length of 9 mm were selected based on previous research and pre-experiments [4]. Therefore, after precise cutting, the direct shear test was performed with the UPEF of 9 mm in length.

2.2. Test Procedure

The first step was to prepare the test specimen. The compaction test was conducted to determine the maximum dry density and the optimum water content of the test soil, which was the target water content of the direct shear test specimen. First, the initial soil was dried, crushed, and dried at 105 °C to completely evaporate the water. Second, the dried soil was sieved through a 0.5 mm sieve for the cylinder test piece of the direct shear test. The dry soil, target moisture content of water, and UPEF were mixed well in a cement mortar mixer made by Jinrui Test Instrument Co., Ltd, Cangzhou, China after 2 min of mixing. The variables for this test were fiber diameters of 0 (plain soil), 0.02, 0.12, and 0.2 mm. Compaction of 0.96 was achieved by using hydrostatic forming, according to the test protocol. Finally, the direct shear specimens were tightly wrapped with cling film to ensure constant moisture.

The prepared specimens were subjected to direct shear tests. The shear stress of the soil at the time of damage was obtained from the direct shear test, and the internal friction angle and cohesion of the soil were calculated based on the Mohr–Coulomb law. The shear strength of four main groups of soils was measured by the direct shear test. The four groups comprised plain soils and fiber-reinforced soils mixed with different UPEF diameters (0.02, 0.12, and 0.2 mm). In the direct shear test, each test group was composed of six test pieces with applied vertical loads of 50, 100, 150, 200, 250, and 300 kPa. During the test, when the horizontal displacement value started to decrease, the specimens were considered damaged, and the shear stress value at this time was recorded as the shear strength. The test procedure was performed according to the relevant code [27].

2.3. Test Results

Direct shear tests were performed in line with the test protocol, and the peak shear stresses obtained were processed. The shear strengths of the specimens with different pressure values were linearly fitted to obtain the shear strength parameters, c and φ, which were calculated by Equation (1):

$$\tau = \sigma \tan \varphi + c \tag{1}$$

The shear strength parameters of fibrous soils with different UPEF diameters are presented in Table 2.

Table 2. Shear strength parameters of fiber soil.

Fiber Diameter	0.02 mm	0.12 mm	0.2 mm	Plain Soil
Cohesion (Pa)	47,540	62,272	60,291	40,747
Internal Friction Angle (°)	21.186	21.653	22.568	22.053

A pattern was observed in the data in Table 2; the variation in fiber diameter had little effect on the internal friction angle of UPEF-reinforced soil, but it could effectively enhance cohesion. The main strengthening mechanism of UPEF was still the reduction of shear interface slip caused by the friction between soil particles and fibers [28]. Therefore, as the fiber diameter increased, the direct contact area between soil particles and fibers increased, and the increasing interfacial friction was the main factor for the increasing cohesion. The cohesion enhancement of the fiber soil was not effective as the fiber diameter exceeded 0.2 mm. Although the interfacial shear strength between a single fiber and the soil increased, the increase in fiber diameter caused a decrease in the number of interfacial shear fibers, because the fiber content was certain [4].

3. Finite Element Analysis

3.1. Computational Model

The numerical model of fiber soil embankment slope was established by the general finite element software ANSYS 12.1. The longitudinal length of the embankment slope is longer than the transverse width, and the disease of the embankment generally occurs mainly on the outward slope. Therefore, the embankment model can be simplified to plane model analysis [19]. In the calculation model, the height of the foundation was 50 m, the top width of the embankment was 36 m, and the fill height and slope angle were the geometric variables of this study. The geometric model was established in ANSYS and meshed; boundary constraints and loads were applied. The finite element model with a slope angle of 30° and an embankment filling height of 30 m is presented in Figure 2.

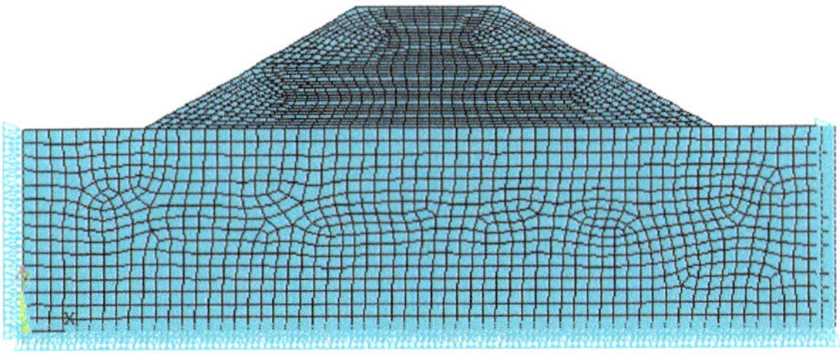

Figure 2. Finite element model of the embankment slope.

Before analyzing the finite element model of an engineering structure, the element type and material parameters of the element must be determined. Since the longitudinal length of the long slope is much larger than the transverse size, the slope stability problem can be classified as a typical plane strain problem. Plane 182 element is a two-dimensional element model with eight nodes; it is a favorable element to reflect the stress state and deformation of soil [19]. The element density was set to 1.725 g/cm^3, which was the maximum dry density of the soil obtained from the compaction test. In addition, the constitutive model of soil was established. The Drucker–Prager (D–P) model was selected as the soil constitutive model of embankment slope, and it is a commonly used constitutive model to simulate slope soil in ANSYS software. There are two important soil shear strength parameters in the D–P model: cohesion (C) and internal friction angle (φ), which were derived from the direct shear test (Table 2).

The finite element geometric model was established based on the geometric parameters. For plane strain problems, Plane182 element has good adaptability. In mesh generation, each side of the model was divided into at least 10 equal parts to ensure sufficient accuracy based on the results of the pretest. By uniform division, the non-convergence of model calculation caused by triangular elements and sharp element angles can be avoided. After the mesh division of the model, the constraint conditions and loads were set. For both sides of the foundation boundary, the horizontal displacement was constrained, and the lower boundary was fully constrained. In addition to setting the self-weight of the embankment slope as the loading condition, a strip load of size 10.5 kN/m was applied on the top of the embankment to simulate the traffic load according to the design specification [29].

The stability of the embankment slope was analyzed by the reduction coefficient method. The shear strength parameters (cohesion and internal friction angle) of the tested slope soil were converted by the reduction coefficient. On the basis of the reduction coefficient method, the stability of the embankment slope was analyzed. The shear strength parameters (cohesion and internal friction angle) of the tested slope were converted by the reduction coefficient method. After selecting an initial reduction coefficient, F, the internal friction angle, c', and cohesion, φ', of the reduced slope soil mass were calculated according to Equations (2) and (3):

$$c' = \frac{c}{F} \tag{2}$$

$$tan\varphi' = \frac{tan\varphi}{F} \quad (3)$$

where c and φ are the initial cohesion and internal friction angle of the soil, respectively.

For slope stability calculation, the discounted shear strength parameters were substituted into the finite element model for calculation. If the calculation results converged, the slope was stable. The reduction coefficient increased for numerical calculations until the results diverged, at which point, F is the stability coefficient of the slope [30].

3.2. Stability Analysis

After the analysis was completed, the results, such as strain and displacement, were viewed through the postprocessing module of ANSYS. The calculated results of the plain soil slope and the UPEF slope with 0.2 mm fiber diameter at a slope angle of 45° were selected for comparison to illustrate the contribution of adding fibers into the soil to the slope stability of the embankment.

After performing stability analysis and solution, viewing the results in ANSYS post-processing revealed that the horizontal displacement in the X-direction changed as the plastic strain developed faster with the increase in the reduction coefficient. When the reduction coefficient of the plain soil slope increased to 1.5, the calculation result showed no convergence, which demonstrated that the slope was unstable. The stability coefficient of the plain soil slope was 1.4, but the stability coefficient of the UPEF slope was 1.7. The plastic strain of the embankment slope under different fills when the reduction coefficient, F, was 1.4 is shown in Figure 3.

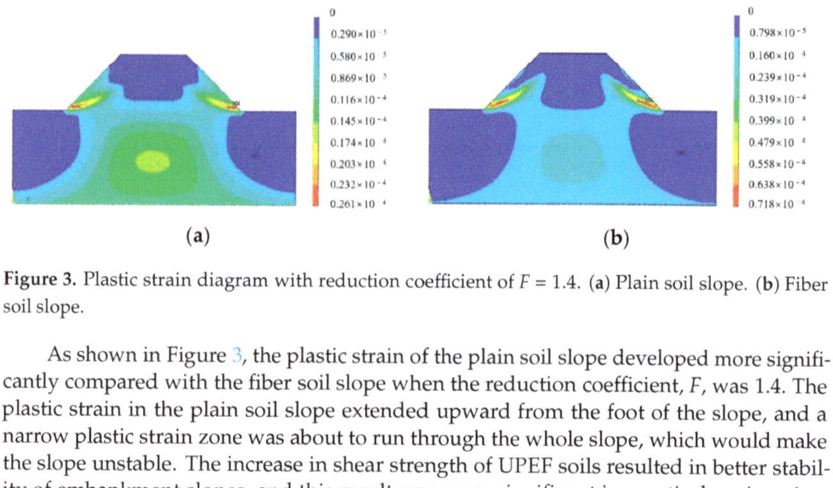

Figure 3. Plastic strain diagram with reduction coefficient of F = 1.4. (**a**) Plain soil slope. (**b**) Fiber soil slope.

As shown in Figure 3, the plastic strain of the plain soil slope developed more significantly compared with the fiber soil slope when the reduction coefficient, F, was 1.4. The plastic strain in the plain soil slope extended upward from the foot of the slope, and a narrow plastic strain zone was about to run through the whole slope, which would make the slope unstable. The increase in shear strength of UPEF soils resulted in better stability of embankment slopes, and this result was more significant in practical engineering, considering the local restraint effect of UPEF on the soil.

The x-directional displacement extremums for each strength reduction factor are shown in Figure 4. The x-directional displacement increased continuously as the reduction coefficient increased and the shear strength decreased. The x-directional displacement extremums of slope increased rapidly when it was close to instability, so there was a critical point to reflect the accelerated increase of the x-direction displacement of the slope at a certain reduction coefficient. The critical point was 1.3 for plain soil slopes and 1.5 for UPEF slopes, and the extreme value of x-direction displacement for UPEF slopes was always smaller than that for plain soil slopes with the same reduction coefficient, which fully demonstrated the high stability of UPEF slopes.

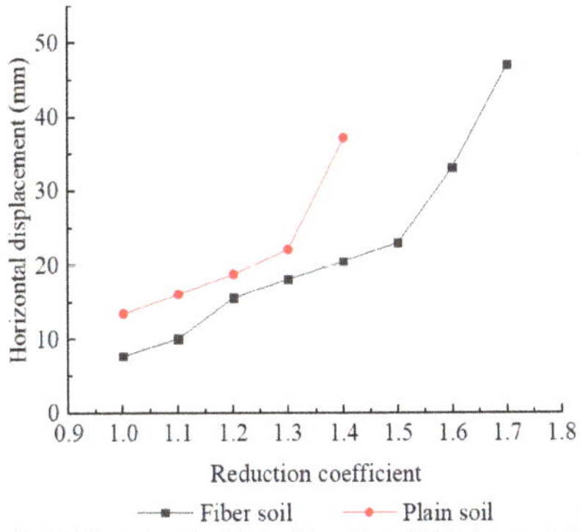

Figure 4. Extreme values of displacement in the x-direction of embankment slope with different reduction coefficients.

4. Response Surface Analysis

4.1. Theory and Method

4.1.1. Response Surface Method

The response surface method is a parameter optimization method for experimental design and statistical analysis proposed by Box and Wilson [31] in 1951 that has good robustness [32]. The response surface method was used for data analysis, which combined the experimental design method with data analysis and statistics for optimization and used explicit polynomial expression to express the implicit function. In the response surface method, the relationship between the imported variable and the response value is explained by the following Equation (4):

$$y = f(x_1, x_2 \cdots x_i) + \varepsilon \tag{4}$$

According to the Taylor formula, the response value, y, can be fitted by the response variables, $x_1, x_2 \ldots x_i$, with polynomial functions, and ε is residual. A common quadratic polynomial equation is represented in Equation (5) [33]:

$$Y = \beta_0 + \sum_{i=1}^{k} \beta_i x_i + \sum_{i=1}^{k} \beta_{ii} {x_{ii}}^2 + \sum_{i=1}^{k-1} \sum_{j=i+1}^{k} \beta_{ij} x_i x_j + \varepsilon \tag{5}$$

where Y represents the shear strength parameters of soil, and β is the undetermined coefficient estimated by the polynomial fitting function; the most commonly used estimation method is the least square method; β_0 is a constant; β_i (i = 1, 2, 3) is the linear coefficient; β_{ii} is the quadratic coefficient of x_{ii}; and β_{ij} is the interaction coefficient.

The accuracy of the response surface was determined by mathematical statistics. Analysis of variance (ANOVA) was used to assess the variance of the test sample data. Response surface models were fitted in accordance with the value coefficients calculated for different embankment slope models. ANOVA and interaction were used to determine the effect of different geometric parameters and fiber incorporation on the slope value coefficients.

4.1.2. Value Engineering Method

The value engineering method is a commonly used method to obtain the best cost–performance ratio. By calculating the value coefficient of the research object, the most cost-effective conditions of use in the study were determined. For embankment slope engineering, the value engineering method mainly considers three aspects for parameter design: cost performance ratio (V), safety stability coefficient (FI), and project cost (CI). The relationship among these aspects is shown in Equation (6) [34].

$$V_i = \frac{FI_i}{CI_i} \tag{6}$$

where V is the value coefficient, FI is the function coefficient, CI is the cost coefficient, and i is the label of the implementation scheme for completing the product.

There are specific requirements for the functional requirements of the project in different engineering environments, so the functional coefficient in value engineering should be specified in advance. In the analysis of the slope, the function coefficient is mainly determined by considering the safety and stability of the slope. The cost coefficient of this study mainly considered the cost of materials used to fill the slope. In this paper, the main influencing factor of the project cost was the amount of fiber. The total cost was obtained by multiplying the cost of fiber per unit area and the total area. The value coefficients were selected to judge the cost effectiveness of different slope-filling solutions, as well as to determine the most cost-effective parameter design solutions based on the value coefficients of slopes with different design parameter conditions.

4.2. Response Surface Test Result

The Box–Behnken Design (BBD) is the most widely and commonly used design in response surface methodology due to its fewer experimental requirements and excellent results [35]. The stability coefficient of the fiber soil embankment slope was analyzed by the value engineering calculation method. A higher value coefficient indicates that it has a relatively high safety coefficient at a lower price cost, which conforms to the principle of safety and economy in engineering design. In this paper, the slope angle, filling height, and fiber diameter were used as the response variables, and the value coefficient was used as the response value to evaluate the comprehensive economy of the UPEF soil embankment slope under the corresponding design parameters. The corresponding response surface fitting equation was established based on the calculation results obtained by the value engineering method.

It is worth noting that the commonly used design parameters are mainly considered as analysis variables when selecting parameters. In this paper, there are three design parameters, namely fill height, slope, and fiber diameter. These three factors affect the geometry and material parameters of the finite element model and are related to the slope stability and cost. The fill height and slope angle were commonly used design values for slopes, and the design value of fiber diameter was also the most commonly used value in the market. Three levels of the three design parameters were set. The slope angle of the embankment slope was designed between 0.3 and 0.6; the fiber diameters of 0.02, 0.12, and 0.2 mm were used. The fiber filling height was 0-layer filling, half filling, and all filling in three ways. The response surface test scheme was designed by using the professional test design software Design Expert. A response surface design with three levels and three factors was obtained by using the BBD experimental design method. The design of the response surface test protocol requires the coding of the influencing factors. The high, medium, and low values of the influencing factors correspond to the codes 1, 0, and −1, respectively, and the coding table is shown in Table 3. The response surface test protocols designed based on Table 3 are listed in Table 4, which also contains the response values (value coefficient) corresponding to the different run numbers.

Table 3. Code of influencing factors.

Coding Levels	Filling Height (h)/m	Fiber Diameter (d)/mm	Slope Angle (s)/°
−1	0	0.02	30
0	15	0.12	45
1	30	0.20	60

Table 4. Calculation table of value coefficients.

Run Number	Influencing Factors			Response Values
	A: h (m)	B: s (°)	C: d (mm)	Value Coefficient
1	0.00	60.00	0.02	7.7381
2	0.00	60.00	0.12	7.7381
3	0.00	30.00	0.12	7.19697
4	0.00	45.00	0.20	7.07071
5	15.00	60.00	0.02	6.92226
6	15.00	30.00	0.02	7.3776
7	15.00	45.00	0.12	7.86218
8	15.00	60.00	0.20	8.4317
9	15.00	30.00	0.20	7.8264
10	30.00	45.00	0.02	7.15488
11	30.00	30.00	0.12	7.24638
12	30.00	60.00	0.12	7.76398
13	30.00	45.00	0.20	7.80533

4.3. Statistical Analysis

The value coefficients obtained from the analysis were fitted, and a trivariate regression fitting formula was obtained as shown in Equation (7):

$$\begin{aligned} F = 7.83 - 0.091 \times A + 0.022 \times B - 11.93 \times C - 2.28 \times 10^{-5} \times AB + 1.1 \times AC \\ -0.027 \times A^2C + 0.21 \times BC - 1.97 \times 10^{-3} \times A^2 - 3.39 \times 10^{-4} \times B^2 - 10.19C^2 \end{aligned} \quad (7)$$

The accuracy of the value coefficient response surface of fiber soil slope was determined by statistical method. ANOVA is a commonly used method to evaluate the difference in value coefficient changes caused by design parameters. A fitted regression model expressed by Equation (9) was constructed based on the value coefficient calculation results provided in Table 4, and the accuracy of the model simulation was determined by ANOVA. The value coefficients in Table 4 were entered into Design Expert for ANOVA, and the ANOVA results are presented in Table 5.

Table 5. ANOVA for response parameters.

Source	Sum of Squares	Degree of Freedom	Mean Square	F-Value	*p*-Value Prob > F
Model	2.38	10	0.24	18.37	0.0010
A-h	0.17	1	0.17	13.39	0.0106
B-s	9.92×10^{-3}	1	9.92×10^{-3}	0.77	0.4153
C-d	0.19	1	0.19	14.90	0.0084
AB	1.09×10^{-4}	1	1.09×10^{-4}	8.42×10^{-3}	0.9299
AC	0.80	1	0.80	61.75	0.0002
BC	0.32	1	0.32	24.55	0.0026
A^2	0.16	1	0.16	12.32	0.0127
B^2	0.026	1	0.026	1.97	0.2099
C^2	0.026	1	0.026	2.02	0.2055
A^2C	0.55	1	0.55	42.42	0.0006
Residual	0.078	6	0.013	-	-
Cor Total	2.46	16	-	-	-

The F-test was performed on the regression fit equation, and the magnitude of the discriminated *p*-value determined the significance of the fitted equation; the smaller the *p*-value, the higher the significance [36]. The Model F-value of 18.37 implied that the model was significant. There was only a 0.10% chance that a "Model F-value" this large could occur due to noise. Values of "Prob > F" greater than 0.10 indicated that the model terms were not significant. Values of "Prob > F" less than 0.05 indicated that model terms were significant. In this case, A, C, AC, BC, A^2, and A^2C were significant model terms.

The above ANOVA on the response surface model of value coefficient showed that the slope angle, filling height, and parameters of fill material had a large influence on the value coefficient of the fiber soil slope. The analysis of the F-value demonstrated that the fiber diameter had the greatest significance on the value coefficient, and the UPEF soil filling height had the lowest F-value; therefore, its influence was relatively less significant The F-value of the multiple-times term was larger, thereby indicating that the influence of several influencing factors on the value coefficient had a non-linear influence.

In the analysis process of the response surface method, the accuracy of the regression equation was verified by testing the residual normality distribution of the response values (value coefficients) and by comparing the predicted values with the analyzed values. These inspections are shown in Figures 5 and 6.

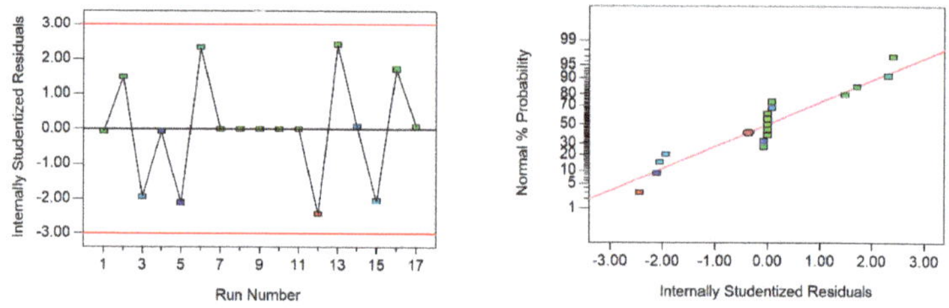

(a) (b)

Figure 5. Residual distribution diagram of the response variable. (**a**) Internally studentized residuals vs. run number (The red line is the residual range line). (**b**) Normal probability vs. internally studentized residuals (The red line is the residual regression line).

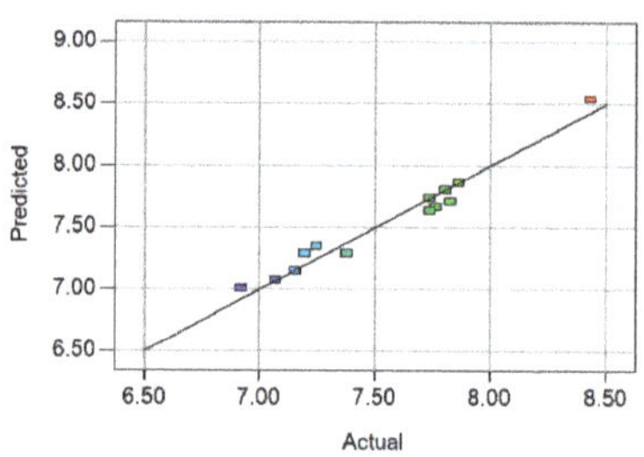

Figure 6. Comparison of calculated and predicted values (The black line is a 45 degree diagonal).

The residual value level is also a powerful condition to reflect the quality of the mathematical model. The closer the residual normal distribution plot is to a straight line, the denser the observations near the regression line are, and the better the fit is [35]. A probability plot of the normal distribution of residuals for the response surface model was plotted (Figure 5). The value coefficients of the fibrous soil slopes were mainly distributed on a sloping straight line, which indicated that the results of the value coefficient analysis showed an approximately normal distribution. In addition, the residual distribution indicated that the residuals were usually random and showed the accuracy of the model [37].

Figure 6 shows a good fit for the response surface model. According to Figure 6, most of the analyzed data points were concentrated above the 45-degree sloping straight line, and only a few calculated points were discrete, which showed that the actual values were very close to the predicted values of the model. Therefore, the fitted equation obtained by regression, using the response surface method, could accurately predict the value coefficient of the fiber soil slope.

In addition, the applicability and significance of the model were checked by various statistical factors in Table 6. The coefficient of variation (CV) is the ratio of the standard deviation (SD) to the mean, which is a normalized measure to reflect the degree of dispersion of the model. In this case, the coefficient of variation was 1.49%, which indicated the high accuracy of the model. R^2 is a common metric used to test predictive models, comparing how well the predicted results match the actual occurrence. The proposed model showed an R^2 of

0.9684, which was a high value indicating a strong agreement between the predicted and true values based on the response surface. "Pred R-Squared" was a negative value, which implied that the overall mean was a better predictor of response than the current model.

Table 6. ANOVA for models.

Statistic Factors	Value	Statistic Factors	Value
SD	0.11	R-Squared	0.9684
Mean	7.62	Adj R-Squared	0.9157
CV %	1.49	Pred R-Squared	−0.1629
PRESS	2.86	Adeq Precision	16.734

The value coefficient was not only related to the safety coefficient but also closely related to the project cost. The three factors were closely related to the above two. The interaction between factors was studied to determine the change in response value under the coupling of multiple factors. The response surface and contour map could intuitively reflect the impact of interaction on the response value. The steeper the surface and the denser the contour, the more significant the impact, and the stronger the interaction between the two factors. The contours of the factor interactions and the response surfaces are shown in Figure 7.

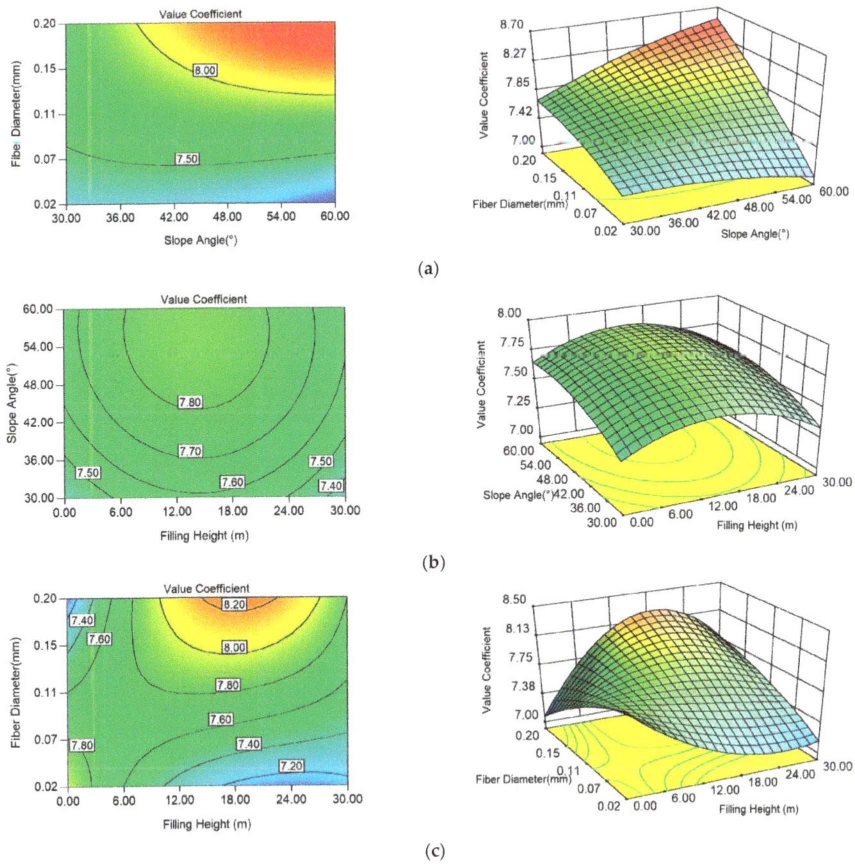

Figure 7. Response surface and contour plots. (**a**) Interaction between fiber diameter and slope angle. (**b**) Interaction between filling height and slope angle. (**c**) Interaction between filling height and fiber diameter.

Figure 7 shows that the three influencing factors of slope angle, filling height, and incorporated fiber diameter had a significant interaction; the interaction between the filling height and fiber diameter was the most obvious. As the fiber diameter increased, the safety coefficient of the soil increased, but so did the cost; the increase in filling height not only changed the value coefficient caused by itself, but it also made the interaction more pronounced due to the fact that the increase in filling height was accompanied with an increase in fibers. The interaction between the fiber diameter and slope angle was very strong, whereas the interaction between the fill height and slope angle was strong to some extent. Thus, each influencing factor did not independently affect the value coefficient of UPEF soil slope, and this result was consistent with the ANOVA results of the interaction term.

In the interaction between fiber diameter and slope angle, the response surface was mainly an oblique upward surface, because the value coefficients all increased with the increase in the influencing factor. Its oblique direction was mainly toward the direction of fiber diameter, so the influence of fiber diameter on the value coefficient was large. The response surface of filling height versus slope angle was spherical, and the contour arrangement was sparser when the slope angle of the side slope was larger. In the interaction between filling height and fiber diameter, the response surface was a complex surface similar to the saddle surface, the contour arrangement was very dense, and the interaction between the two was very significant. The value coefficient increased and then decreased as the filling height increased, and this indicated that the interaction of the two factors had a complex effect on the value coefficient.

The above contour map and response surface map showed that the contour map under the interaction of the three influencing factors in the whole boundary range was in the shape of opening, thus showing that the extreme value of the UPEF slope value coefficient may not appear in the selected parameter design range. On the basis of the obtained response surface equation, combined with the optional conditions for the design value coefficient of fiber soil slope, the optimal design parameters were obtained. The optional conditions were set, as shown in Equation (8):

$$\begin{aligned} \mathrm{MaxFT} = \; & \mathrm{f}\,(\mathrm{A},\ \mathrm{B},\ \mathrm{C}) \\ & \mathrm{s.t.} \\ & 30 \geq \mathrm{A} \geq 0 \\ & 60 \geq \mathrm{B} \geq 30 \\ & 0.2 \geq \mathrm{C} \geq 0.02 \end{aligned} \tag{8}$$

In the above equation, FT is the fitted response value; A, B, and C are the filling height, slope angle, and fiber diameter, respectively. As shown in Equations (9) and (10), the optimal design parameters for determining the soil slope based on the value coefficient of UPEF soil slope were obtained as follows: the filling height of 18.52 m, the fiber slope angle of 56.11°, and the fiber diameter of 0.19 mm (value coefficient of 8.48).

5. Conclusions

The shear strength parameters of UPEF soil were measured through direct shear tests, and the slope stability was analyzed by using the finite element analysis model. The value coefficient of fiber soil slope was calculated based on the value coefficient method. The following conclusions were obtained by analyzing the value coefficient of the fiber soil embankment slope based on the response surface method:

(1) Compared with plain soil, the internal friction angle of fiber-reinforced soil showed no obvious change, but the cohesion was significantly enhanced. When analyzing the slope with ANSYS, the stability of the fiber-reinforced soil embankment slope showed a significant improvement.
(2) The results of the ANOVA indicated that the fiber diameter had the most significant effect on the value coefficient; the response surface illustrated the interaction between

the three influencing factors, and the interaction between the fiber diameter and filling height is the most significant.

(3) The cost-effectiveness of the fiber soil slope was obtained by studying the strength reduction coefficient and value engineering method, and a regression polynomial with significant fitting effect was obtained to provide a reference for the actual project.

Author Contributions: Conceptualization, Y.H.; methodology, Y.G. and J.S.; formal analysis and investigation, Y.G. and J.S.; writing—original draft preparation, Y.G. and J.S.; writing—review and editing, Y.H.; funding acquisition, Y.G.; resources, G.M.; supervision, G.M. All authors have read and agreed to the published version of the manuscript.

Funding: The research was funded by the National Key R&D Program of China (2021YFB2600604, 2021YFB2600600), the Technological Developing Scheme Program of Jilin Province (20200403157SF), the Key Project of Department of Transportation of Heilongjiang Province of China (2022-1, 2022-2), the Transportation Technology Program of Jilin Province of China (2021-1-1), and the Transportation Technology Program of Jilin Province of China (2022-1-8).

Institutional Review Board Statement: Not applicable.

Informed Consent Statement: Not applicable.

Data Availability Statement: The data presented in this study are available upon request from the corresponding author.

Acknowledgments: The authors would like to thank anonymous reviewers for their constructive suggestions and comments that improved the quality of the paper.

Conflicts of Interest: The authors declare no conflict of interest.

References

1. DeJong, J.T.; Mortensen, B.M.; Martinez, B.C.; Nelson, D.C. Bio-mediated soil improvement. *Ecol. Eng.* **2010**, *36*, 197–210. [CrossRef]
2. Winterkorn, H.F.; Pamukcu, S. *Soil Stabilization and Grouting*; Springer: Boston, MA, USA, 1991; pp. 317–378.
3. Chen, C.; Wei, K.; Gu, J.; Huang, X.; Dai, X.; Liu, Q. Combined Effect of Biopolymer and Fiber Inclusions on Unconfined Compressive Strength of Soft Soil. *Polymers* **2022**, *14*, 787. [CrossRef]
4. Han, C.; He, Y.; Tian, J.; Zhang, J.; Li, J.; Wang, S. Shear strength of polypropylene fibre reinforced clay. *Road Mater. Pavement Des.* **2020**, *22*, 2783–2800. [CrossRef]
5. Sharma, V.; Vinayak, H.K.; Marwaha, B.M. Enhancing compressive strength of soil using natural fibers. *Constr. Build. Mater.* **2015**, *93*, 943–949. [CrossRef]
6. Ibrahim, M.; Dufresne, A.; El-Zawawy, W.K.; Agblevor, F.A. Banana fibers and microfibrils as lignocellulosic reinforcements in polymer composites. *Carbohydr. Polym.* **2010**, *81*, 811–819. [CrossRef]
7. Babu, G.L.S.; Vasudevan, A.K. Strength and Stiffness Response of Coir Fiber-Reinforced Tropical Soil. *J. Mater. Civ. Eng.* **2008**, *20*, 571–577. [CrossRef]
8. Babu, G.L.S.; Vasudevan, A.K. Seepage Velocity and Piping Resistance of Coir Fiber Mixed Soils. *J. Irrig. Drain. Eng.* **2008**, *134*, 485–492. [CrossRef]
9. Güllü, H.; Khudir, A. Effect of freeze–thaw cycles on unconfined compressive strength of fine-grained soil treated with jute fibre, steel fibre and lime. *Cold Reg. Sci. Technol.* **2014**, *106–107*, 55–65. [CrossRef]
10. Yixian, W.; Panpan, G.; Shengbiao, S.; Haiping, Y.; Binxiang, Y. Study on Strength Influence Mechanism of Fiber-Reinforced Expansive Soil Using Jute. *Geotech. Geol. Eng.* **2016**, *34*, 1079–1088. [CrossRef]
11. Ahmad, F.; Bateni, F.; Azmi, M. Performance evaluation of silty sand reinforced with fibres. *Geotext. Geomembr.* **2010**, *28*, 93–99. [CrossRef]
12. Wu, Y.K.; Li, Y.B.; Niu, B. Investigation of mechanical properties of randomly distributed sisal fibre reinforced soil. *Mater. Res. Innov.* **2014**, *18* (Suppl. S2), 953–959. [CrossRef]
13. Yong, L.L.; Emmanuel, E. Effects of Corn Starch and Corn Silk Binary Blends on the Strength Properties of A Soft Clay: A Preliminary Study. *J. Nat. Fibers* **2021**, *39*, 1–15. [CrossRef]
14. Yarbasi, N. Effect of Freezing-Thawing on Clayey Soils Reinforced with Human Hair Fibers. *J. Nat. Fibers* **2019**, *17*, 921–931. [CrossRef]
15. Yarbasi, N. Effect of Freeze-Thaw on Compressive Strength of Clayey Soils Reinforced with Wool. *J. Nat. Fibers* **2021**, *19*, 382–393. [CrossRef]
16. Gowthaman, S.; Nakashima, K.; Kawasaki, S. A State-of-the-Art Review on Soil Reinforcement Technology Using Natural Plant Fiber Materials: Past Findings, Present Trends and Future Directions. *Materials* **2018**, *11*, 553. [CrossRef]

17. Kutanaei, S.S.; Choobbasti, A.J. Triaxial behavior of fiber-reinforced cemented sand. *J. Adhes. Sci. Technol.* **2015**, *30*, 579–593. [CrossRef]
18. Diambra, A.; Ibraim, E.; Wood, D.M.; Russell, A. Fibre reinforced sands: Experiments and modelling. *Geotext. Geomembr.* **2010**, *28*, 238–250. [CrossRef]
19. Gong, Y.; He, Y.; Han, C.; Shen, Y.; Tan, G. Stability Analysis of Soil Embankment Slope Reinforced with Polypropylene Fiber under Freeze-Thaw Cycles. *Adv. Mater. Sci. Eng.* **2019**, *2019*, 5725708. [CrossRef]
20. Bouteben, Y.M.; Zeineddine, B. Mechanical behaviour of sandy soils embankments treated with cement and reinforced with discrete elements (fibres). *Frattura ed Integrità Strutturale* **2022**, *16*, 174–186. [CrossRef]
21. Li, Y.; Su, L.; Ling, X.; Wang, J.; Yang, Y. Model Studies on Load-Settlement Characteristics of Coarse-Grained Soil Treated with Geofiber and Cement. *Polymers* **2018**, *10*, 621. [CrossRef]
22. Akbulut, S.; Arasan, S.; Kalkan, E. Modification of clayey soils using scrap tire rubber and synthetic fibers. *Appl. Clay Sci.* **2007**, *38*, 23–32. [CrossRef]
23. Consoli, N.C.; Montardo, J.P.; Prietto, P.D.M.; Pasa, G.S. Engineering Behavior of a Sand Reinforced with Plastic Waste. *J. Geotech. Geoenviron. Eng.* **2002**, *128*, 462–472. [CrossRef]
24. Al-Adili, A.; Azzam, R.; Spagnoli, G.; Schrader, J. Strength of soil reinforced with fiber materials (Papyrus). *Soil Mech. Found. Eng.* **2012**, *48*, 241–247. [CrossRef]
25. Cavey, J.K.; Krizek, R.J.; Sobhan, K.; Baker, W.H. Waste fibers in cement-stabilized recycled aggregate base course material. *Transp. Res. Rec.* **1995**, *1468*, 97–107.
26. Wang, W.; Cao, G.; Li, Y.; Zhou, Y.; Lu, T.; Zheng, B.; Geng, W. Effects of Freeze–Thaw Cycles on Strength and Wave Velocity of Lime-Stabilized Basalt Fiber-Reinforced Loess. *Polymers* **2022**, *14*, 1465. [CrossRef] [PubMed]
27. Ministry of Transport of the People's Republic of China. *Test Methods of Soils for Highway Engineering*; Ministry of Transport of the People's Republic of China: Beijing, China, 2017. (In Chinese)
28. Tang, C.-S.; Shi, B.; Zhao, L.-Z. Interfacial shear strength of fiber reinforced soil. *Geotext. Geomembr.* **2010**, *28*, 54–62. [CrossRef]
29. Ministry of Transport of the People's Republic of China. *Technical Standard of Highway Engineering*; Ministry of Transport of the People's Republic of China: Beijing, China, 2014. (In Chinese)
30. Zienkiewicz, O.C.; Taylor, R.L. *The Finite Element Method*, 4th ed.; McGraw-Hill: New York, NY, USA; pp. 319–355.
31. Box, G.E.P.; Wilson, G. On the Experimental Attainment of Optimum Conditions. *J. R. Stat. Society. Ser. B Methodol.* **1951**, *13*, 1–45. [CrossRef]
32. Fan, W.L.; Zhang, C.T.; Li, Z.L.; Feng, H. An adaptive response surface method with cross terms. *Eng. Mech.* **2013**, *30*, 68–72. [CrossRef]
33. Singh, K.P.; Gupta, S.; Singh, A.K.; Sinha, S. Optimizing adsorption of crystal violet dye from water by magnetic nanocomposite using response surface modeling approach. *J. Hazard. Mater.* **2011**, *186*, 1462–1473. [CrossRef]
34. *GB/T 8223-1987*; Value Engineering-General Terms and Work Program. Standards Press of China: Beijing, China, 1988. (In Chinese)
35. Guo, L.; Guo, Y.; Zhong, L. Research on the back analysis and failure mechanism of recycled concrete aggregate meso-parameters based on Box-Behnken Design response surface model. *J. Build. Eng.* **2022**, *51*, 104317. [CrossRef]
36. Yetilmezsoy, K.; Demirel, S.; Vanderbei, R.J. Response surface modeling of Pb(II) removal from aqueous solution by Pistacia vera L.: Box–Behnken experimental design. *J. Hazard. Mater.* **2009**, *171*, 551–562. [CrossRef] [PubMed]
37. Scrivener, K.L.; Crumbie, A.K.; Laugesen, P. The Interfacial Transition Zone (ITZ) Between Cement Paste and Aggregate in Concrete. *Interface Sci.* **2004**, *12*, 411–421. [CrossRef]

polymers

MDPI

Article

Laboratory Experiments and Numerical Simulation Study of Composite-Material-Modified Loess Improving High-Speed Railway Subgrade

Li Luo, Xingang Wang *, Chen Xue, Daozheng Wang and Baoqin Lian

State Key Laboratory of Continental Dynamics, Department of Geology, Northwest University, Xi'an 710069, China
* Correspondence: xgwang@nwu.edu.cn

Abstract: Construction of high-speed railway subgrade on loess soils in the Loess Plateau is risky because such soil is susceptible to differential settlements. Various soil-improvement methods have been used to enhance the mechanical properties of loess. Lime-ash soil and cement-lime soil are the most commonly used methods in the improvement of loess subgrade, while few studies have been found on loess subgrade improvement by using composite material consisting of traditional materials and new materials. A series of direct shear tests and unconfined compressive tests were conducted on the loess specimen with the addition of three kinds of composite materials: traditional material cement, new material polypropylene fiber and SCA-2 soil curing agent. The numerical simulation was conducted on loess subgrade in an actual engineering practice. The experimental results show that cement, polypropylene fiber and SCA-2 soil curing agent can effectively improve the shear strength and compressive strength of loess, and the influence degree is cement > fiber > curing agent. Additionally, based on the relative strength characteristics of the improved loess, an optimal improvement scheme for the composite-material-modified loess was obtained: 16% cement content + 0.5% fiber content + 4% curing agent content. The numerical simulation results revealed that the compressive strength index of the improved loess has a significant impact on the subgrade settlement, and the optimal improvement scheme obtained from comprehensive analysis can effectively improve the settlement of high-speed railway subgrade under vibration load.

Keywords: loess; composite material; improvement; high-speed rail; subgrade settlement

Citation: Luo, L.; Wang, X.; Xue, C.; Wang, D.; Lian, B. Laboratory Experiments and Numerical Simulation Study of Composite-Material-Modified Loess Improving High-Speed Railway Subgrade. *Polymers* **2022**, *14*, 3215. https://doi.org/10.3390/polym14153215

Academic Editors: Wensheng Wang, Yongchun Cheng, Heping (Fred) Chen and Guojin Tan

Received: 9 July 2022
Accepted: 2 August 2022
Published: 8 August 2022

1. Introduction

High-speed railway subgrade has been identified as one type of an important foundation of bearing track structures and trains [1–3]. It is known that excessive settlement of the subgrade will cause serious disasters such as subgrade collapse and train derailment (Figure 1); however, the millimeter-level settlement-control standard of high-speed railway lines poses unprecedented challenges to subgrade engineering, which is a special issue for loess soils in loess areas of China [4].

Loess soils are characterized by a metastable structure, high porosity, water sensitivity and being loosely compacted [5–7], which leads to a high risk of disaster under external forces [8–10]. Thus, constructions of high-speed railway in loess areas are prone to uneven settlement of high fill embankment, subgrade slope instability and other engineering disasters [11,12]. Therefore, effectively improving the loess and controlling the settlement and deformation of the subgrade has become a significant problem that needs to be solved urgently in the current high-speed railway construction. At present, it is recognized that the improved subgrade has been widely used in high-speed railway engineering [13,14]. Japan, Germany and other countries have clearly stipulated that the railway subgrade could be improved by cement, lime and other materials. For the improvements of soil subgrade, the common treatment methods include physical and chemical methods. Generally, the physical improvement method is a way to change soil gradation by adding coarse particles such as sand and gravel [15], and the chemical improvement is usually employed by mixing cement [16], lime [17,18] and polypropylene fiber [19,20] into the soil.

Figure 1. Typical high-speed railway foundation settlement disasters in China: (**a**) Xiaoyong railway collapse disaster; (**b**) Kunming railway collapse disaster; (**c**) Longhai railway collapse disaster; (**d**) South Tongpu railway collapse disaster.

In recent years, research scholars have carried out a series of studies on the improvement of subgrade in loess areas by using traditional materials including cement, lime, etc. and by using new improvement materials such as polypropylene fiber, curing agent, etc. By conducting field compaction tests on loess with different lime contents, Zhang et al. [21] investigated the effects of moisture content, lime content and compaction strength on the compaction characteristics of loess subgrade. The results show that the improvement effect of lime on loess increases with water content. Jiang et al. [22] studied the influence of cement content, compaction coefficient and curing time on the mechanical strength index of cement-modified loess (CML) compacted by vertical vibration compaction (VVCM), and finally established a model to predict the growth law of strength characteristics of CML. With the general improvement of environmental protection awareness, new environmentally friendly improvement materials have been used in soil improvement [23–25]. By extracting lignin from waste paper to improve loess with different contents, and carrying out triaxial tests and SEM observations on the modified loess, Wang et al. [26] found that when the lignin content was 4%, the modified loess had the best liquefaction resistance and compactness. Wang et al. [20] studied the role of polypropylene fiber played as a reinforcing material to enhance the mechanical properties of loess. Through ring shear tests, the optimum content of polypropylene fiber to improve loess properties was found to be 0.5%.

Previous studies revealed that the mechanical properties such as compressive strength [27,28] and shear strength [29] of the improved soil are significantly enhanced, which thus can effectively decrease the settlement deformation of high-speed railway subgrade under external load. Additionally, it has been widely recognized that the improvement of soil mass by adding a variety of composite materials with a certain mixing ratio is better than that by adding a single material [30,31]. At present, the addition of lime-ash or cement-lime mixes to soil is the most-used method for the improvement of loess subgrade, while few studies were found on the improved materials improve loess subgrade by combining new materials with traditional materials [32].

In this study, to explore a new method of loess subgrade improvement, traditional improvement materials (cement) and new improvement materials (polypropylene fiber, SCA-2 soil curing agent) were selected to form composite materials with different contents to improve loess. By conducting a series of laboratory test and orthogonal analysis, the influence of composite material dosage on the strength index of improved loess was studied. Additionally, the influence of different strength indexes of the improved loess on subgrade settlement was also discussed by conducting numerical simulation. Finally, the best scheme of the composite material to improve loess was obtained, which can provide new ideas and experimental reference for the improvement of high-speed railway subgrade in loess areas.

2. Materials and Methods

2.1. Materials

The loess samples were collected from a high-speed railway subgrade in Xi'an, Shaanxi Province of China (Figure 2). The grain-size distribution curve of soil sample was determined by conducting sieve analysis (Figure 3a). The basic physical indexes including dry density, water content, specific gravity, liquid limit (WL) and plastic limit (Wp) shown in Table 1, were determined according to the Chinese National Standards (CNS) GB/T50123–1999 [33]. The results reveal that the soil sample has a relatively low water content and low dry density. According to the Casagrande plasticity chart shown in Figure 3b, the loess is classified as a low-plastic clay. Three improvement materials used in the test are as follows: the ordinary portland cement with a strength grade of 42.5 [17], the chemical composition of which is shown in Table 2; the bundle monofilament polypropylene fiber with 6 mm length [31], the main physical and mechanical properties of which are shown in Table 3; and the SCA-2 soil curing agent (dark gray) is produced in Qingdao, China, which is a polymer organic liquid curing agent with pH value less than 1, strong cohesiveness, high-concentration liquid and density ranging from 1.30 to 1.35 g/cm^3.

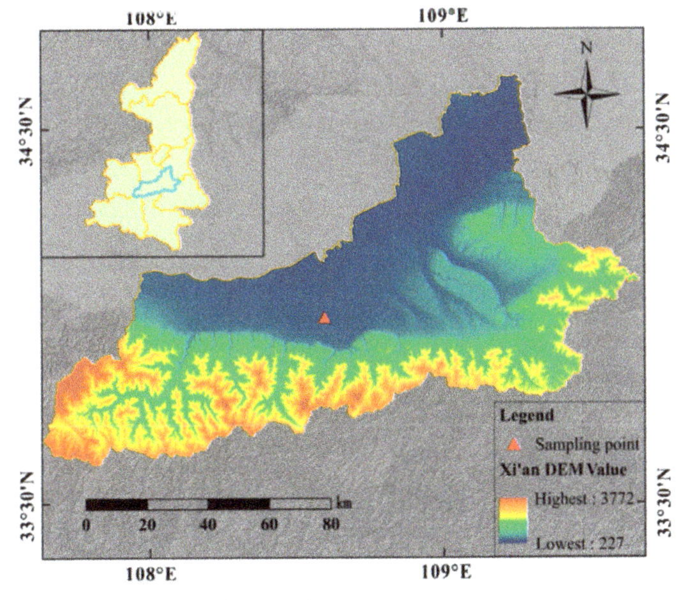

Figure 2. Location map of the sampling site.

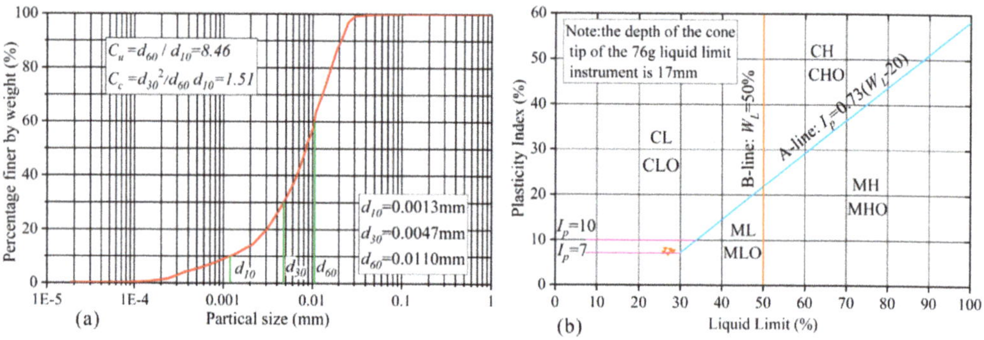

Figure 3. Physical properties of loess specimens showing: (**a**) grain-size distribution curve; (**b**) Casagrande plasticity chart.

Table 1. Basic physical indexes of loess.

ρ_d	W	ρ	G_S	W_L	W_p	Grain Size Fractions (%)		
						<0.005 mm	0.005–0.075 mm	0.075–0.05 mm
1.35	12	1.51	2.70	27.1	17.6	26.4	71.3	2.3

Notes: ρ_d = dry density (g/cm^3); W = natural water content (%); ρ = natural density (g/cm^3); G_S = specific gravity; W_L = liquid limit (%); W_p = plastic limit (%).

Table 2. The main chemical composition of cement.

Raw Material	Chemical Composition (%)						
	CaO	Fe_2O_3	SiO_2	Al_2O_3	MgO	SO_3	LOI
Cement	63.89	4.53	22.78	5.46	0.79	1.31	1.24

Table 3. Physicomechanical characteristics of polypropylene fiber.

Type	Density (g/cm^3)	Diameter (mm)	Tensile Strength (MPa)	Elastic Modulus (MPa)	Melting Point (°C)
Bundle monofilament	0.91	0.045~0.18	≥350	≥3600	165~175

2.2. Experimental Program and Procedure

In this study, the effects of the composite material composition on the shear strength and compressive strength of loess soil were investigated by conducting direct shear and unconfined compression tests on soil samples added with different amounts of improving materials, i.e., cement, polypropylene fibres and curing agent SCA-2. The contents of added materials were expressed on a dry basis of loess soil and varied from 0 to 16% for cement, from 0 to 1% for fibres and from 0 to 16% for the curing agent [24]. The content of materials is controlled by the ratio of the materials amount to the dry loess amount. The orthogonal test method was used to design the test scheme, which can discuss the influence of each test factor on the strength index under the premise of lower test quantity [34]. L_{25} (5^6) orthogonal table was adopted in this test, and there were 25 combination ratio schemes (Table 4).

Table 4. Test scheme.

Sample No.	Cement Content (%)	Fiber Content (%)	Curing Agent Content (%)	Sample No.	Cement Content (%)	Fiber Content (%)	Curing Agent Content (%)
1	0	0	0	14	8	0.75	0
2	0	0.25	4	15	8	1	4
3	0	0.5	8	16	12	0	12
4	0	0.75	12	17	12	0.25	16
5	0	1	16	18	12	0.5	0
6	4	0	4	19	12	0.75	4
7	4	0.25	8	20	12	1	8
8	4	0.5	12	21	16	0	16
9	4	0.75	16	22	16	0.25	0
10	4	1	0	23	16	0.5	4
11	8	0	8	24	16	0.75	8
12	8	0.25	12	25	16	1	12
13	8	0.5	16	—	—	—	—

During sample preparation, the cement, fiber and curing agent were mixed according to the established ratios listed in Table 4. To be more specific, the mixture was added in sequence (first cement, then fiber, then curing agent). After that, water was added into the mixed soil samples according to the optimal moisture content of loess (14%) by compaction test. Finally, different samples were prepared by sample preparation instrument. Each group sample includes two kinds of samples: four ring-knife samples with a diameter of 61.8 mm and height of 20 mm were prepared, and a cylindrical sample (i.e., approximately 50 mm in diameter and 100 mm in height) was prepared (Figure 4). The compaction coefficient K of the prepared samples was controlled to be 0.93. The prepared samples were wrapped in plastic and stored in a sealed container that prevents evaporation for about 7 days before conducting tests to obtain uniform water content within samples [22].

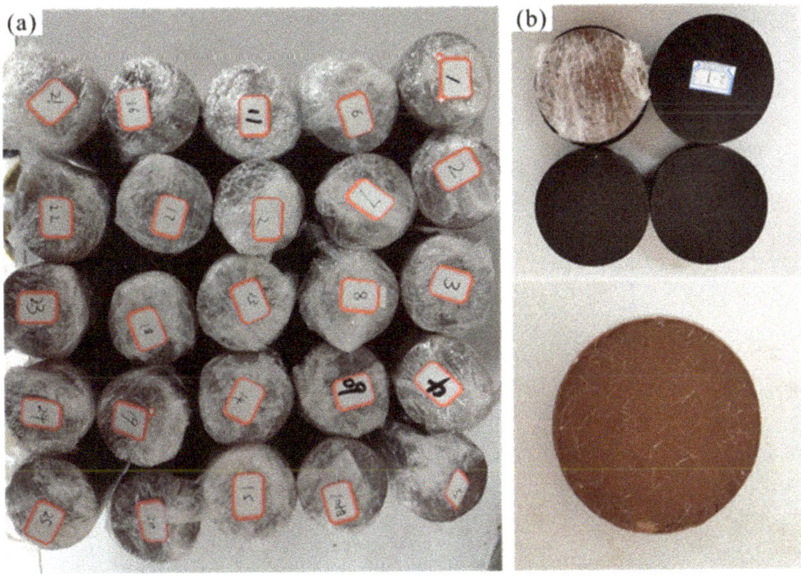

Figure 4. The improved loess samples: (**a**) cylindrical samples; (**b**) ring knife samples.

3. Results

To study the influence of the various test factors on the strength indexes of loess, the extreme difference analysis and the variance analysis methods were applied to the results of direct shear stress and unconfined compressive tests. In addition, the influences of various factors on the internal friction angle, cohesion and compressive strength of loess were also discussed.

3.1. Test Results

Due to the limited space, the test results of some samples were selected in Figure 5. The peak shear stress of improved loess in the direct shear tests was identified as the failure shear strength when the peak shear stress emerged. Conversely, without the emergence of the peak shear stress, the shear stress corresponding to 15% strain was determined to be the shear strength. Then, the internal friction angle and cohesion of modified loess can be obtained by Mohr–Coulomb theory. The axial stress–strain curve of the unconfined compressive strength test of some samples is shown in the Figure 5b; the stress–strain curve showed a strain-softening trend, so the maximum axial pressure was identified as the unconfined compressive strength of the improved loess. Additionally, if the stress–strain curve showed a strain-hardening trend, the stress was taken as the unconfined compressive strength of the sample when the axial strain was 15%.

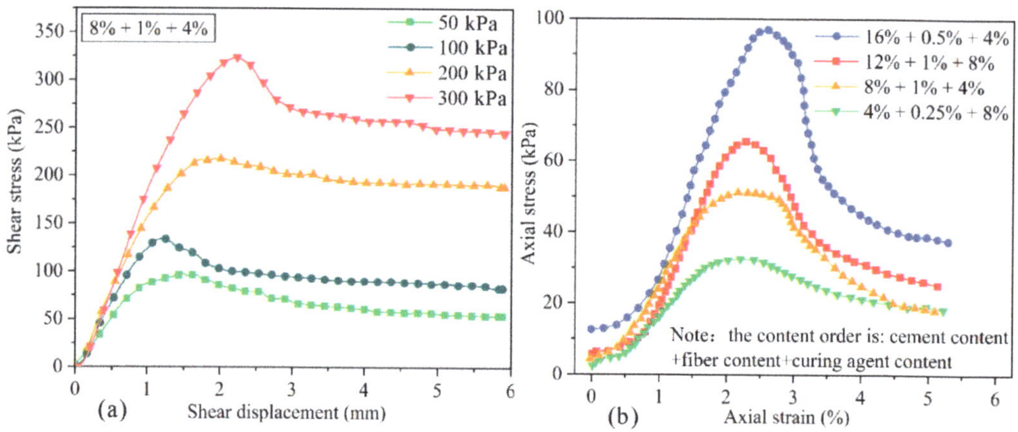

Figure 5. Test results of some samples: (**a**) shear stress–shear displacement curve of direct shear tests; (**b**) axial stress–strain curve of unconfined compressive tests.

3.2. The Extreme Difference Analysis Results

The extreme difference analysis is an important method in the analysis of orthogonal test results [35], which can represent the average value of the test index (i.e., friction angle, cohesion and compressive strength) corresponding to a single factor at a specific level. The change under the same conditions reflects the different levels of influence impacting the test indexes. By comparing the extreme difference average value R (difference between maximum and minimum average values under different contents) of test indexes, the influence degree of each factor on the test index was directly obtained. It is worth noting that the greater R value demonstrates the greater influence of this factor on the test index.

To more intuitively analyze the influence level of various factors on test indexes, the content of cement, fiber and curing agent was taken as the abscissa, and the average value of the test index corresponding to the addition of cement, fiber and curing agent with different content was set as the ordinate; the influence trend diagram of the test result index under different levels of each factor is shown in Figures 6–8.

The influence of various factors on the relevant mechanical properties of the improved loess is shown in Figures 6–8. By comparing the average value R, the level of influence of cement, polypropylene fiber and curing agent on the internal friction angle, cohesion and compressive strength was obtained as follows: cement > polypropylene fiber > SCA-2 soil curing agent. Additionally, it can be found that the cement content is positively correlated with the average value of the internal friction angle, cohesion and compressive strength of improved loess, and with the increase in curing-agent content, the average strength index of improved loess increases at first and then decreases, reaching the maximum value when the curing agent content is about 4%. However, the variations of internal friction angle, cohesion and compressive strength of loess soils with polypropylene fibers are quite different; the average value of internal friction angle increases continuously with the increase in fiber content, while the average cohesion and compressive strength increase first and then decrease with the increase in fiber content, and reach the maximum when the fiber content is approximately 0.25% and 0.5%, respectively.

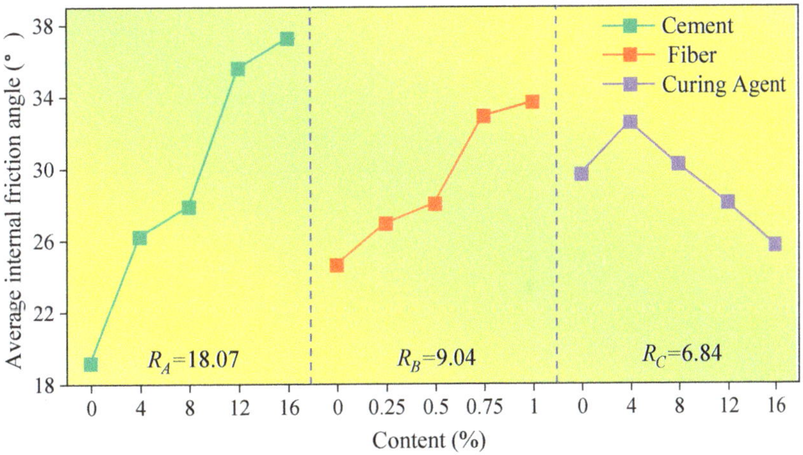

Figure 6. The influence of various factors on the internal friction angle.

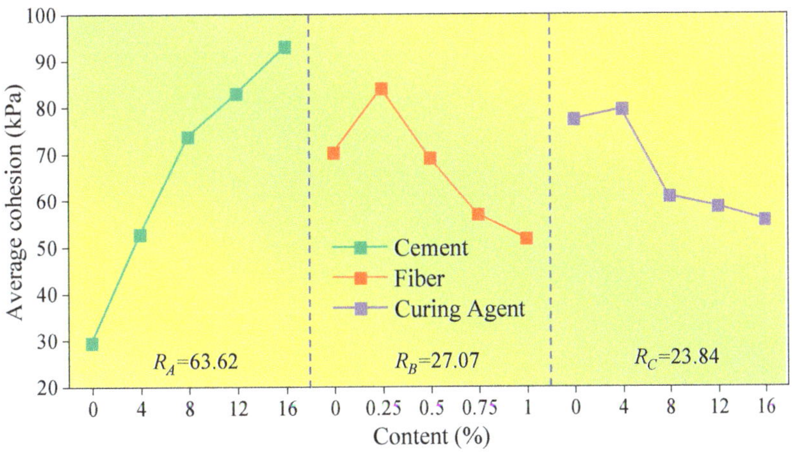

Figure 7. The influence of various factors on the cohesion.

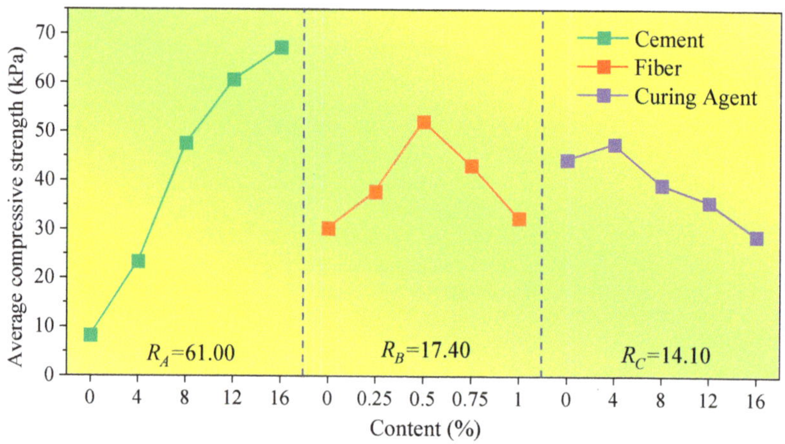

Figure 8. The influence of various factors on the compressive strength. **Notes**: R_A, R_B and R_C represent the extreme difference of each strength index under different contents of cement, fiber and curing agent, respectively.

3.3. The Variance Analysis Results

It must be considered that range analysis can directly analyze the impact of test factors on the results, but ignores the impact of error [36]. To better analyze test results, data-analysis software SPSS was employed to conduct variance analysis on the results. In this software, p (when $p < 0.05$, it means that the factor has a significant influence on the dependent variable) was analyzed. Therefore, the significance of the influence factors on the test index was investigated, which can provide reference for the study of the optimal content of each experimental factor.

The variance analysis results (Tables 5–7) show that the order of influence of cement, polypropylene fiber and curing agent on internal friction angle, cohesion and compressive strength is as follows: cement > fiber > curing agent, which is consistent with the results obtained from the extreme difference analysis (Figures 6–8). By comparing the significance p value, it can be found that cement has a great influence on the internal friction angle, cohesion and compressive strength of improved loess, while polypropylene fiber has a greater influence on the internal friction angle and compressive strength compared with cohesion. Additionally, little effect of SCA-2 curing agent on the internal friction angle, cohesion and compressive strength of improved loess was found.

Table 5. Results of variance analysis of internal friction angle.

	Class III Sum of Squares	Degrees of Freedom	Mean Square	F	p
Modified model	1492.3	12	124.4	5.0	0.005
Intercept	20,793.1	1	20,793.4	833.1	0.000
Cement	918.4	4	229.64	9.2	0.001
Fiber	490.2	4	122.6	4.9	0.014
Curing agent	83.6	4	20.9	0.8	0.527
Error	299.5	12	25.0		

Table 6. Results of variance analysis of the cohesion.

	Class III Sum of Squares	Degrees of Freedom	Mean Square	F	p
Modified model	23,171.4	12.0	1930.9	1.8	0.162
Intercept	106,268.4	1.0	106,268.4	98.7	0.000
Cement	13,730.7	4.0	3432.7	3.2	0.043
Fiber	3239.1	4.0	809.8	0.8	0.575
Curing agent	6201.6	4.0	1550.4	1.4	0.280
Error	12,918.0	12.0	1076.5		

Table 7. Results of variance analysis of the compressive strength.

	Class III Sum of Squares	Degrees of Freedom	Mean Square	F	p
Modified model	15,366.6	12.0	1280.6	10.0	0.000
Intercept	38,181.2	1.0	38,181.2	299.6	0.000
Cement	12,511.4	4.0	3127.9	24.5	0.001
Fiber	1606.6	4.0	401.7	3.2	0.048
Curing agent	1248.5	4.0	312.1	2.4	0.103
Error	1529.2	12.0	127.4		

According to the extreme difference analysis results and the variance analysis results, it is found that the maximum value of internal friction angle, cohesion and compressive strength of the improved loess was achieved by the addition of 16% cement, 4% SCA-2 soil curing agent and 1–0.25% polypropylene fiber.

4. Numerical Simulation of Settlement of High Railway Foundation

In this paper, MIDAS/GTS finite-element-analysis software was used to simulate the settlement of high railway foundation in actual engineering practice. The improvement scheme obtained from the test results was used to improve the subgrade, and the settlement of the improved subgrade under the vibration load of high-speed railway was compared, then the best improvement scheme for improving the railway foundation in the loess area was obtained and the feasibility of the loess improvement scheme was also tested.

4.1. Model Description

According to the Chinese national technical standard "Standard of Railway Subgrade Desiga" (TB10001–2016) [37], the finite element model of train dynamic subgrade is established; the section length and height of the model are 70 m and 38.3 m, respectively (Figure 9a), and the extension length of subgrade is 50 m (Figure 9b). The numerical simulation model is divided into 6920 units and the boundary is viscoelastic. The embankment thickness of the high-speed railway subgrade is 3 m, which is the main part of subgrade improvement. The way to improve the embankment was to use the construction method of layered filling; the embankment was divided into six layers, with each layer being 50 cm high, and the improved loess in the embankment was sandwiched by loess.

To simulate the vibration load caused by the train running, the moving axle load of periodic loading and unloading was applied to the high railway foundation model (Figure 10). The calculated high-speed rail speed is 300 km/h. The train time history analysis module in Midas numerical simulation software was used, the simulation results data were assessed over an interval of 0.03 s and the variation in the settlement in the duration of 3 s was analyzed when the train passed by.

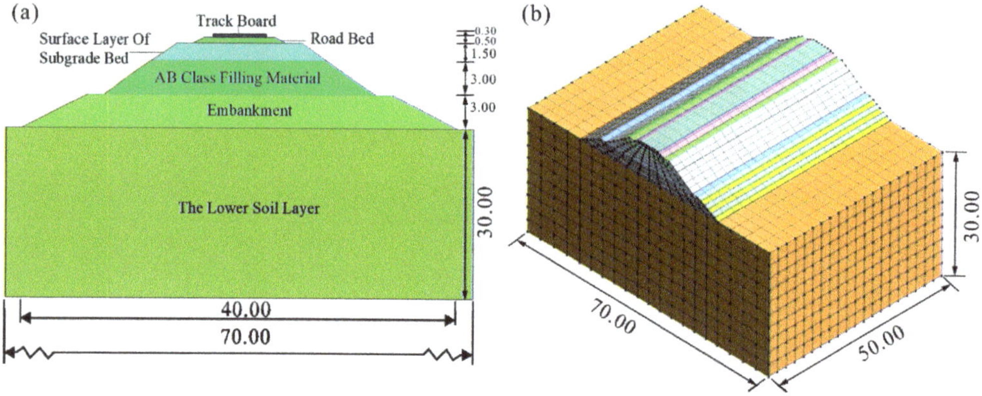

Figure 9. Model diagram of high-speed railway subgrade. (**a**) Schematic diagram of the standard section of subgrade (unit: m); (**b**) 3D model of high-speed railway subgrade (unit: m).

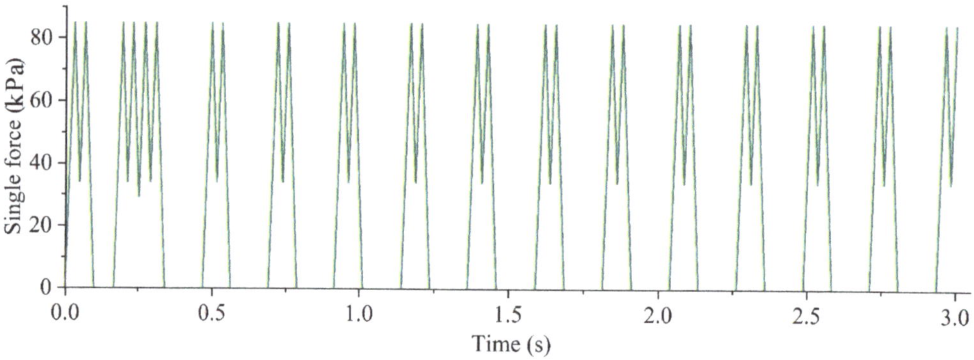

Figure 10. Time-history curve of excitation force load.

The high-speed railway subgrade is mainly divided into the following parts: track slab, road bed, surface layer of subgrade bad (graded gravel), bottom layer of subgrade bad (A B fillers), embankment (improved part) and lower soil (loess). According to the results of the orthogonal test, three improvement schemes were obtained, which gave priority to the influence of the internal friction angle (16% + 1% + 4%), cohesion (16% + 0.25% + 4%) and compressive strength (16% + 0.5% + 4%), respectively, and these schemes were denoted by schemes X, Y and Z, respectively (the content order of the improved scheme is: cement content + fiber content + curing agent content). To acquire the optimal fiber content, the modified scheme W (16% + 0.75% + 4%) was added. The relevant numerical simulation parameters of the above materials were obtained through laboratory tests and literature review [38]. The cohesion and internal friction angle of loess and improved loess were obtained by using the Coulomb–Mohr theory through direct shear tests, and the elastic modulus was obtained from the results of unconfined compressive strength tests. The calculation parameters of the finite element model are shown in Table 8.

Table 8. Calculation parameters of the finite element model.

Parameter	h	c	φ	E	μ	γ
Track board	0.3	—	—	325,00	0.18	26
Ballast bed	0.5	40	—	220	0.23	22
Surface layer of subgrade	1.5	30	20	200	0.32	22
Bottom layer of subgrade	3	42	28	150	0.32	21
Embankment (Loess)	3	26.21	13.7	21.07	0.4	17
(Scheme X)	—	117.35	35.78	64.87	0.37	18
(Scheme Y)	—	140.86	27.42	71.92	0.37	18
(Scheme Z)	—	135.66	30.43	85.37	0.37	18
(Scheme W)	—	131.17	32.36	79.46	0.37	18

Notes: h = thickness (m); c = cohesion (kPa); φ = internal friction angle (°); E = elastic modulus (kPa); μ = Poisson ratio; γ = bulk density (kN/m^3).

4.2. Numerical Simulation Results

Settlement variation of unimproved high railway subgrade is shown in Figure 11. Using the model section interception method (Figure 11b), the settlement contour map of the section with the greatest subgrade settlement was obtained (Figure 11c). As seen in Figure 11, the settlement phenomenon of the high-speed railway subgrade without improvement is obvious when a single high-speed railway train passes by, with the maximum settlement value of the subgrade reaching about 1.48 mm.

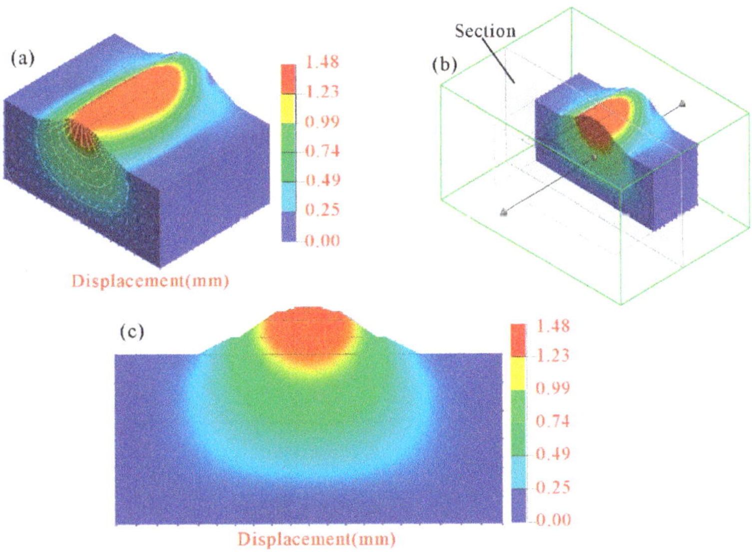

Figure 11. Settlement contour map of unimproved high-speed railway subgrade. (**a**) Three-dimensional contour map; (**b**) schematic diagram of section position; (**c**) section diagram of maximum settlement.

The changes in settlement of high-speed railway subgrade after improvement are shown in Figure 12. After the improvement of the scheme X, the settlement of high-speed railway subgrade is greatly decreased, and the maximum settlement drops to 1.09 mm, which is 0.37 mm lower than that of the unimproved high-speed railway subgrade. The maximum settlement of the high-speed railway subgrade improved by the scheme Y is 0.92 mm, which is 0.54 mm lower than that of the unimproved subgrade. The improvement effect of scheme Z is pronounced, with the maximum settlement only 0.72 mm, which is

0.74 mm less than that of the unimproved high-speed railway subgrade. The maximum settlement of the high-speed railway subgrade improved by the scheme W is 0.79 mm, which is 0.67 mm less than the unimproved subgrade. Clearly, the improvement scheme Z has the best effect on the improvement of high-speed railway subgrade, thus the best improvement scheme for improving the high-speed railway foundation with composite materials is obtained: 16% cement content + 0.5% fiber content + 4% curing agent content.

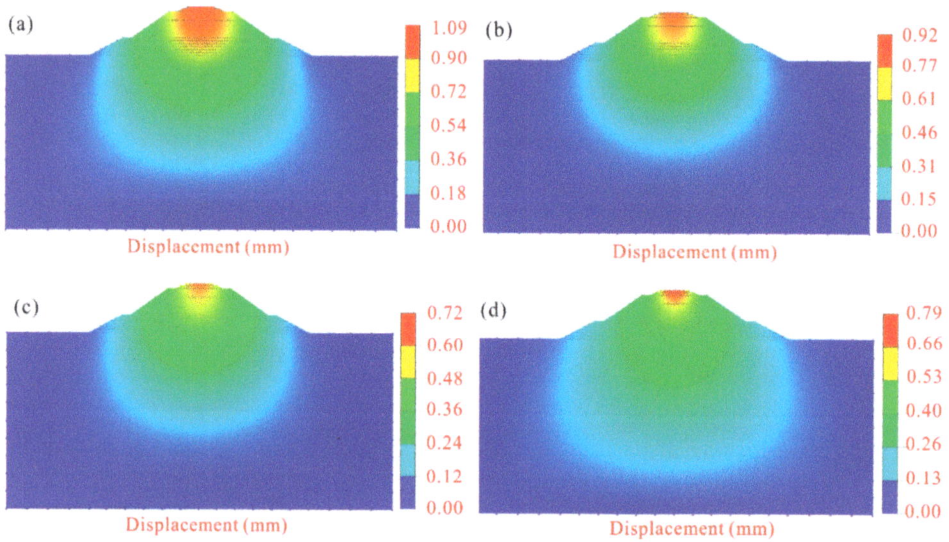

Figure 12. Settlement contour map of high-speed railway subgrade after improvement. (**a**) Scheme X; (**b**) scheme Y; (**c**) scheme Z; (**d**) scheme W.

5. Discussion

5.1. Influence Mechanism of Different Materials on the Strength Properties of Modified Loess

Through direct shear test and unconfined compressive strength test, it can be found that the improvement effect of cement, polypropylene fiber and SCA-2 soil curing agent on the internal friction angle, cohesion and compressive strength of modified loess is quite different. The reason is that the three modifying materials have different ways of interacting with loess [39], and the influence degree on the relevant strength characteristics of loess is also quite different (Figure 13).

The addition of cement will react with water to produce cement hydrate [40] that forms a cement-soil skeleton between soil particles [29], which makes the soil particles more closely connected and increases the friction resistance on the surface of the particles. With the increase in cement content, cement hydrate even generates gel to wrap soil particles to form aggregates (Figure 13a,c), which makes the soil cemented and hardened. Therefore, this can further explain the phenomenon in Figures 6–8 that the internal friction angle, cohesion and compressive strength of the modified loess increase continually with the increase in cement content.

As shown in Figure 13a,b, the random distribution of polypropylene fibers in soil can form a three-dimensional reinforcement network that is similar to the antislide pile, which can increase the interface friction of soil and restrict the relative movement of soil particles [39,41]. Therefore, the internal friction angle of the modified loess can be greatly increased, and this can further explain the phenomenon shown in Figure 6 that the fiber content and the internal friction angle is positively correlated. However, for the cohesion of modified loess, the addition of a small amount of polypropylene fiber causes friction and interlocking between fibers and between fibers and soil particles [42]. An increase in the fiber content will cause obvious cracks and pores in the soil [20], resulting in a negative effect. Consequently, as seen in Figure 7, when the fiber content is 0.25%, the improved loess has the greatest cohesion. For the compressive strength of improved loess, cement enhances the compressive capacity of loess, and thus the improved loess is prone to brittle fracture. The addition of an appropriate amount of fiber enhances the ductility of the soil [31], and this can further explain the phenomenon illustrated in Figure 8 that the compressive strength of the modified loess is the maximum when the fiber content is 0.5%.

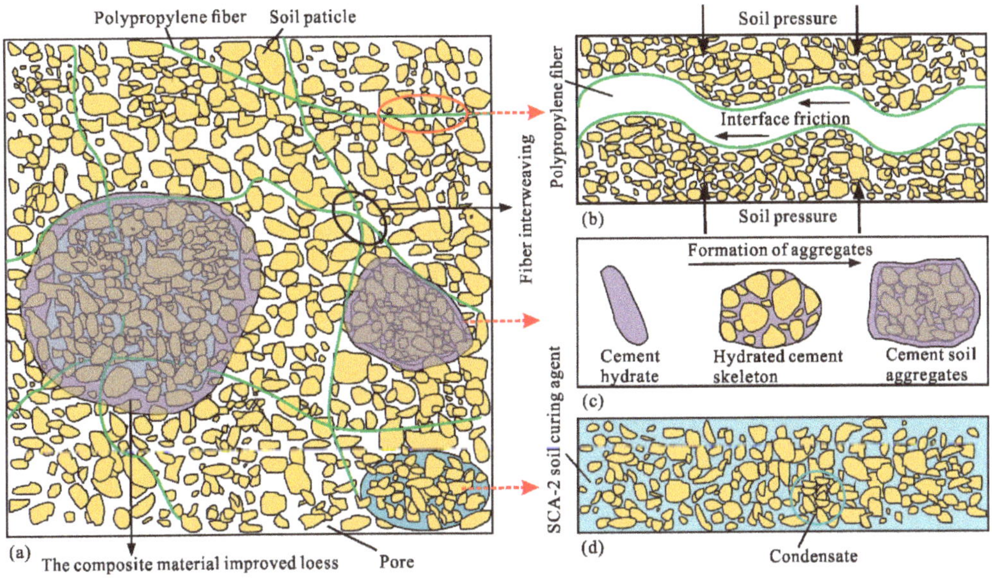

Figure 13. Microscopic schematic diagram of the mechanism of composite materials improving loess (modified from Yan et al. 2021). (**a**) General schematic diagram; (**b**) polypropylene fiber; (**c**) cement; (**d**) SCA-2 soil curing agent.

SCA-2 soil curing agent is mainly composed of sodium silicate and polyacrylamide, can chemically react with cement to produce gel, which can wrap soil particles, reduce pores in soil, and thus increase the cohesion and compressive strength of modified loess [43]. However, when the content of curing agent exceeds 4%, the soil produces a saturation reaction to it (Figures 6–8), and the excess liquid curing agent fills the pores of the soil and acts as a lubricant, which reduces the strength index of the soil (Figure 13a,d), and this can further explain the phenomenon shown in Figures 6–8 that the internal friction angle, cohesion and compressive strength of the modified loess increased firstly with the curing agent content and then decreased from the maximum strength at the content of 4%.

5.2. Influence of Loess Strength Characteristics on Subgrade Settlement

The influence of the strength characteristics of modified loess on subgrade settlement is complicated. From the experimental results and numerical simulation results, it can be found that the improvement scheme can effectively reduce the settlement of high-speed railway subgrade. Through analyzing the settlement change curve of subgrade after the train passes by (Figure 14), three stages of subgrade settlement were identified: the instantaneous settlement stage, in which the settlement of high railway subgrade under train dynamic load reaches the maximum value in a very short time; the settlement attenuation stage, due to the influence of the change of train dynamic load; and the settlement stable stage, in which the settlement attenuation decreased to a certain value [44].

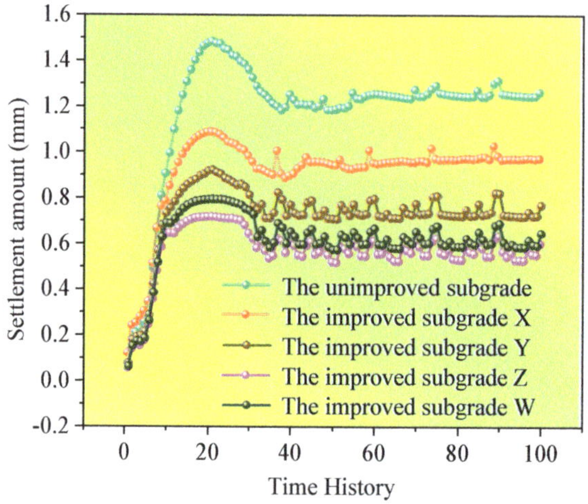

Figure 14. Time-history curve of subgrade settlement.

The subgrade settlement and its variation under different improvement schemes were compared (Figure 15), and the results show that the compressive strength of improved loess has the most influence on the settlement of high-speed railway subgrade, followed by cohesion, and the influence of the internal friction angle is relatively little, indicating that the settlement of subgrade is more affected by the compressive capacity of soil compared with the shear strength of loess. Therefore, when considering the improvement of high railway foundation filler in loess area, the compressive strength of modified loess should be given priority under the premise of comprehensively considering the improvement of loess strength indexes.

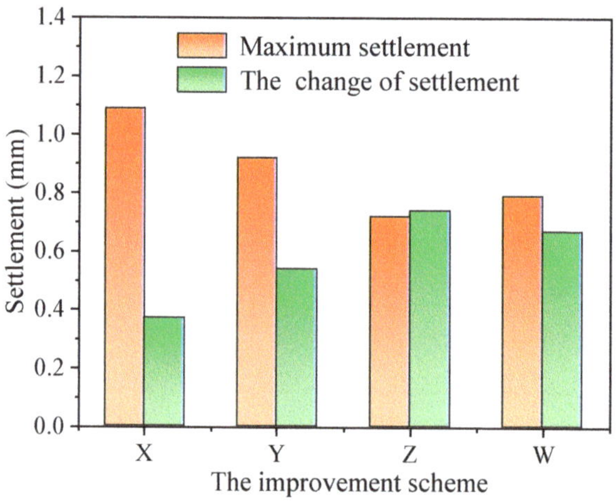

Figure 15. Changes in the maximum settlement of each improvement scheme.

Using the direct shear test, unconfined compressive test and numerical simulation analysis, the influence of single train passing on subgrade settlement was investigated. Considering that long-term settlement of loess subgrade was frequently found, which is attributed to the influence of long-term vibration load, the influence of long-term train vibration load on loess subgrade settlement would be studied in future research.

6. Conclusions

In this paper, a series of direct shear tests and unconfined compressive tests were conducted on the loess specimens improved by the composite material consisting of cement, polypropylene fiber and SCA-2 soil curing agent. The influence of the improvement scheme obtained from the test on the subgrade settlement was studied by numerical simulation. The following conclusions can be drawn:

(1) Cement, polypropylene fiber and SCA-2 soil curing agent can effectively improve the shear strength and compressive strength of loess, and the degree of influence of cement, polypropylene fiber and curing agent on the internal friction angle, cohesion and compressive strength is cement > fiber > curing agent.

(2) Based on the orthogonal test results and MIDAS numerical simulation method, the optimal improvement scheme for composite materials to improve loess was obtained: 16% cement content + 0.5% fiber content + 4% curing agent content.

(3) Through the numerical simulation calculation method, it is found that the compressive strength index of the improved loess has a significant impact on the subgrade settlement, and the optimal improvement scheme obtained can effectively decrease the settlement of high-speed railway subgrade under vibration load, and can effectively avoid safety accidents caused by cumulative subgrade settlement.

Author Contributions: Writing—original draft preparation, L.L.; conceptualization, X.W.; writing—review and editing, D.W. and B.L.; visualization, C.X. All authors have read and agreed to the published version of the manuscript.

Funding: This work was supported by The National Natural Science Foundation of China (No.41902268), Nature Science Basic Research Plan in Shaanxi Province of China (No. 2022JQ-253) and China Postdoctoral Science Foundation (No. 2021M702648) are gratefully acknowledged.

Institutional Review Board Statement: Not applicable.

Informed Consent Statement: Not applicable.

Data Availability Statement: The data presented in this study are available upon request from the corresponding author.

Conflicts of Interest: The authors declare no conflict of interest.

References

1. Radampola, S.S.; Gurung, N.; Mcsweeney, T.; Dhanasekar, M. Evaluation of the properties of railway capping layer soil. *Comput. Geotech.* **2008**, *35*, 719–728. [CrossRef]
2. Liu, J.; Xiao, J. Experimental study on the stability of railroad silt subgrade with increasing train speed. *J. Geotech. Geoenviron. Eng.* **2010**, *136*, 833–841. [CrossRef]
3. Lazorenko, G.; Kasprzhitskii, A.; Khakiev, Z.; Yavna, V. Dynamic behavior and stability of soil foundation in heavy haul railway tracks: A review. *Constr. Build. Mater.* **2019**, *205*, 111–136. [CrossRef]
4. Ma, X.; Zhang, Z.; Zhang, P.; Wang, X. Long-term dynamic stability of improved loess subgrade for high-speed railways. *P. I. Civil. Eng-Geotec.* **2019**, *173*, 217–227. [CrossRef]
5. Xu, Z.; Lin, Z.; Zhang, M. Loess in China and loess landslides. *Chin. J. Rock Mech. Eng.* **2007**, *26*, 1297–1312.
6. Li, Y. A review of shear and tensile strengths of the Malan Loess in China. *Eng. Geol.* **2018**, *236*, 4–10. [CrossRef]
7. Assallay, A.M.; Rogers, C.D.F.; Smalley, I.J. Formation and collapse of metastable particle packings and open structures in loess deposits. *Eng. Geol.* **1997**, *48*, 101–115. [CrossRef]
8. Peng, J.; Lin, H.; Wang, Q.; Zhuang, J.; Cheng, Y.; Zhu, X. The critical issues and creative concepts in mitigation research of loess geological hazards. *Eng. Geol.* **2014**, *22*, 684–691.
9. Juang, C.H.; Dijkstra, T.; Wasowski, J.; Meng, X. Loess geohazards research in China: Advances and challenges for mega engineering projects. *Eng. Geol.* **2019**, *251*, 1–10. [CrossRef]
10. Wu, Z.; Zhang, D.; Wang, S.; Liang, C.; Zhao, D. Dynamic-response characteristics and deformation evolution of loess slopes under seismic loads. *Eng. Geol.* **2020**, *267*, 105507. [CrossRef]
11. Li, G.; Ma, W.; Mu, Y.; Wang, F.; Fan, S.; Wu, Y. Effects of freeze-thaw cycle on engineering properties of loess used as road fills in seasonally frozen ground regions, North China. *J. Mt. Sci.* **2017**, *14*, 356–368. [CrossRef]
12. Zhang, J.; Li, J.; Wang, J.; Xu, S. Characteristics of the Interface between Bamboo Grids and Reinforced Soil of High-Filled Embankments in Loess Areas. *Adv. Civ. Eng.* **2021**, *2021*, 5135756. [CrossRef]

13. Mariri, M.; Ziaie Moayed, R.; Kordnaeij, A. Stress–strain behavior of loess soil stabilized with cement, zeolite, and recycled polyester fiber. *J. Mater. Civ. Eng.* **2019**, *31*, 04019291. [CrossRef]
14. Chen, R.; Cai, G.; Dong, X.; Duan, W. Mechanical properties and micro-mechanism of loess roadbed filling using by-product red mud as a partial alternative. *Constr. Build. Mater.* **2019**, *216*, 188–201. [CrossRef]
15. Salehi, M.; Bayat, M.; Saadat, M.; Nasri, M. Experimental study on mechanical properties of cement-stabilized soil blended with crushed stone waste. *KSCE J. Civ. Eng.* **2021**, *25*, 1974–1984. [CrossRef]
16. Wang, F.; Pang, W.; Qin, X.; Han, L.; Jiang, Y. Durability-Aimed Design Criteria of Cement-Stabilized Loess Subgrade for Railway. *Appl. Sci.* **2021**, *11*, 5061. [CrossRef]
17. Gu, K.; Chen, B. Loess stabilization using cement, waste phosphogypsum, fly ash and quicklime for self-compacting rammed earth construction. *Constr. Build. Mater.* **2020**, *231*, 117195. [CrossRef]
18. Wang, Z.; Zhang, W.; Jiang, P.; Li, C. The Elastic Modulus and Damage Stress–Strain Model of Polypropylene Fiber and Nano Clay Modified Lime Treated Soil under Axial Load. *Polymers.* **2022**, *14*, 2606. [CrossRef]
19. Tang, C.; Li, H.; Pan, X.; Yin, L.; Cheng, L.; Cheng, Q. Coupling effect of biocementation-fiber reinforcement on mechanical behavior of calcareous sand for ocean engineering. *Bull. Eng. Geol. Environ.* **2022**, *81*, 1–15. [CrossRef]
20. Wang, X.; Liu, K.; Lian, B. Experimental study on ring shear properties of fiber-reinforced loess. *Bull. Eng. Geol. Environ.* **2021**, *80*, 5021–5029. [CrossRef]
21. Zhang, C.; Jiang, G.; Su, L.; Zhou, G. Effect of cement on the stabilization of loess. *J. Mt. Sci.* **2017**, *14*, 2325–2336. [CrossRef]
22. Jiang, Y.; Li, Q.; Yi, Y.; Yuan, K.; Deng, C.; Tian, T. Cement-modified loess base for intercity railways: Mechanical strength and influencing factors based on the vertical vibration compaction method. *Materials.* **2020**, *13*, 3643. [CrossRef]
23. Tabarsa, A.; Latifi, N.; Meehan, C.L.; Manahiloh, K.N. Laboratory investigation and field evaluation of loess improvement using nanoclay–A sustainable material for construction. *Constr. Build. Mater.* **2018**, *158*, 454–463. [CrossRef]
24. Puppala, A.J.; Musenda, C. Effects of fiber reinforcement on strength and volume change in expansive soils. *Transp. Res. Rec.* **2000**, *1736*, 134–140. [CrossRef]
25. Gu, L.; Lv, Q.; Wang, S.; Xiang, J.; Guo, L.; Jiang, J. Effect of sodium silicate on the properties of loess stabilized with alkali-activated fly ash-based. *Constr. Build. Mater.* **2021**, *280*, 122515. [CrossRef]
26. Wang, Q.; Zhong, X.; Ma, H.; Wang, S.; Liu, Z.; Guo, P. Microstructure and reinforcement mechanism of lignin-modified loess. *J. Mater. Civ. Eng.* **2020**, *32*, 04020319. [CrossRef]
27. Zhang, Y.; Ye, W.; Wang, Z. Study on the compaction effect factors of lime-treated loess highway embankments. *Civ. Eng. J.* **2017**, *3*, 1008–1019. [CrossRef]
28. Wei, L.; Chai, S.; Zhang, H.; Shi, Q. Mechanical properties of soil reinforced with both lime and four kinds of fiber. *Constr. Build. Mater.* **2018**, *172*, 300–308. [CrossRef]
29. Liu, H.; Wang, L.; Gao, P. The mechanical properties of cement reinforced loess and pore microstructure characteristics. Applied Mechanics and Materials. *Trans. Tech. Publications. Ltd.* **2014**, *527*, 25–30.
30. Mahedi, M.; Cetin, B.; Cetin, K.S. Freeze-thaw performance of phase change material (PCM) incorporated pavement subgrade soil. *Constr. Build. Mater.* **2019**, *202*, 449–464. [CrossRef]
31. Yang, B.; Weng, X.; Liu, J.; Kou, Y.; Jiang, L.; Li, H.; Yan, X. Strength characteristics of modified polypropylene fiber and cement-reinforced loess. *J. Cent. South Univ.* **2017**, *24*, 560–568. [CrossRef]
32. Chen, C.; Wei, K.; Gu, J.; Huang, X.; Dai, X.; Liu, Q. Combined Effect of Biopolymer and Fiber Inclusions on Unconfined Compressive Strength of Soft Soil. *Polymers.* **2022**, *14*, 787. [CrossRef] [PubMed]
33. *GB/T 50123—1999*; Standard for soil test method. The State Bureau of Quality and Technical Supervision; Ministry of Construction: Beijing, China, 1999.
34. Ren, W.; Yang, T.; Huang, M.; Zhang, A.; Wei, H.; Mi, W.; Wang, Y.; Hu, J. Optimal mixing ratio and SWCC fitting of lightweight soil with cotton stalk fibres. *Soils. Found.* **2021**, *61*, 453–464. [CrossRef]
35. Wang, L. Orthogonal Test and Multi-Element Linear Regression Analysis of Compressive Strength of Nanometer Silicon Cemented Soil. *Adv. Mater. Res.* **2011**, *317*, 34–41. [CrossRef]
36. Sun, Y.; Xu, Z.; Li, A.; Wang, C. Analysis of influence factors on end soil reinforcement effect of rectangular pipe jacking. *P. I. Civil. Eng-Geotec.* **2021**, 1–10. [CrossRef]
37. *TB10001—2016;* China Railway Press. Standard of Railway Subgrade Design. State Railway Administration: Beijing, China, 2016.
38. Tai, B.; Yue, Z.; Sun, T.; Qi, S.; Li, L.; Yang, Z. Novel anti-frost subgrade bed structures a high speed railways in deep seasonally frozen ground regions: Experimental and numerical studies. *Constr. Build. Mater.* **2020**, *269*, 121266. [CrossRef]
39. Xu, J.; Wu, Z.; Chen, H.; Shao, L.; Zhou, X.; Wang, S. Study on strength behavior of basalt fiber-reinforced loess by digital image technology (DIT) and scanning electron microscope (SEM). *Arabian J. Sci. Eng.* **2021**, *46*, 11319–11338. [CrossRef]
40. Wahab, N.A.; Roshan, M.J.; Rashid, A.S.A.; Hezmi, M.A.; Jusoh, S.N.; Nik, N.N.D.; Tamassoki, S. Strength and Durability of Cement-Treated Lateritic Soil. *Sustainability.* **2021**, *13*, 6430. [CrossRef]
41. Ruan, B.; Ding, H.; Teng, J.; Deng, W.; Zheng, S.; Ruan, C. Mechanical properties and microstructure of polypropylene fiber reinforced cement mortar soil. *KSCE J. Civ. Eng.* **2021**, *25*, 1985–1997. [CrossRef]
42. Yan, C.; An, N.; Wang, Y.; Sun, W. Effect of dry-wet cycles and freeze-thaw cycles on the antierosion ability of fiber-reinforced loess. *Adv. Mater. Sci. Eng.* **2021**, *2021*, 8834598. [CrossRef]

43. Zhang, B.; Jiang, W.; Xu, Q.; Yuan, D.; Shan, J.; Lu, R. Experimental feasibility study of ethylene-vinyl acetate copolymer (EVA) as cement stabilized soil curing agent. *Road Mater. Pavement Des.* **2022**, *23*, 617–638. [CrossRef]
44. Ma, X.; Wang, X. Comparative study on settlement prediction of railway subgrade on collapsible loess area. *Appl. Mech. Mater.* **2012**, *188*, 49–53.

polymers

MDPI

Article

Mechanical Characteristics of Lime-Treated Subgrade Soil Improved by Polypropylene Fiber and Class F Fly Ash

Wei Wang [1,2], Beifeng Lv [1], Chen Zhang [3,*], Na Li [1] and Shaoyun Pu [1]

1 School of Civil Engineering, Shaoxing University, Shaoxing 312000, China; wellswang@usx.edu.cn (W.W.); 20020852047@usx.edu.cn (B.L.); lina@usx.edu.cn (N.L.); pushaoyun@seu.edu.cn (S.P.)
2 Department of Civil & Environmental Engineering, National University of Singapore, Singapore 117576, Singapore
3 School of Transportation, Southeast University, Nanjing 211189, China
* Correspondence: zhangchenedu@163.com

Abstract: To improve the limitations of lime-treated subgrade soil (LS), a series of unconsolidated and undrained triaxial tests were conducted to investigate the improvement effect of fiber modified lime-treated soil (PLS) and fly ash modified lime-treated soil (FLS). The test results showed that (1) The deviatoric stress-strain curves of LS, PLS, and FLS were basically of the softening type. (2) The addition of fiber and fly ash improved the ductility and stiffness of LS. The ductility of PLS increased by 134% compared with LS, while the mechanical strength of FLS increased by 53%. (3) The microscopic tests showed that a denser skeleton structure was generated inside LS with the addition of fiber and fly ash. (4) The deviatoric stress-strain curves of LS, PLS, and FLS under different confining pressures were better characterized with the CES curve model. The above results indicate that fiber and fly ash can effectively improve the mechanical characteristics of lime-treated subgrade soil.

Keywords: polypropylene fiber; fly ash; lime-treated subgrade soil; mechanical characteristics; microscopic test; curve model

Citation: Wang, W.; Lv, B.; Zhang, C.; Li, N.; Pu, S. Mechanical Characteristics of Lime-Treated Subgrade Soil Improved by Polypropylene Fiber and Class F Fly Ash. *Polymers* **2022**, *14*, 2921. https://doi.org/10.3390/polym14142921

Academic Editor: Wensheng Wang

Received: 22 June 2022
Accepted: 12 July 2022
Published: 19 July 2022

1. Introduction

In the subgrade project, the uneven settlement of subgrade often occurs under the influence of a soft foundation, leading to significant cracks in the pavement and affecting its comfort and stability [1,2]. Therefore, it is the focus of research to improve the soft soil subgrade with low strength and high compressibility in engineering construction [3–5]. Currently, a series of solidifying materials (e.g., cement, lime, and fly ash) are usually used to improve the poor mechanical characteristics of soft soils, and finally realize the stability and safety of the subgrade project [6–8].

Because of its advantages of high compressive strength, strong water, ability, and low cost, lime-treated soil was commonly used in subgrade construction [9]. However, some research showed that the application of lime-treated soil caused tensile failure and serious deformation problems in some projects [10–12]. Hence, a great deal of materials were used by researchers to improve the limitations of lime-treated soil, and found that fibers and fly ash were better modification materials [13–17]. For example, some studies showed that fibers had a good effect on the tensile strength, ductility, crack resistance, and brittle failure of lime-treated soils to some extent, while that of fly ash could enhance its shear strength, bearing capacity, and deformation resistance [18,19]. Rudramurthy et al. [20] studied the effects of different fiber contents on the mechanical characteristics of lime-treated clay by a series of unconfined compressive strength tests, and found that the ductility and brittle failure of soil samples were improved by adding 1% fibers. Dhar et al. [21] proposed that the mechanical characteristics of lime-treated clay were improved after adding fiber. Turan et al. [22] used unconfined compression tests to investigate the improvement effect of fly ash on the lime-treated clay, and their results indicated that the compressive strength, brittleness index, and secant modulus of soil samples were enhanced. Li et al. [23] studied lime-treated clay modified with fiber and fly ash with an unconfined compression test, and suggested that the addition of fly ash and fiber increased the stiffness and ductility of soil samples. Ghosh et al. [24] investigated the modification effect of fly ash on lime soil under the conditions of soaking and non-soaking, and pointed out that fly ash could strengthen the hydration reaction of lime and enhance the mechanical strength of soil samples. The above-mentioned literature reviews show that fibers and fly ash can be used as a modifying material to improve the limitations of lime-treated soils. Moreover, the influence of confining pressure is rarely considered in the current research. In practical engineering, the soil reinforcement layer is usually buried underground, thus the influence of confining pressure on soil mechanical characteristics cannot be ignored [25].

In summary, a series of unconsolidated and undrained triaxial (UU) tests and SEM tests were carried out to investigate the mechanical characteristics and micro-structure of modified soils. Meanwhile, a CSE curve model is proposed to analyze the stress and strain characteristics of modified soils, providing help for the application of FLS and PLS in the subgrade engineering construction, design, and numerical simulation.

2. Experimental Scheme

2.1. Test Materials and Scheme

The subgrade soil was taken from a construction site in Shaoxing City, Zhejiang Province, China. Its physical property indexes shown in Table 1, which is from the research of Wang et al. [25].

Table 1. Physical property indexes of subgrade soil [25].

Density ($g \cdot cm^{-3}$)	Pore Ratio	Water Content (%)	Liquid Limit (%)	Plastic Limit (%)	Liquidity Index	Plastic Index
1.65	1.64	33.0	46.2	26.4	1.7	19.8

The length of polypropylene (PP) fiber used in the test was 6 mm obtained from Shaoxing City, Zhejiang Province, China. Its appearance is shown in Figure 1. In addition, the main technical indexes shown in Table 2, which is from the research of Wang et al. [25].

The lime was produced in Xinyu City, Jiangxi Province, China. The main component contents of lime were 89.4% CaO, 1.8% MgO and 8.8% other components by oxide composition analysis [13].

Figure 1. Polypropylene fibers.

Table 2. Main technical indexes of polypropylene fiber [25].

Fiber Type	Diameter (μm)	Length (mm)	Tensile Strength (MPa)	Elasticity Modulus (GPa)	Stretch Limits (%)
Bunchy monofilament	18–48	6	>358	>3.50	>15

The fly ash was produced in Shaoxing City, Zhejiang Province, China. The main component contents of fly ash were 8.9% CaO, 25.3% Al_2O_3, 12.4% Fe_2O_3, 35.6% SiO_2 and 17.8% other components by oxide composition analysis. Due to the content of CaO in fly ash being 8.9%, lower than 10%, it thus belonged to class F fly ash as per the Standard Specification for Coal Fly Ash (ASTM C618, 2019) [26].

The UU test instrument used in the test was the TKA-TTS-3S, produced by Nanjing TKA Technology Co., Ltd. [25].

Table 3 shows the mass dosing scheme of different modified samples. According to previous works [13], the optimum values of lime content and water content were used. Meanwhile, the 1% PP fiber content and 12% fly ash content were determined as per the research results of Wang et al. [25] and Zhou et al. [27].

Table 3. Mass dosing scheme of different samples.

Group	Water Content (%)	Lime Content (%)	Fiber Content (%)	Fly Ash Content (%)
	$\frac{m_{water}}{m_{flyash}+m_{lime}+m_{drysoil}} \times 100\%$	$\frac{m_{lime}}{m_{drysoil}} \times 100\%$	$\frac{m_{fiber}}{m_{lime}+m_{drysoil}} \times 100\%$	$\frac{m_{flyash}}{m_{lime}+m_{drysoil}} \times 100\%$
LS	17.5	6	0	0
PLS	17.5	6	1	0
FLS	17.5	6	0	12

2.2. Sample Preparation

According to the Chinese National Geotechnical Test Standard (GB/T 50123 1999) [28] and the test mix proportion designed in Table 3, the sample preparation steps are divided into the following steps:

(1) Place the subgrade soil in an oven with the constant temperature for 24 h, set the temperature to 105 °C, and then fully crush the subgrade soil.

(2) The fully crushed subgrade soil is sieved with a 2 mm standard sieve in order to remove soil particles and impurities with a particle size greater than 2 mm in the subgrade soil.

(3) According to the mix proportion designed in the test scheme, weigh a certain quality of subgrade soil, lime, PP fiber, fly ash, and water, and let stand for 24 h after mixing evenly. When the lime-treated soil is initially hydrated, the triaxial sample is made with a three-valve saturator. The height of the sample is 80 mm and its diameter is 39.1 mm.

(4) After the samples are prepared, they are put in a standard curing box for curing for seven days. The curing temperature and humidity are 20 °C and 95%, respectively.

3. Test Results and Discussion

3.1. Mechanical Characteristics

3.1.1. Deviatoric Stress and Strain Characteristics

Through a series of triaxial UU tests, the deviatoric stress-strain curves of modified soils are obtained and shown in Figure 2. It can be seen from Figure 2 that the stress-strain curves of LS, PLS, and FLS samples are basically softening types. According to the ASTM D2850-15 [29], the axial strain (ε = 15%) is considered to be the maximum strain level. In addition, the peak stress (q_p), peak strain (ε_p) and residual stress (q_r) of the modified soils obtained from the deviatoric stress-strain curves are summarized in Table 4. Among them, the peak stress and residual stress reflect the resistance to shear damage of soil samples and its residual strength after damage.

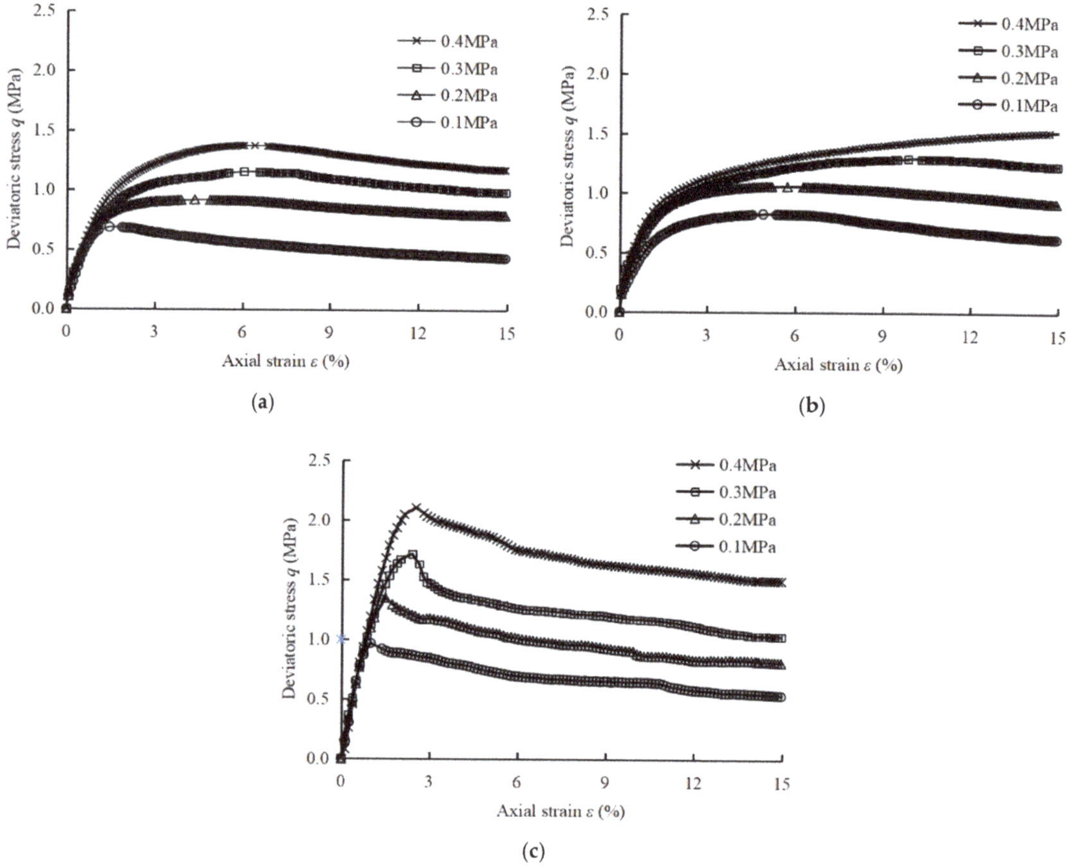

Figure 2. Stress-strain curves. (**a**) LS sample; (**b**) PLS sample; (**c**) FLS sample.

Table 4. Mechanical parameters of modified samples.

Soil Samples	Confining Pressure (MPa)	Peak Stress q_p (MPa)	Peak Strain ε_p (%)	Residual Stress q_r (MPa)
LS	0.1	0.7	2.5	0.4
	0.2	0.9	4.4	0.7
	0.3	1.2	6.0	1.0
	0.4	1.4	6.4	1.2
PLS	0.1	0.8	4.9	0.6
	0.2	1.1	5.7	0.9
	0.3	1.3	9.8	1.2
	0.4	1.5	15.0	1.5
FLS	0.1	1.0	1.0	0.5
	0.2	1.4	1.5	0.8
	0.3	1.7	2.4	1.1
	0.4	2.1	2.5	1.5

From Table 4, it can be seen that when the confining pressure is 0.4 MPa, compared with 0.1 MPa confining pressure, the peak stress of LS, PLS, and FLS increase by 98, 85 and 117%, the strain at peak stress increase by 156%, 206%, and 144%, and the residual stress increase by 168, 145 and 177%, respectively. Compared with LS, the peak stress of PLS and FLS increase by 10–19% and 40–53%, the strain at peak stress increase by 31–134% and 58–65%, and the residual stress increase by 26–42% and 12–27%, respectively. The above results indicate that with the increase of confining pressure, fiber perform a good effect on improving the ductility and secondary damage resistance of LS, while fly ash is mainly used to enhance its bearing capacity and strength.

3.1.2. Shear Strength Curve

When the normal stress and shear stress are the abscissa and ordinate, the Mohr's circle of modified soils are drawn in the τ–σ stress plan with $(\sigma_1 + \sigma_3)/2$ and $(\sigma_1 - \sigma_3)/2$ as the center and radius [25] and shown in Figure 3. Meanwhile, the shear strength parameters c and ϕ of soil samples are obtained and listed in Table 5.

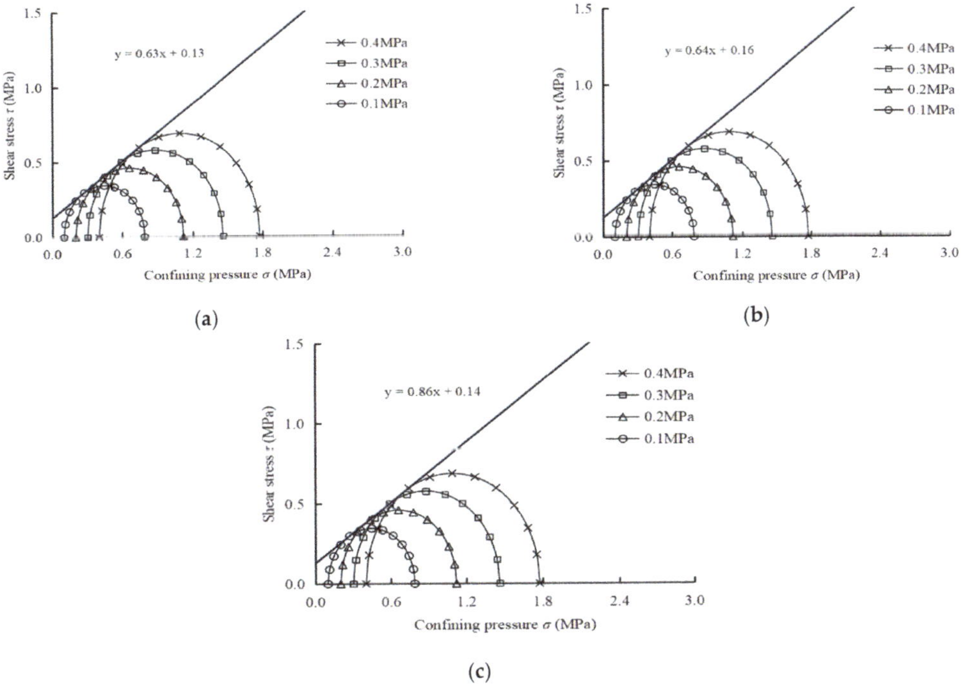

Figure 3. Mohr's circle. (**a**) LS sample; (**b**) PLS sample; (**c**) FLS sample.

Table 5. Strength parameters.

Group	Strength Equation	c (MPa)	ϕ (°)
LS	$\tau = 0.63\sigma + 0.13$	0.13	32.3
PLS	$\tau = 0.64\sigma + 0.16$	0.16	32.6
FLS	$\tau = 0.86\sigma + 0.14$	0.14	40.8

As shown in Table 5 and Figure 3, the cohesion c of LS, PLS and FLS samples are 0.13 MPa, 0.16 MPa and 0.14 MPa, the internal friction angles ϕ are 32.3, 32.6 and 40.8°, respectively. Compared with LS samples, the c value of PLS and FLS samples increased by 23.1 and 7.7%, and the ϕ value of FLS samples increased by 26.3%. It can be found that the addition of fiber and fly ash improved the shear strength of LS to a certain extent. The effect of fiber and fly ash are realized by increasing the cohesion and internal friction angle of soil sample. The test results are similar to the existing literature [30,31], in which the mechanical characteristics and micro-structure of fiber and fly ash modified lime soft soil had been studied. The main reason was that lime generated a large amount of gelling matrix during the hydration process, which combine with fiber and soil particles to form more compact particle gels, thus improving the ductility and cohesion of LS. Meanwhile, the addition of fly ash promotes the hydration reaction of lime and fill its internal pores, thus increasing the bearing capacity and internal friction angle of sample. As shown in Tables 4 and 5, with the addition of fiber and fly ash, the bearing capacity, ductility, and shear strength of LS is improved to some extent.

3.2. Failure Characteristics

3.2.1. Stress Softening Coefficient

The peak stress q_p and residual stress q_r are the traditional characteristic points of stress-strain curve. For further analyzing the softening characteristics of samples, the stress relative softening coefficient k is defined as follows [32]:

$$k = \frac{q_p - q_r}{q_p} \times 100\% \tag{1}$$

where that the smaller k is, the closer the q_p and q_r values of the stress-strain curve are, the less obvious the softening characteristics are, and the more difficult the sample is to break; when $k = 0$, the curve shows hardening curve. By calculating the data in Table 6 using Equation (1), the softening coefficients of LS, PLS and FLS samples can be obtained and are shown in Table 6.

Table 6. Failure characteristics of modified samples.

Soil Samples	Confining Pressure (MPa)	Softening Coefficient k (%)	Brittleness Index I (MPa)	Secant Modulus E_{50} (MPa)
LS	0.1	36.7	0.6	0.5
	0.4	14.5	0.2	0.4
PLS	0.1	24.4	0.3	0.3
	0.4	0.0	0.0	0.2
FLS	0.1	44.3	0.8	1.3
	0.4	28.9	0.4	1.2

Comparing the data in Table 6, it can be found that the softening coefficients of LS, PLS, and FLS samples under the confining pressure of 0.4 MPa are reduced by 60, 100, and 35%, respectively, compared with that of 0.1 MPa. Compared with LS samples, the softening coefficient of PLS sample is reduced by 100%. In Figure 4b, it can also be found that the deviatoric stress-strain curve of the PLS sample shows a hardening type under high confining pressure. On the contrary, the softening coefficient of FLS samples increased by 130%. The results showed that the addition of fiber could significantly improve the softening characteristics of LS samples.

(a) (b) (c)

Figure 4. SEM images of modified soil samples. (**a**) LS sample; (**b**) PLS sample; (**c**) FLS sample.

3.2.2. Brittleness Index

To further investigate the brittle failure characteristics of soil samples in the shear failure process, Consoli et al. [33] proposed an evaluation index called brittle index I, and its calculation formula is shown in Equation (2):

$$I = qp/qr - 1 \quad (2)$$

where the greater the I value, the more obvious the brittle failure of samples. The brittleness index of samples calculated by Equation (2) are shown in Table 6. When the confining pressure is 0.4 MPa, compared with 0.1 MPa confining pressure, the brittleness index of LS, PLS, and FLS samples decrease by 71, 100 and 50%, respectively. Meanwhile, Compared with LS samples, the brittleness index of PLS samples decrease from 44 to 100%, and that of FLS samples increase from 27 to 203%. It indicates that when the confining pressure increases, fiber have the optimum improvement effect on the brittleness of LS samples.

3.2.3. Secant Modulus

For studying the ability of samples to resist deformation, Kutanai et al. [34] used the secant modulus E_{50} as the evaluation criteria and the calculation equation is as follows:

$$E_{50} = \frac{q_{50}}{\varepsilon_{50}} \quad (3)$$

where ε_{50} represents the strain at 50% peak stress, and q_{50} represents the stress value against the strain at 50% peak stress. The secant modulus of samples obtained by Equation (3) is plotted in Table 6. It can be seen that when the confining pressure is 0.4 MPa, compared with 0.1 MPa confining pressure, the secant modulus of LS, PLS, and FLS samples decrease by 28, 41 and 11%, respectively. Compared with LS samples, the secant modulus of PLS samples decreases from 44 to 53%, and that of FLS samples increase from 140 to 198%. It indicates that the addition of fly ash has a greater improvement effect on the stiffness of LS samples. The secant modulus of samples decreases with increasing confining pressure, because the improvement effect of confining pressure on the brittleness of samples is better than that of the rigidity. For example, the brittleness index of LS, PLS, and FLS samples decreased by 71, 100 and 50%, respectively. The brittleness of the samples is improved significantly, which hinders the rigidity increase of the samples to a certain extent.

3.3. Microscopic Analysis

To analyze the micro-structure of modified soils, the JSM-6360LV type high vacuum and low vacuum scanning electron microscope (SEM) was used, which was produced by Tokyo, Japan, Electronics Co., Ltd. The damage samples after the triaxial tests of modified soils were put into the oven to dry for 24 h and then SEM tests were performed. Figure 4 is the SEM images of modified soil samples.

As can be seen from Figure 4a, the LS sample has different sizes of gelling particles, its overall structure is poorly compacted, and there are more pores between the gelling particles. With the admixture of PP fibers and fly ash, the overall structure of LS sample is improved to some extent. In Figure 4b, PP fibers are closely bonded to the gelling particles. When the soil sample is stressed, interfacial friction is generated between the fiber and gelling particles, improving the damage resistance of soil sample, while in Figure 4c, fly ash mainly plays a role in promoting lime hydration and pore filling, which makes the internal structure of LS more compact, improving the mechanical strength of LS sample. The mechanical characteristics are expressed as follows: the brittle failure of LS samples is improved by adding fiber, while the addition of fly ash improves its rigidity, as shown in Table 6. Jiang et al. [35] investigated the improvement effect of PP fibers on the micro-structure of LS by microscopic tests and suggested that fibers mainly played a bonding role in LS. Zhou et al. [27] proposed that fly ash would promote the hydration reaction of lime and form large gelling particles with soil particles to fill the pores of the sample, resulting in a significant increase for its mechanical strength.

3.4. Discussion

In summary, PP fibers and class F fly ash modified lime-treated soils are innovative and feasible. The results of triaxial tests and SEM tests indicated that the mechanical characteristics and micro-structure of lime-treated soils were improved by adding PP fiber and class F fly ash. Among them, the fiber forms a close bond with the lime gelling particles and soil particles, thus improving the ductility and brittleness of LS. While the fly ash mainly promotes the hydration reaction of lime and fills the pores of soil sample, thus increasing the strength, toughness, and stiffness of LS. Jiang et al. [34] proposed that the compressive strength and tensile strength of lime-treated soils were significantly enhanced due to the better bonding structure between the 1% fiber and lime gelling particles. Abdi et al. [36] concluded that incorporating fiber into lime-treated soil could significantly increase the shear strength, compressive strength, and ductility of soil samples. Meanwhile, the fiber significantly reduced the cracks of soil in subgrade engineering and landfills. Eskisar [37] investigated the application prospects of fly ash modified lime in subgrade engineering and proposed that fly ash modified lime could significantly improve the compressive strength, bearing capacity, and stability of subgrade soil. Moreover, although good triaxial test results are obtained for fly ash and fiber modified lime soils, further modeling developments are needed for their application in engineering design.

4. CSE Curve Model

In subgrade engineering, the study of soil constitutive model is related to the reliability of numerical calculation results, and the relationship between stress and strain is the core issue to accurately describe the mechanical characteristics of soils. In this study, the CSE model proposed by Wang et al. [38] was used to analyze the stress-strain characteristics of treated soils in Equation (4).

$$q = a\sin[b(1-\exp(-c\varepsilon))] \tag{4}$$

In Equation (4), deviatoric stress $q = \sigma_1 - \sigma_3$; ε represents the axial strain. a, b and c are the non-negative undetermined parameters, obtained by fitting to the test data of soil samples. The simulation results are shown in Table 7. It can be seen from Table 7 that there is a certain correlation between the fitting parameters of samples under different confining pressures, which can be analyzed through quadratic function, as shown in Equation (5). Table 8 shows the fitting parameters and the resulting formulas.

$$y_{(abc)} = j\sigma^2 + m\sigma + n \tag{5}$$

where $y_{(abc)}$ represents the corresponding value of parameters a, b, and c, respectively, σ represents different confining pressures, and j, m, and n are fitting parameters. To obtain the CSE prediction models for the three samples, the formulas in Table 8 are substituted into Equation (4), and the results are shown as follows:

1. LS samples:

$$q = (2.09\sigma + 0.52)\sin[(11.75\sigma^2 - 7.01\sigma + 2.99)(1-\exp((8.75\sigma^2 - 3.75\sigma - 0.24)\varepsilon))] \tag{6}$$

2. PLS samples:

$$q = (2.89\sigma + 0.54)\sin[(-4.65\sigma + 2.71)(1-\exp((8\sigma^2 - 4.72\sigma + 0.09)\varepsilon))] \tag{7}$$

3. FLS samples:

$$q = (3.42\sigma + 0.61)\sin[(-0.53\sigma + 2.51)(1-\exp((0.72\sigma - 0.66)\varepsilon))] \tag{8}$$

Table 7. Fitting results of CSE model.

Group	σ (MPa)	a	b	c	R^2
LS	0.1	0.73	2.42	0.56	0.99
	0.2	0.93	2.01	0.55	0.93
	0.3	1.13	1.99	0.67	0.95
	0.4	1.36	2.05	0.31	0.96
PLS	0.1	0.85	2.28	0.31	0.93
	0.2	1.04	1.89	0.51	0.96
	0.3	1.50	0.99	0.63	0.96
	0.4	1.66	1.03	0.51	0.93
FLS	0.1	0.96	2.45	0.59	0.92
	0.2	1.28	2.4	0.49	0.95
	0.3	1.55	2.35	0.46	0.95
	0.4	2.01	2.29	0.36	0.98

To sum up, the accuracy of CSE model is verified by comparison with the measured stress-strain curves of modified soil samples, and the results are shown in Figure 5, where PC is the predicted curve and TR is the measured value. It can be seen from Figure 5 that the prediction results of CSE model are in good agreement with the measured data, thus the CSE model can better characterize the stress-strain characteristic of modified soil samples under different confining pressures. Moreover, the feasibility of CSE model was verified by the indoor tests and mathematical derivation in the previous works of the authors [38]. When it is similar to the research background in this study, the stress-strain curve of corresponding samples can be predicted and analyzed by the CSE model, providing assistance for the application of CSE model in the subgrade design and numerical simulation.

Table 8. Fitting parameters and formulas.

	Group	*j*	*m*	*n*	*R2*	Formula
a	LS	0	2.09	0.52	0.99	$y = 2.09\sigma + 0.52$
	PLS	0	2.89	0.54	0.96	$y = 2.89\sigma + 0.54$
	FLS	0	3.42	0.61	0.99	$y = 3.42\sigma + 0.61$
b	LS	11.75	−7.01	2.99	0.95	$y = 11.75\sigma^2 - 7.01\sigma + 2.99$
	PLS	0	−4.65	2.71	0.91	$y = -4.65\sigma + 2.71$
	FLS	0	−0.53	2.51	0.99	$y = -0.53\sigma + 2.51$
c	LS	−8.75	3.75	0.24	0.85	$y = -8.75\sigma^2 + 3.75\sigma + 0.24$
	PLS	−8	4.72	−0.09	0.98	$y = -8\sigma^2 + 4.72\sigma - 0.09$
	FLS	0	−0.72	0.66	0.96	$y = -0.72\sigma + 0.66$

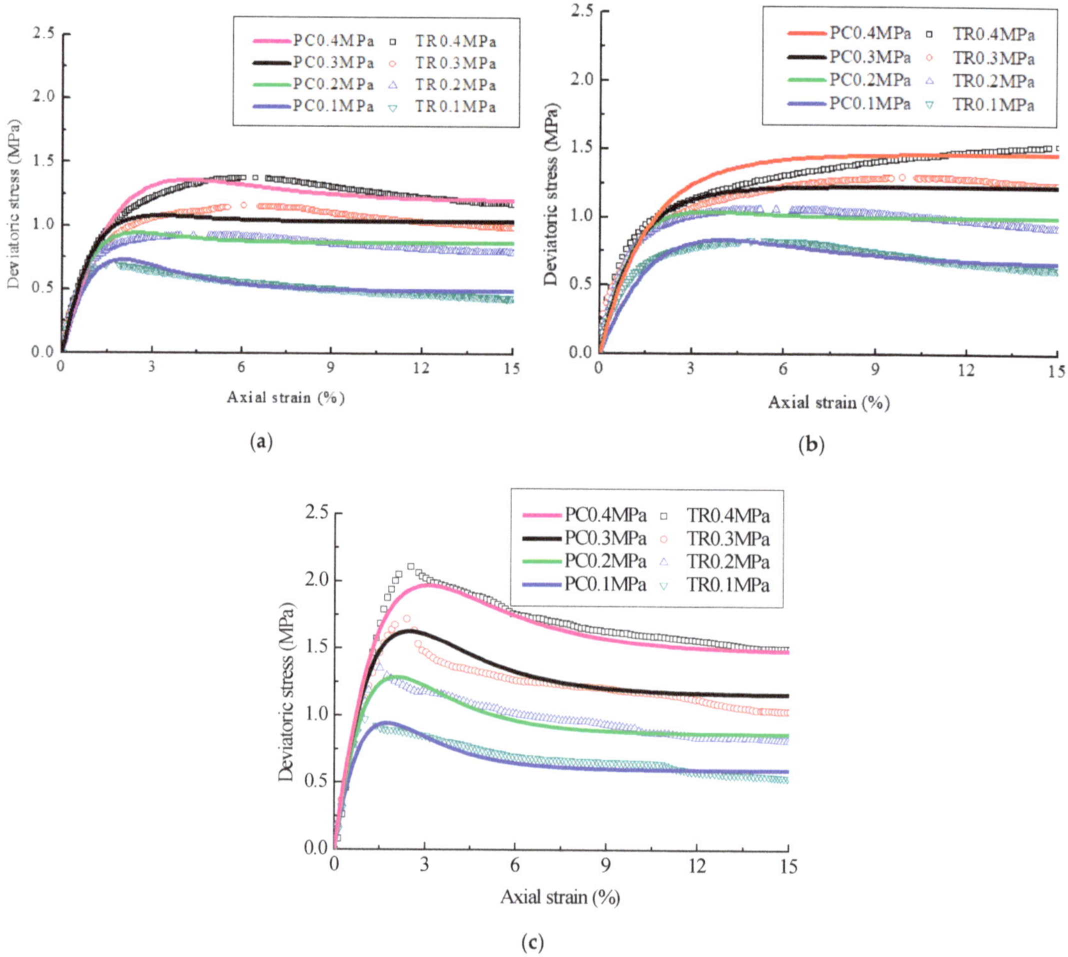

Figure 5. CSE prediction model verification. (**a**) LS sample; (**b**) PLS sample; (**c**) FLS sample.

5. Conclusions

In this study, PP fibers and fly ash were used to improve the triaxial mechanical characteristics of lime-treated subgrade soil. Through a series of triaxial UU and SEM tests, the mechanical characteristics and micro-structure of modified soils were investigated. Meanwhile, the CSE model was proposed to analyze the stress-strain characteristics of modified soils. The relevant results are as follows:

(1) The stress-strain curves of LS, PLS, and FLS samples are both of the weak softening type, which can be better fit by the proposed CSE model.
(2) Fly ash has a good lifting effect on the mechanical strength of lime-treated subgrade soil, while fiber mainly acts on the ductility lifting thereof. For example, when the confining pressure is 0.4 MPa, compared with LS samples, the peak stress, peak strain, cohesion, internal friction angle, and secant modulus of FLS increased by 53, 65, 23, 26 and 53%, and PLS increased by 19, 134, 8, 0 and 198%, respectively.
(3) The addition of fiber and fly ash improve the overall structure of LS samples to a certain extent and make its overall skeleton compact and denser.

It is worth noting that the effects of impurities in the sample, the sample size, and the optimum mass content of fibers and fly ash are not considered in this study, which is worthy for further research. Meanwhile, for the applicability of the CSE model, it is necessary to establish a relevant experimental database for its further study in the future.

Author Contributions: Conceptualization, W.W. and C.Z.; formal analysis, B.L.; investigation, N.L.; writing—original draft preparation, C.Z.; writing—review and editing, W.W. and S.P. All authors have read and agreed to the published version of the manuscript.

Funding: This research was funded by the National Natural Science Foundation of China (Grant No. 52179107) and China Scholarship Council (201607910002).

Institutional Review Board Statement: Not applicable.

Informed Consent Statement: Not applicable.

Data Availability Statement: Not applicable.

Conflicts of Interest: The authors declare no conflict of interest.

References

1. Wang, W.; Kang, H.; Li, N.; Liu, Y. Experimental investigations on the mechanical and microscopic behavior of cement-treated clay modified by Nano-MgO and fibers. *Int. J. Geomech.* **2022**, *22*, 04022059. [CrossRef]
2. Karkush, M.-O.; Yassin, S. Using sustainable material in improvement the geotechnical properties of soft clayey soil. *J. Eng. Sci. Technol.* **2020**, *15*, 2208–2222.
3. Onyelowe, K.C. Review on the role of solid waste materials in soft soils reengineering. *Mater. Sci. Energy Technol.* **2019**, *2*, 46–51. [CrossRef]
4. Peter, L.; Jayasree, P.K.; Balan, K. Optimization, quantification of Strength and Design Benefits of Flexible Pavements Resting on Soft Soil Subgrades Treated with Coir Waste and Lime. *J. Nat. Fibers* **2022**, *19*, 1805–1818. [CrossRef]
5. Zhang, C.; Wang, W.; Zhu, Z.-D.; Li, N.; Pu, S.-Y.; Wan, Y.; Huo, W.-W. Triaxial Mechanical Characteristics and Microscopic Mechanism of Graphene-Modified Cement Stabilized Expansive Soil. *KSCE J. Civ. Eng.* **2022**, *26*, 96–106. [CrossRef]
6. Xiao, H.; Wang, W.; Goh, S.H. Effectiveness study for fly ash cement improved marine clay. *Constr. Build. Mater.* **2017**, *157*, 1053–1064. [CrossRef]
7. Ojuri, O.O.; Adavi, A.A.; Oluwatuyi, O.E. Geotechnical and environmental evaluation of lime–cement stabilized soil–mine tailing mixtures for highway construction. *Transp. Geotech.* **2017**, *10*, 1–12. [CrossRef]
8. Zimar, Z.; Robert, D.; Zhou, A.; Giustozzi, F.; Setunge, S.; Kodikara, J. Application of coal fly ash in pavement subgrade stabilisation: A review. *J. Environ. Manag.* **2022**, *312*, 114926. [CrossRef] [PubMed]
9. Auler, A.C.; Pires, L.F.; Dos Santos, J.A.B.; Caires, E.; Borges, J.A.R.; Giarola, N.F.B. Effects of surface-applied and soil-incorporated lime on some physical attributes of a Dystrudept soil. *Soil Use Manag.* **2017**, *33*, 129–140. [CrossRef]
10. Wang, Z.; Zhang, W.; Jiang, P.; Li, C. The Elastic Modulus and Damage Stress–Strain Model of Polypropylene Fiber and Nano Clay Modified Lime Treated Soil under Axial Load. *Polymers* **2022**, *14*, 2606. [CrossRef]
11. Kafodya, I.; Okonta, F. Density control method for compression test of compacted lime-flyash stabilised fiber-soil mixtures. *MethodsX* **2018**, *5*, 848–856. [CrossRef] [PubMed]

12. Bozbey, I.; Kamal, N.A.; Abut, Y. Effects of soil pulverisation level and freeze and thaw cycles on fly-ash- and lime-stabilised high plasticity clay: Implications on pavement design and performance. *Road Mater. Pavement Des.* **2017**, *18*, 1098–1116. [CrossRef]
13. Jiang, P.; Zhou, X.; Qian, J.; Li, N. Experimental Study on the Influence of Dry–Wet Cycles on the Static and Dynamic Characteristics of Fiber-Modified Lime and Fly Ash-Stabilized Iron Tailings at Early Curing Age. *Crystals* **2022**, *12*, 568. [CrossRef]
14. Madarvoi, S.; Rama, P.-S. Mechanical characterization of graphene-hexagonal boron nitride-based kevlar-carbon hybrid fabric nanocomposites. *Polymers* **2022**, *14*, 2559. [CrossRef]
15. Lisuzzo, L.; Caruso, M.R.; Cavallaro, G.; Milioto, S.; Lazzara, G. Hydroxypropyl Cellulose Films Filled with Halloysite Nanotubes/Wax Hybrid Microspheres. *Ind. Eng. Chem. Res.* **2021**, *60*, 1656–1665. [CrossRef]
16. Lisuzzo, L.; Cavallaro, G.; Milioto, S.; Lazzara, G. Halloysite nanotubes filled with MgO for paper reinforcement and deacidification. *Appl. Clay Sci.* **2021**, *213*, 106231. [CrossRef]
17. Linh, N.; Chuong, B.; Liem, T.-N. Dioctyl phthalate-modified graphene nanoplatelets: An effective additive for enhanced mechanical properties of natural rubber. *Polymers* **2022**, *14*, 2541.
18. Zhang, J.; Li, C. Experimental Study on Lime and Fly Ash–Stabilized Sintered Red Mud in Road Base. *J. Test. Eval.* **2018**, *46*, 1539–1547. [CrossRef]
19. Wei, L.; Chai, S.X.; Zhang, H.Y.; Shi, Q. Mechanical properties of soil reinforced with both lime and four kinds of fiber. *Constr. Build. Mater.* **2018**, *172*, 300–308. [CrossRef]
20. Rudramurthy, G.; Ramasamy, P.; Rajendran, A. Stabilization of Clayey Soil Using Lime and Prosopis Fibers. In *Conference of the Arabian Journal of Geosciences*; Springer: Cham, Switzerland, 2018; pp. 259–261. [CrossRef]
21. Dhar, S.; Hussain, M. The strength behaviour of lime-stabilised plastic fibre-reinforced clayey soil. *Road Mater. Pavement Des.* **2019**, *20*, 1757–1778. [CrossRef]
22. Turan, C.; Javadi, A.; Consoli, N.C.; Turan, C.; Vinai, R.; Cuisinier, O.; Russo, G. Mechanical Properties of Calcareous Fly Ash Stabilized Soil. In Proceedings of the EuroCoalAsh, Dundee, UK, 10–12 June 2019; pp. 184–194.
23. Li, L.; Zhang, J.; Xiao, H.; Hu, Z.; Wang, Z. Experimental Investigation of Mechanical Behaviors of Fiber-Reinforced Fly Ash-Soil Mixture. *Adv. Mater. Sci. Eng.* **2019**, *2019*, 1–10. [CrossRef]
24. Ghosh, A.; Subbarao, C. Strength Characteristics of Class F Fly Ash Modified with Lime and Gypsum. *J. Geotech. Geoenvironmental Eng.* **2007**, *133*, 757–766. [CrossRef]
25. Wang, W.; Zhang, C.; Guo, J.; Li, N.; Li, Y.; Zhou, H.; Liu, Y. Investigation on the Triaxial Mechanical Characteristics of Cement-Treated Subgrade Soil Admixed with Polypropylene Fiber. *Appl. Sci.* **2019**, *9*, 4557. [CrossRef]
26. *C618-19*; Standard Specification for Coal Fly Ash and Raw or Calcined Natural Pozzolan for Use in Concrete. ASTM International: West Conshohocken, PA, USA, 2019.
27. Zhou, S.-Q.; Zhou, D.-W.; Zhang, Y.-F.; Wang, W.-J. Study on Physical-Mechanical Properties and Microstructure of Expansive Soil Stabilized with Fly Ash and Lime. *Adv. Civ. Eng.* **2019**, *2019*, 1–15. [CrossRef]
28. *GB/T 50123-1999*; Chinese National Geotechnical Test Standard: Beijing, China, 1999.
29. *D2850-15*; Standard Test Method for Unconsolidated-Undrained Triaxial Compression Test on Cohesive Soils. ASTM International: West Conshohocken, PA, USA, 2015.
30. Wang, Y.; Guo, P.; Li, X.; Lin, H.; Liu, Y.; Yuan, H. Behavior of Fiber-Reinforced and Lime-Stabilized Clayey Soil in Triaxial Tests. *Appl. Sci.* **2019**, *9*, 900. [CrossRef]
31. Ghosh, A.; Subbarao, C. Microstructural Development in Fly Ash Modified with Lime and Gypsum. *J. Mater. Civ. Eng.* **2001**, *13*, 65–70. [CrossRef]
32. Wang, W.; Song, X.; Zhou, G. Composite exponential-hyperbolic model for stress-strain curve of seashore soft soil. *Chin. J. Geotech. Eng.* **2010**, *32*, 1455.
33. Consoli, N.C.; Prietto, P.D.M.; Ulbrich, L.A. Influence of Fiber and Cement Addition on Behavior of Sandy Soil. *J. Geotech. Geoenvironmental Eng.* **1998**, *124*, 1211–1214. [CrossRef]
34. Kutanaei, S.S.; Choobbasti, A.J. Triaxial behavior of fiber-reinforced cemented sand. *J. Adhes. Sci. Technol.* **2016**, *30*, 579–593. [CrossRef]
35. Jiang, P.; Zhou, L.; Zhang, W.; Wang, W.; Li, N. Unconfined Compressive Strength and Splitting Tensile Strength of Lime Soil Modified by Nano Clay and Polypropylene Fiber. *Crystals* **2022**, *12*, 285. [CrossRef]
36. Abdi, M.R.; Ghalandarzadeh, A.; Chafi, L.S. An investigation into the effects of lime on compressive and shear strength characteristics of fiber-reinforced clays. *J. Rock Mech. Geotech. Eng.* **2021**, *13*, 885–898. [CrossRef]
37. Eskisar, T. The role of carbide lime and fly ash blends on the geotechnical properties of clay soils. *Bull. Eng. Geol. Environ.* **2021**, *80*, 6343–6357. [CrossRef]
38. Wang, W.; Lin, H.; Sun, B.-X. Composite exponential-sine model for dynamic stress-strain curve of soft soil. *J. Hydroelectr. Eng.* **2011**, *30*, 11–15.

polymers

MDPI

Article

The Elastic Modulus and Damage Stress–Strain Model of Polypropylene Fiber and Nano Clay Modified Lime Treated Soil under Axial Load

Zhichao Wang, Weiqing Zhang, Ping Jiang and Cuihong Li *

School of Civil Engineering, Shaoxing University, Shaoxing 312000, China; 21020859083@usx.edu.cn (Z.W.); zhangweiqingusx@163.com (W.Z.); jiangping@usx.edu.cn (P.J.)
* Correspondence: lich@usx.edu.cn

Abstract: Using polypropylene fiber (PPF) and nano clay modified lime treated soil (LS), the static and dynamic properties of fiber modified lime treated soil (FLS), nano clay modified lime treated soil (NLS), and fiber nano clay composite modified lime treated soil (NFLS) were studied. Through the unconfined compressive strength (UCS) test and dynamic triaxial test of FLS, NLS, and NFLS, the static and dynamic elastic modulus characteristics at 7 day curing age were explored, and the damage stress–strain model was established. The results show that: (1) Polypropylene fiber and nano clay can significantly enhance the mechanical properties of NFLS. Nano clay can promote the reaction between lime and soil to produce calcium silicate hydrate (C-S-H) and calcium aluminate hydrate (C-A-H), thus improving the strength of NFLS, and UCS can be increased by up to 103%. Polypropylene fiber can enhance the ductility of NFLS and increase the residual ductility strength, and the residual strength can be increased by 827%. (2) Nano clay can enhance the static and dynamic elastic modulus of modified lime treated soil. The static and dynamic elastic modulus of NLS, FLS, and NFLS are linear with the change of polypropylene fiber and nano clay content. The static and dynamic elastic modulus of NLS, FLS, and NFLS are linear, exponential, and logarithmic, respectively. (3) The mesoscopic random damage model can characterize the stress–strain relationship of NFLS. Polypropylene fiber and nano clay can improve the ductility and strength of modified LS, and the composite addition of polypropylene fiber and nano clay can improve the ability of modified LS to resist damage.

Keywords: lime treated soil; polypropylene fiber; nano clay; mechanical properties; damage model

Citation: Wang, Z.; Zhang, W.; Jiang, P.; Li, C. The Elastic Modulus and Damage Stress–Strain Model of Polypropylene Fiber and Nano Clay Modified Lime Treated Soil under Axial Load. *Polymers* **2022**, *14*, 2606. https://doi.org/10.3390/polym14132606

Academic Editors: Wensheng Wang, Yongchun Cheng, Heping (Fred) Chen and Guojin Tan

Received: 9 June 2022
Accepted: 23 June 2022
Published: 27 June 2022

1. Introduction

Due to the demand of economic and urbanization development, the quality requirements of road infrastructure construction are increasing year by year. Using lime treated construction waste soil for road base is an effective measure to reduce construction waste. At the same time, it can replace sand, stone, and other materials, so as to reduce the irreversible damage to the ecological environment caused by mountain excavation and river excavation. The traffic load is mainly axial load. Due to the increase of road load, higher requirements are put forward for the mechanical properties of road base. Therefore, it is necessary to deeply explore the properties of lime treated soil and the main technical means to improve its mechanical properties.

Liu et al. [1] proved that after adding lime into the soil, the soil properties change, which can improve its unconfined compressive strength (UCS) to meet the actual engineering needs. Lime can not only significantly improve the UCS of the improved soil and reduce the expansion and contraction of the soil, but also react with the pozzolanic ash to form cementitious materials, which is an excellent material for stabilizing the soil [2,3]. Lemaire et al. [4] conducted UCS and X-ray diffraction (XRD) tests on lime treated soil. The results show that lime can make the soil form a "honeycomb" structure and significantly improve the mechanical properties of the soil. Lime treated soil can improve the mechanical properties of soil to a certain extent, but its brittle failure mode will affect the strength and durability of road base. Therefore, further study of the mechanical properties of LS modified by adding polypropylene fiber and nano clay is required [5,6]. Jiang et al. [7] studied the interfacial action and energy dissipation of polypropylene fiber and glass fiber modified iron tailings through single fiber drawing test, and established a quantitative relationship model between dissipated energy, effective fiber length, and freeze–thaw cycle. The UCS, CBR, and TCT tests of polypropylene fiber modified soil show that the addition of polypropylene fiber will improve the strength, ductility, and residual strength of soil [8,9]. With the progress of science and technology, nano clay can significantly improve the mechanical properties of soil due to its unique physical and chemical properties, and has begun to become a new stable material in the field of construction engineering [10]. Niu et al. [11] studied the mechanical properties, durability, shrinkage, hydration, and microstructure of cement-based materials modified by nano clay. The results show that an appropriate amount of nano clay is beneficial to improve the properties of cement-based materials. Gao et al. [12] conducted triaxial undrained and SEM tests on nano MgO modified soil, and found that nano MgO can effectively improve the strength and cohesion of engineering waste silt and reduce the porosity of soil. Many scholars have studied the modification characteristics of nano materials, and the results show that nano materials can improve the compressive strength of LS and improve the compression characteristics of LS [13–15]. Lu et al. [16] studied the effects of moisture content, compactness, cyclic deviatoric stress, and confining pressure on the dynamic elastic modulus of lime treated expansive soil by using dynamic triaxial test, and verified the applicability of UT Austin model. Zhao et al. [17] measured the stress–strain curve of solidified soft soil under different lime content and confining pressure through dynamic triaxial test, and analyzed the relationship between dynamic and static strength and the influence of lime content on dynamic strength. Therefore, fiber materials and nano materials play a positive role in the mechanical properties of modified LS.

Fiber materials and nano materials have good engineering practical significance for modified LS. Many scholars have deeply studied the stress–strain model of modified materials. Zhou et al. [18] discussed the relationship between the failure mechanism, deformation characteristics, and strength of lime treated expansive soil, and constructed a nonlinear elastic model. Poinard et al. [19] revealed the main differences between damage and failure mechanisms by studying the meso damage mechanism of concrete under hydrostatic pressure and triaxial pressure, using X-ray computed tomography instrument. Jiang et al. [20] established UCS and STS prediction models of NFLS by studying the effects of nano clay and polypropylene fiber on the unconfined compressive strength and splitting performance of LS, taking the content of nano clay, pore volume, and lime volume as independent variables. Wang et al. [21] studied the stress relaxation of lime treated unsaturated soil according to a series of triaxial tests, and established a structural model considering time effect to characterize the cumulative damage of LS.

Polypropylene fiber and nano clay are used to modify LS. The mechanical properties of FLS, NLS, and NFLS were studied through UCS test and dynamic triaxial test. Based on the meso random damage theory, the stress–strain damage relationship of NFLS is established, which provides a theoretical reference for the application of modified LS in road engineering.

2. Materials and Methods

2.1. Materials

The raw materials used in the test were soil, quicklime powder, polypropylene fiber, and nano clay. The soil was from the foundation pit construction site in Shaoxing City, Zhejiang Province, China. The soil is grayish brown, and its main physical and mechanical properties are shown in Table 1. The quicklime powder was produced by Jiangxi Xinyu Liangliang Trading Co., Ltd., Xinyu City, Jiangxi Province, China, with CaO content of

88% and purity of 98%. The polypropylene fiber was a bundle monofilament, which was produced by Shaoxing Fiber High Tech Co., Ltd., Shaoxing City, Zhejiang Province, China, with a diameter of 18–46 μm and length of 6 mm, which has the characteristics of light weight and good stability, and can significantly improve the ductility and bending resistance of soil. The nano clay used was produced by Hubei Jinxi montmorillonite Co., Ltd., Wuhan City, Hubei Province, China, the density is 0.45 g/cm^3, 99.9% of nano clay can pass the 200 mesh screen, and the content of montmorillonite is 96–98%, which is beige powder.

Table 1. Physical properties of soil.

Natural Moisture Content (%)	Specific Gravity	Uniformity Coefficient Cu	Curvature Coefficient Cc	Liquid Limit (%)	Plastic Limit (%)	Plasticity Index (%)
33	2.53	12.2	1.24	37.8	17.2	20.6

2.2. Test Scheme

LS was modified by nano clay, polypropylene fiber, composite nano clay and polypropylene fiber. The static and dynamic properties of the modified LS were studied by UCS test and dynamic triaxial test. The content of lime was 6%, and, on this basis, nano clay with the content of 0%, 2%, 4%, and 6% and polypropylene fiber with the content of 0%, 0.25%, 0.5%, and 0.75% was added respectively. The content of lime, nano clay, and polypropylene fiber was the mass ratio of dry soil. The mechanical properties of NFLS were investigated by adding 2%, 4%, and 6% nano clay in the optimal FLS and 0.25%, 0.5%, and 0.75% polypropylene fiber in the optimal NLS [20].

According to the Chinese test code for Inorganic Binder Stabilized Materials in Highway Engineering [22], the undisturbed soil of the required soil sample is crushed and dried in a 105 °C oven. After crushing, a 2 mm sieve is used to screen the soil. According to the test scheme, after weighing the corresponding quality of dry soil, quicklime, nano clay, and polypropylene fiber with an electronic scale, add water and stir evenly according to the optimal moisture content to make each admixture disperse and not agglomerate. Put the uniformly stirred aggregate into the sealing stuffs of plastic bags for 24 h, and make samples by static compaction, as shown in Figure 1. After demoulding, it should be cured for 7 days. In the first 6 days, it should be cured in a standard curing box with a temperature of 20 ± 2 °C and a relative humidity of more than 95%. On the last day, the sample should be taken out of the curing box and soaked in water at 20 ± 2 °C for curing. The water surface is approximately 2.5 cm above the sample. The test scheme is shown in Table 2.

Table 2. Mix ratio design.

NO	Lime Content (%)	Nano Clay Content (%)	PP Fiber Content (%)	Maximum Dry Density (g/cm^3)	Optimum Moisture Content (%)	Compactness (%)	Curing Time (d)
LS	6	0	0	1.762	17.5	98	7
NLS	6	2, 4, 6	0				
FLS	6	0	0.25, 0.5, 0.75				
NFLS	6	2, 4, 6	0.25, 0.5, 0.75				

Note, L, N, and F respectively represent lime, nano clay, and polypropylene fiber. Lime content, nano clay content, and polypropylene content are the percentage of dry soil mass, and water content is the percentage of all dry mixtures.

Figure 1. Sample preparation process.

The UCS test sample is a cylinder with a diameter of 50 mm and a height of 50 mm. After curing, TKA-WCY-1F automatic unconfined compressive strength tester is selected for testing, as shown in Figure 2. Referring to the Chinese code for Highway Geotechnical Test [23], the loading rate of the instrument was 1 mm/min, and the stress–strain curve of the sample is obtained through the computer data acquisition system.

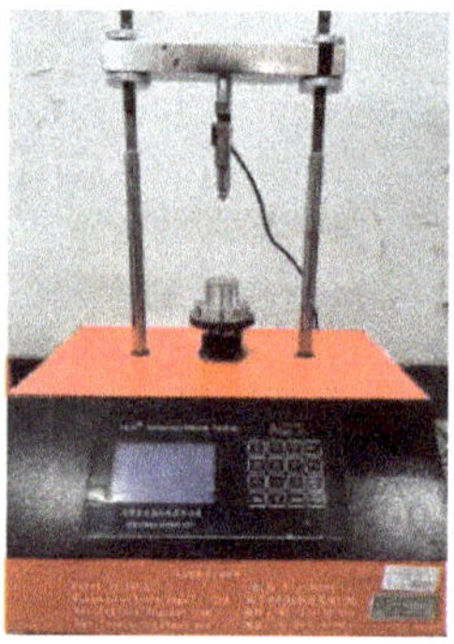

Figure 2. UCS test system.

Because the traffic load has a long-term load impact on the road base, the dynamic elastic modulus of NFLS is studied by dynamic triaxial test to deeply explore the long-term traffic load bearing capacity of modified LS. After curing, GDS servo motor control system is selected for dynamic triaxial test. The size of the dynamic triaxial test sample is a cylinder with a diameter of 39.1 mm and a height of 80 mm. Because the surrounding earth pressure has little impact on the road base, the confining pressure is 0 kPa, and the displacement control method is used to vibrate for 50 times. The test is carried out with a sine wave with a frequency of 1 Hz and an amplitude of 0.1 mm. Data acquisition is carried out through microcomputer system, as shown in Figure 3.

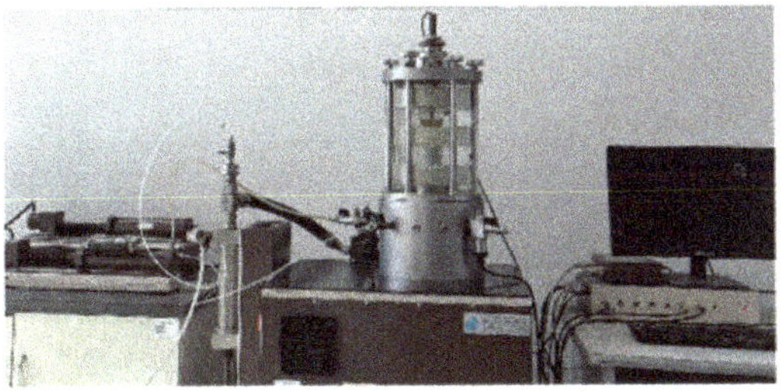

Figure 3. Dynamic triaxial test loading system.

For SEM test, a Nissan high-low vacuum scanning electron microscope (JSM-6360LV, JEOL, Tokyo, Japan) was used. The electro-adhesive was evenly spread on the aluminum tray, and then the powder of the sample to be tested was glued to the electro-adhesive. After the required test sample was glued, a layer of platinum film was uniformly sprayed on the surface of the sample. After confirming uniform spraying, the aluminum tray was put into the electron microscope for SEM test.

FT-IR was tested by KBr compression method. The instrument used in the experiment was the Fourier Infrared spectrometer (NEXUS, Thermo Nicolet Corporation, American, Waltham, MA, USA), with a wavelength range of 4000–400 cm^{-1}. The test sample was ground into powder and then added into potassium bromide powder. Grinding continued so that the sample powder and potassium bromide powder was fully mixed and ground into a fine powder. Then the powder was pressed into sheets by KBr method and put into the instrument for infrared spectrum analysis.

2.3. Stress–Strain Damage Model

According to the random damage mechanics of concrete, Li et al. [24,25] abstracted the concrete sample as a micro spring random damage system from the meso point of view, so as to establish the random damage constitutive model of concrete. According to the meso random damage theory, the damage stress–strain relationship of materials under compression is shown in Equation (1).

$$\sigma = (1 - D)E\varepsilon \tag{1}$$

where E is the elastic modulus, D is the damage variable, and the mean and variance of D are shown in Equations (2) and (3), respectively.

$$\mu_D = \frac{(1-\beta)\mu_{D_S}}{1-\beta\mu_{D_S}} \tag{2}$$

$$V_D{}^2 = \frac{(1-\beta)\mu_{D_S}}{(1-\beta)^2}\left(2\int_0^1 (1-\eta)F_S(\gamma,\gamma,\eta)d\eta - \mu_{D_S}\right) \tag{3}$$

The meso random damage model is mainly related to parameters λ, ζ, and ω [26,27]. E can be obtained from the dynamic triaxial test of NFLS, and then the parameters of λ, ζ, and ω can be identified according to the UCS test results, so as to establish the stress–strain damage model of NFLS.

3. Results and Discussion

3.1. Static Characteristics

3.1.1. UCS Test Results

UCS and residual strength are important mechanical indexes for the study of NFLS. The peak value of stress–strain curve is taken as the UCS, and the stress corresponding to peak strain plus 5% is selected as the residual strength. The UCS test results of NFLS are shown in Table 3.

Table 3. Test results of UCS test.

NO	UCS (kPa)	Residual Strength (kPa)	Elastic Modulus (MPa)
L6	1239	192	81
L6-N2	1425	321	89
L6-N4	1635	301	117
L6-N6	1947	246	122
L6-F0.25	1439	381	66
L6-F0.5	1913	1228	69
L6-F0.75	1794	1315	55
L6-N2-F0.75	2037	1498	66
L6-N4-F0.75	2057	1596	65
L6-N6-F0.25	2007	840	113
L6-N6-F0.5	2193	1245	98
L6-N6-F0.75	2513	1762	90

Figure 4a,b show the stress–strain curve of NFLS with nano clay content N = 6%. The stress–strain curve of NFLS is a softening curve, that is, the stress first increases to the stress peak with strain, then decreases, and then tends to be gentle. Figure 4c,d show the UCS and residual strength of NFLS with nano clay content N = 6%. The increase of nano clay and fiber content can significantly improve the UCS of LS. The content of nano clay increases from 0% to 6%. The UCS of NFLS increases with the increase of nano clay content. When the content of nano clay is 6%, the UCS of NLS is 1947 kPa and the relative growth rate of UCS is 57%. When the optimal nano clay content is 6%, the UCS of NFLS increases with the increase of polypropylene fiber content, and when the fiber content is 0.75%, the UCS of NFLS is 2513 kPa, the growth rate relative to NLS is 29%, and the growth rate of UCS is the fastest. It shows that the strength of NFLS can be significantly improved by adding nano clay and fiber, and the optimal contents of nano clay and polypropylene fiber are 6% and 0.75% respectively. The residual strength of NLS increases first and then decreases with the increase of nano clay content, which is due to the increase of brittleness of NLS and the decrease of residual strength. When the content of nano clay is 6%, the residual strength of NFLS increases linearly with the increase of the content of polypropylene fiber. This is because the fiber can slow down the development of failure cracks of NFLS. With the increase of the content of fiber, the cracks develop more slowly and the bearing capacity is greater, indicating that polypropylene fiber can significantly enhance the ductility of NFLS and slow down its brittle failure. The interfacial force generated between the polypropylene fiber and soil can play a better role in the bonding of cracks. As shown in Figure 4e, under the same strain, with the increase of polypropylene fiber content, the fibers emerging from the crack of the sample increase significantly, the pull effect is significant, and the crack development slows down [28,29].

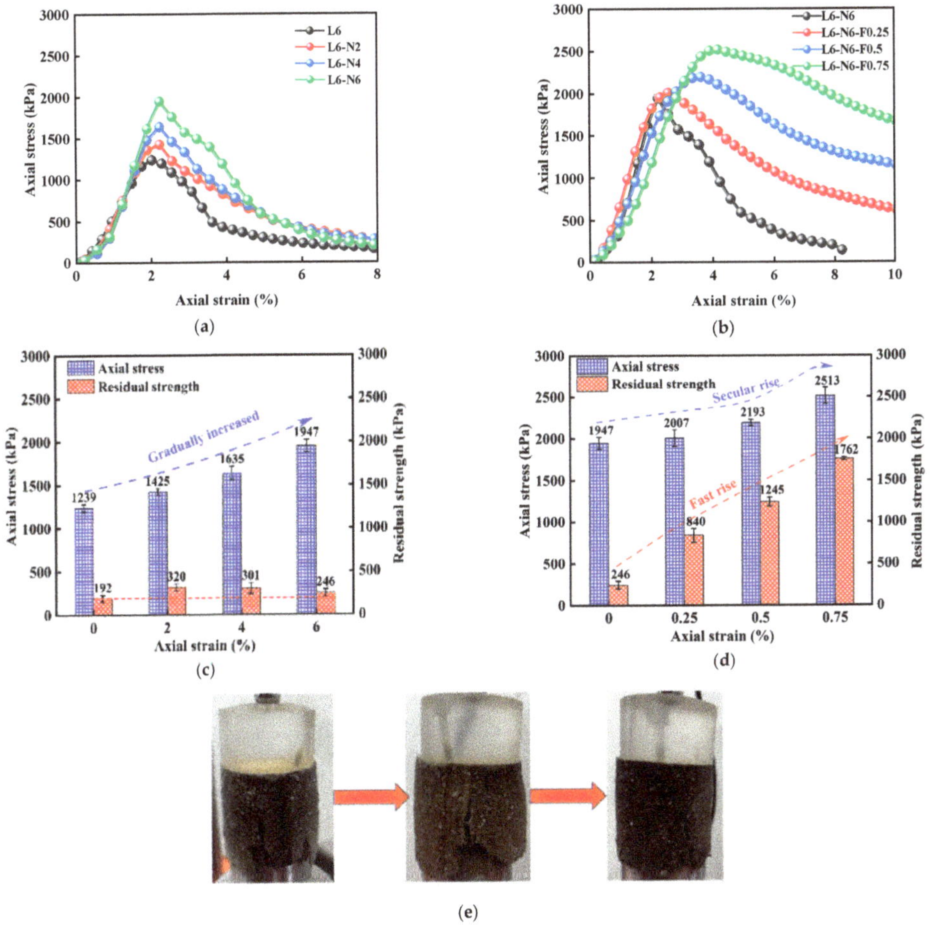

Figure 4. Effect of PF on the statics of NFLS: (**a**) stress–strain curve of NLS; (**b**) stress–strain curve of NFLS (N = 6%); (**c**) UCS and residual strength of NLS; (**d**) UCS and residual strength of NFLS (N = 6%); and (**e**) the variation of cracks with the content of polypropylene fiber under the same strain.

Figure 5a,b shows the stress–strain curve of FLS and NFLS with fiber content F = 0.75%, and Figure 5c,d shows the UCS and residual strength diagram of NFLS with fiber content F = 0.75%. With the increase of polypropylene fiber content, the UCS of FLS first increases and then decreases. When the polypropylene fiber content is 0.75%, the UCS of FLS shows a downward trend, which is due to the overhead phenomenon when the network structure is formed between the soil mass due to the excessive fiber content, and the relative compactness of the soil mass decreases, leading to a decrease in UCS growth. When the fiber content is 0.75%, the UCS of NFLS increases with the increase of nano clay content, and when the nano clay content is 6%, the UCS of NFLS increases the fastest. This is because nano clay can fill the pores between LS particles and make its internal structure more dense. Secondly, nano clay can promote the further reaction of LS, and on the other hand, it can have pozzolanic reaction with LS [30–33]. The volcanic ash reaction generates a sheet structure, as shown in Figure 6a. The sheet structure is cementitious substances such as hydrated calcium silicate C-S-H and hydrated calcium aluminate C-A-H, as shown in Figure 6b,c. The generated cementitious material can fill the pores between the LS skeleton and make the LS particles agglomerate better. At the same time, the interface between fiber and soil particles is closer, the interfacial friction is greater, the damage degree of external force to the sample is reduced, and the strength of NFLS is further increased. However, the residual strength of NFLS increases slowly with the increase of nano clay content, indicating that polypropylene fiber contributes more to the residual strength of NFLS than nano clay.

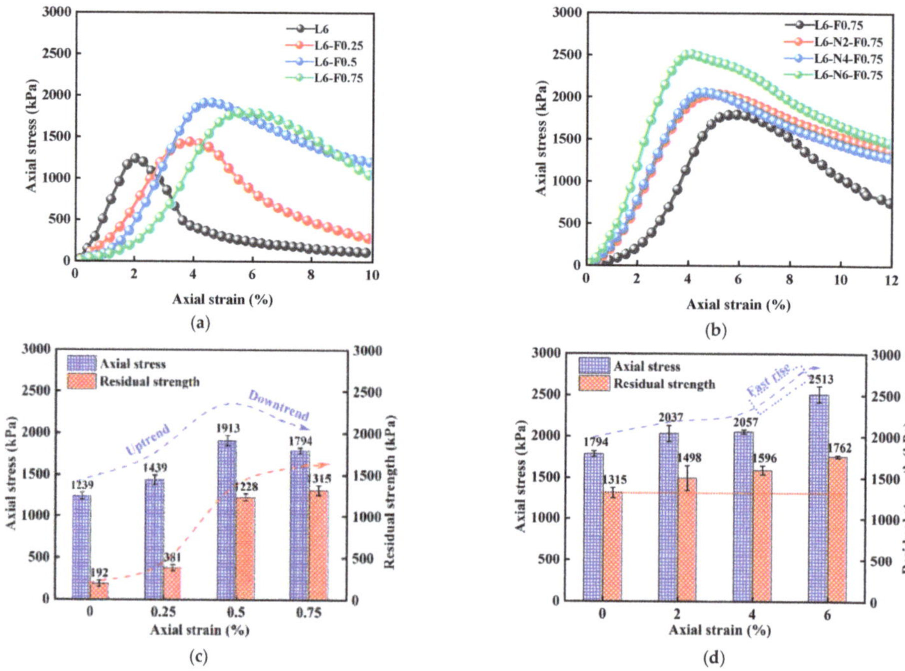

Figure 5. Effect of nano clay on the statics of NFLS: (**a**) stress–strain curve of FLS; (**b**) stress–strain curve of NFLS (F = 0.75%); (**c**) UCS and residual strength of FLS; (**d**) UCS and residual strength of NFLS (F = 0.75%).

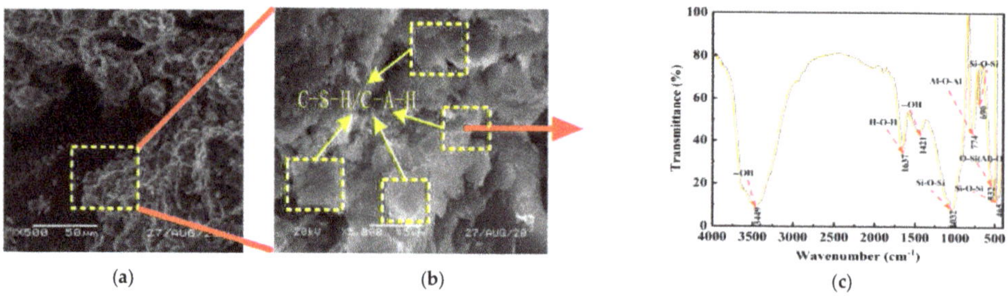

Figure 6. Microscopic analysis graphs of NFLS: (**a**) 500−fold SEM; (**b**) 5000−fold SEM; (**c**) FT−IR.

The above phenomena show that nano clay mainly provides strength for NFLS, but will enhance the brittle failure of LS. Polypropylene fiber can enhance the ductility of NFLS and improve its residual strength.

3.1.2. Variation Law of Static Elastic Modulus

The static elastic modulus is the ratio of stress to strain in the elastic deformation stage under compression. Through the unconfined compressive strength test data, the stress–strain curve of NFLS in the elastic stage under compression is fitted, and the straight-line slope is the static elastic modulus. Figure 7a–d shows the variation curve of static elastic modulus under different content of nano clay and polypropylene fiber. Figure 7a shows the variation law of the static elastic modulus of NLS with the increase of nano clay content. When the content of nano clay is 6%, the static elastic modulus of NLS increases to 122 MPa, and the growth rate is 51% higher than that of LS, indicating that the addition of nano clay can improve the static elastic modulus of LS and enhance its deformation resistance. Figure 7b shows that when the content of nano clay is 6%, the static elastic modulus of NFLS decreases with the increase of the content of polypropylene fiber, because the stress of NFLS in the elastic stage decreases with the growth rate of strain. In Figure 7c, it can be found that the content of polypropylene fiber increases from 0.25% to 0.75%, and the static elastic modulus of FLS first increases and then decreases. When the content of polypropylene fiber is 0.5%, the elastic modulus of FLS is 69 MPa, while when the content of polypropylene increases by 0.75%, the static elastic modulus of FLS decreases to 55 MPa, indicating that on the one hand, this is due to the large amount of C-S-H and C-A-H produced by lime soil reaction attached to the surface of polypropylene fiber, so polypropylene fiber can enhance the adhesion of the interface between soil and fiber and make the samples bond together better, as shown in Figure 8. On the other hand, with the increase of the content of polypropylene fiber, the porosity between soil particles and fiber may increase, resulting in the decrease of its static elastic modulus. Figure 7d shows that when the content of polypropylene fiber is 0.75%, the static elastic modulus of NFLS first increases, then decreases, and finally significantly increases with the increase of nano clay content. When the content of nano clay is 6%, the static elastic modulus of NFLS is significantly increased to 90 MPa. This is because nano clay can fill the pores between fiber and soil, enhance the stiffness of NFLS, and reduce elastic deformation.

The relationship is fitted between the static elastic modulus and the content of nano clay and polypropylene fiber of NLS, FLS, and NFLS. The fitting formulas are shown in Equations (4)–(7).

$$E_{(\mathrm{NLS})} = 6.877y + 80.9, R^2 = 0.92 \tag{4}$$

$$E_{(\mathrm{NFLS,\ N=6\%})} = -44.16x + 122.46, R^2 = 0.98 \tag{5}$$

$$E_{(\mathrm{FLS})} = -28.216x + 78.26, R^2 = 0.90 \tag{6}$$

$$E_{(\mathrm{NFLS,F=0.75\%})} = 0.8188y^2 + 0.3525y + 56.755, R^2 = 0.89 \tag{7}$$

where E is the static elastic modulus (MPa); x is the content of polypropylene fiber (%); and y is the content of nano clay (%).

3.2. *Dynamic Characteristics*

3.2.1. Hysteretic Curve

The dynamic elastic modulus of the material can be calculated from the slope of the hysteretic curve to characterize the stiffness of the soil [34]. Figure 9a–d shows the variation law of dynamic stress–strain hysteretic curves of NLS, FLS, and NFLS.

Figure 9a,b shows that the hysteresis curves of NLS gradually approaches the stress axis with the increase of nano clay content. When the content of nano clay is 6%, different content of polypropylene fiber has a significant effect on the hysteretic curve. With the increase of polypropylene fiber content, the angle between hysteretic curve and stress axis increases. The dynamic elastic modulus of NFLS decreases with the increase of polypropylene fiber content. Figure 9c,d shows that the hysteresis curves of FLS fluctuates with the increase of polypropylene fiber content and its area and angle with the stress axis. The FLS area is the largest and the included angle with the stress axis is the smallest when the content of polypropylene fiber is 0.5%. When the content of polypropylene fiber is 0.75%, the inclination angles of hysteresis curves increase gradually with the content of nano clay.

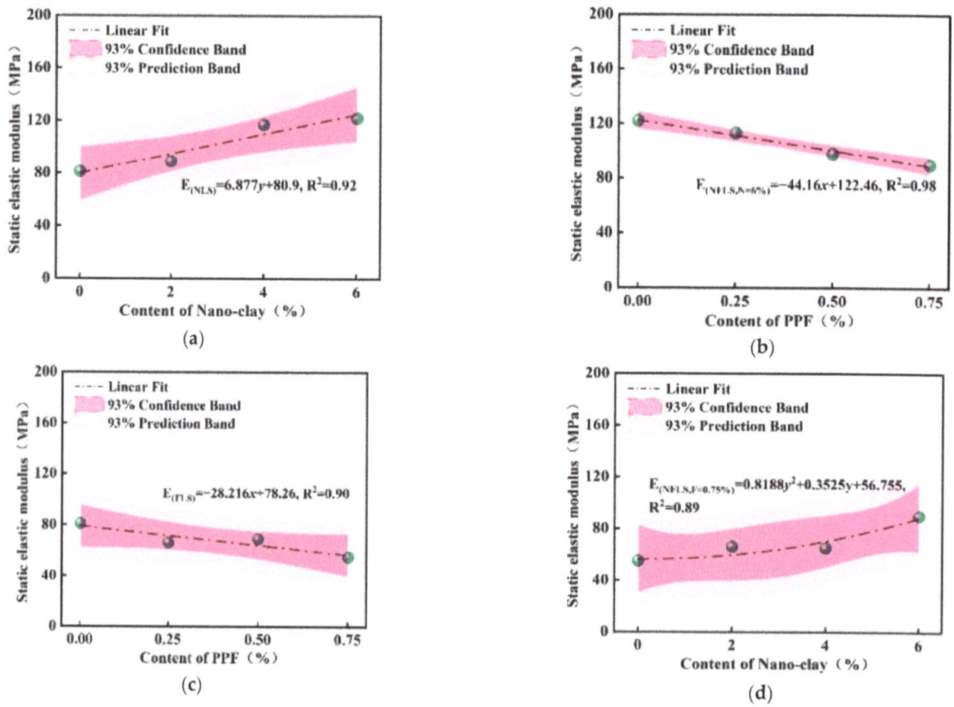

Figure 7. Static elastic modulus: (**a**) static elastic modulus curve of NLS; (**b**) static elastic modulus curve of NFLS (N = 6%); (**c**) static elastic modulus curve of FLS; (**d**) static elastic modulus curve of NFLS (F = 0.75%).

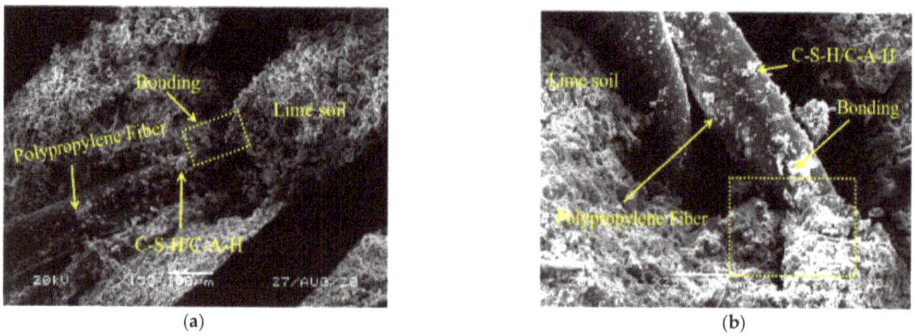

Figure 8. Microcosmic diagram of the interface between PPF and lime-soil: (**a**) 150-fold SEM; (**b**) 250-fold SEM.

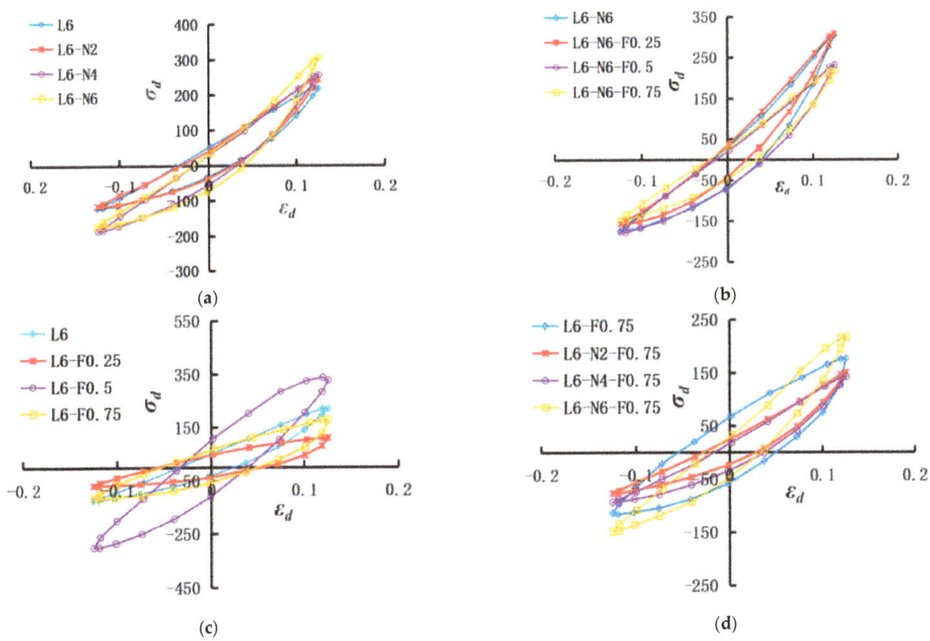

Figure 9. Dynamic stress–strain hysteresis curve: (**a**) change of NLS with nano clay content; (**b**) changes of NFLS (N= 6%) with PPF content; (**c**) changes of FLS with PPF content; (**d**) change of NFLS (F = 0.75%) with nano clay content.

3.2.2. Dynamic Elastic Modulus

According to the hysteretic curves, the dynamic elastic modulus of NFLS with different nano clay and polypropylene fiber content can be obtained according to Equation (8), as shown in Figure 10.

$$E_d = \frac{\sigma_d}{\varepsilon_d} \tag{8}$$

Figure 10a,b shows that the dynamic elastic modulus of NLS increases with the increase of nano clay content. The nano clay content increases from 0% to 6%, the dynamic elastic modulus of NLS increases to 189 MPa, and nano clay can significantly enhance the elastic modulus of LS. When the content of nano clay is 6%, the dynamic elastic modulus of NFLS decreases with the increase of the content of polypropylene fiber. Nano clay can enhance the dynamic elastic modulus of LS because nano clay can produce pozzolanic reaction by modifying LS to produce more cementitious materials and make its structure more compact. When the sample is pressed and does not break, the strain produced is very small, so the force on the fiber can be decomposed into a compressive force and a flexural force, which has almost no effect on the sample and mainly plays a role of reinforcement [35–37].

Figure 10c,d shows that the dynamic elastic modulus of FLS first decreases, then increases and then decreases with the increase of polypropylene fiber content. When the polypropylene fiber content is 0.5%, the dynamic elastic modulus of FLS is 120 MPa, while when the polypropylene fiber content increases to 0.75%, the dynamic elastic modulus decreases to 112 MPa. When the polypropylene fiber content is large, it is easy to agglomerate and interweave, reducing the compactness of soil. The capillary action of fiber will lead to soil erosion and reduce the dynamic elastic modulus of LS. When the content of polypropylene fiber is certain, the dynamic elastic modulus of NFLS decreases first and then increases significantly with the increase in the content of nano clay. When the content of nano clay is 6%, the dynamic elastic modulus of NFLS is 150 MPa.

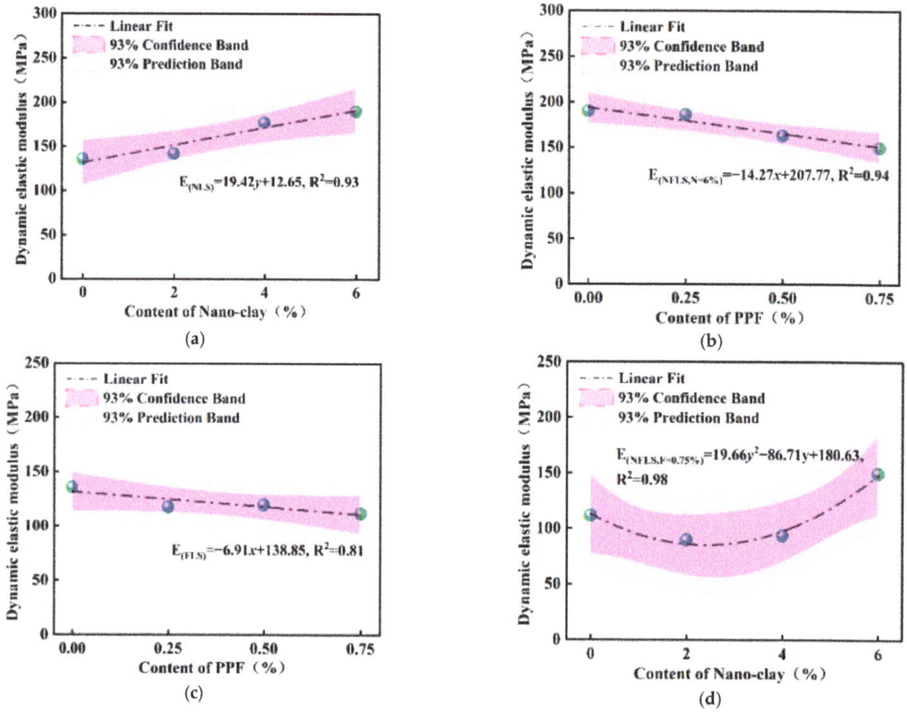

Figure 10. Dynamic elastic modulus: (**a**) dynamic elastic modulus curve of NLS; (**b**) dynamic elastic modulus curve of NFLS (N = 6%); (**c**) dynamic elastic modulus curve of FLS; (**d**) dynamic elastic modulus curve of NFLS (F = 0.75%).

The relationship between dynamic elastic modulus of NLS, FLS, and NFLS and content of nano clay and polypropylene fiber was fitted. The fitting formulas are shown in Equations (9)–(12) respectively.

$$E_{d(\mathrm{NLS})} = 19.42y + 112.65, R^2 = 0.93 \tag{9}$$

$$E_{d(\mathrm{NFLS,N=6\%})} = -14.27x + 207.77, R^2 = 0.94 \tag{10}$$

$$E_{d(\mathrm{FLS})} = -6.91x + 138.85, R^2 = 0.81 \tag{11}$$

$$E_{d(\mathrm{NFLS,F=0.75\%})} = 19.66y^2 - 86.71y + 180.63, R^2 = 0.98 \tag{12}$$

where E_d is dynamic elastic modulus (MPa).

3.2.3. Relationship between Static and Dynamic Elastic Modulus

Comparing Figures 9 and 10, it can be found that the content of polypropylene fiber and nano clay has a significant impact on the static and dynamic elastic modulus of NFLS. Therefore, the prediction model of the impact of different content of polypropylene fiber and nano clay on the static and dynamic elastic modulus of modified LS can be established. Figure 11a,b shows the prediction model of the relationship between the content of polypropylene fiber, nano clay, and static and dynamic elastic modulus. It reflects the relationship between the static and dynamic elastic modulus of NFLS and the content of polypropylene fiber and nano clay. Under the same nano clay content, the predicted model surface shows a downward convex shape with the increase of polypropylene fiber content, indicating that the static and dynamic elastic modulus of NFLS first decreases and then increases with the change of polypropylene fiber content. Under the same polypropylene fiber content, the prediction model surface increases with the increase of nano clay content, indicating that the static and dynamic elastic modulus of NFLS increases linearly with the increase of nano clay content. The fitting formulas of static and dynamic elastic modulus of NFLS are shown in Equations (13) and (14):

$$E = 86.9 - 42.39x + 0.99y^2, R^2 = 0.92 \tag{13}$$

$$E_d = 138.71 - 61.699x + 1.526y^2, R^2 = 0.85 \tag{14}$$

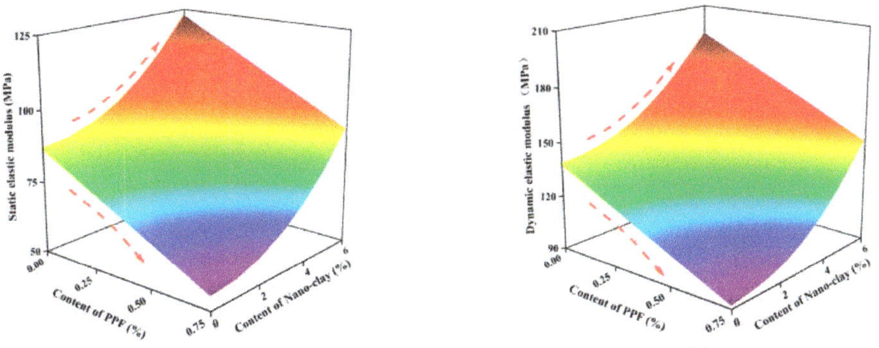

Figure 11. Influence of PPF and nano clay content variation on elastic modulus prediction model. (**a**) static elastic modulus; (**b**) dynamic elastic modulus.

From the comparison of static and dynamic elastic modulus prediction models according to Figure 11, it can be found that the change trend between the static and dynamic elastic modulus of NFLS and the content of nano clay and polypropylene fiber is basically the same. In the actual road engineering construction, the static elastic modulus is easy to obtain, but in the complex engineering environment, it is difficult to obtain the dynamic elastic modulus through the precise dynamic triaxial test. The static and dynamic elastic modulus test data of NFLS are fitted to establish the variation relationship of them, as shown in Figure 12. The static and dynamic elastic modulus of NLS, FLS, and NFLS conforms to the linear, exponential, and logarithmic relations respectively, and the fitting formula is shown in Equations (15)–(17).

$$E_{d(\text{NLS})} = 0.6557E + 75.325, R^2 = 0.99 \tag{15}$$

$$E_{d(\text{FLS})} = 29.382e^{0.0208E}, R^2 = 0.95 \tag{16}$$

$$E_{d(\text{NFLS})} = 165.66\ln(E) - 597.22, R^2 = 0.96 \tag{17}$$

3.3. Stress–Strain Damage Model and Damage Evolution Law of NFLS

According to the UCS test data, the elastic modulus of NFLS is substituted into the objective function [25,26], and the parameters λ and ζ are identified by particle swarm optimization algorithm. The identification results are shown in Table 4. The mean value of damage variables μ_D is calculated according to the identification results, and the stress–strain random damage model is established by using λ and ζ.

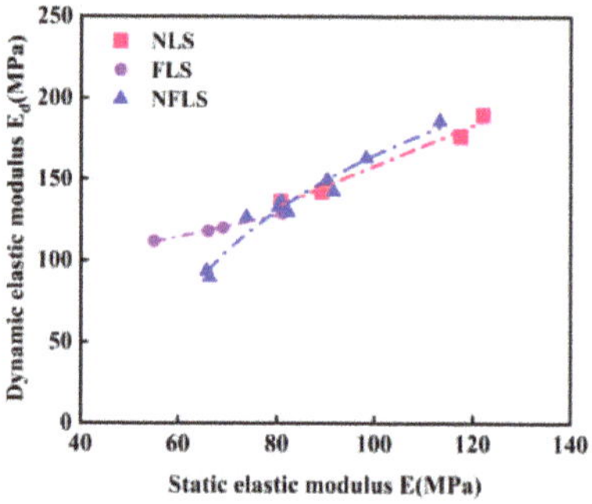

Figure 12. Relationship between static and dynamic elastic modulus.

Table 4. Calculation results of λ and ζ.

NO	λ	ζ	η (%)
L6	−3.5443	0.6054	1.0
L6-N2	−3.2778	0.4326	1.9
L6-N4	−3.1601	0.3403	2.5
L6-N6	−3.1448	0.2955	1.9
L6-F0.25	−3.2111	0.7107	3.3
L6-F0.5	−2.9546	0.7141	2.7
L6-F0.75	−2.8662	0.7064	3.6
L6-N2-F0.75	−2.8381	0.7826	0.5
L6-N4-F0.75	−2.9024	0.7730	1.5
L6-N6-F0.25	−2.9897	0.5031	1.6
L6-N6-F0.5	−2.6872	0.5547	1.3
L6-N6-F0.75	−2.6819	0.6176	0.8

Note, $\eta = q/\text{UCS}$, $q = \frac{\sqrt{\sum_{i}^{m}(\sigma_i - \sigma(\varepsilon_i))}}{m}$, where σ_i and σ (ε_i) is the measured stress and model calculated stress with the strain is ε_i, respectively, and m is the number of measured data groups.

Figure 13a–d shows the damage evolution curves of NFLS. When D = 0, the sample is not damaged. When D = 1, the sample is completely destroyed. When the strain is less than 1%, because the last day of curing is immersion curing, the early stage is mainly the compression process of sample pores, and there is no damage. With the increase of strain, the specimen is gradually compressed and destroyed until the damage variable D = 1 and the specimen is completely destroyed.

Figure 13a shows that in the early stage of NLS test, before the strain reaches 1%, the sample is in the compaction stage, and the stress changes little with the strain. When the strain increases from 1% to 5%, the specimen is mainly in the linear elastic stage and plastic stage, the stress changes obviously with the strain, and the damage accumulates gradually. When the strain is 5%, the damage variable D = 0.8, the sample begins to crack and plastic failure. When the strain is greater than 5%, the damage variable D tends to be stable with the increase of strain until the specimen is completely destroyed. Figure 13b shows the change of damage variable D of NFLS with the content of polypropylene fiber when the content of nano clay is 6%. When the strain is 8%, the damage variable of NFLS without fiber is D = 1 and the sample is completely destroyed. Then, with the increase of fiber content, the destruction speed of the sample slows down. Figure 13c reflects the change of damage variable D of FLS with polypropylene content. When the strain of the sample without fiber is 10%, the damage variable D = 1, and the sample is completely destroyed. With the increase of the content of polypropylene fiber, the failure rate of FLS slows down, and the sample is not completely destroyed until the strain is 12%. Figure 13d shows the curves of the damage variable *D* of NFLS with the content of nano clay when the content of fiber is fixed at 0.75%. With the increase of nano clay content, the damage curve is dense and the plastic damage is obvious. Due to the addition of polypropylene fiber, the damage of the sample slows down. When the strain is 10%, the sample is not completely damaged. On the one hand, nano clay can fill the pores between soil particles and delay the development of cracks. On the other hand, the fiber plays a reinforcing role, provides tension, and inhibits its plastic deformation.

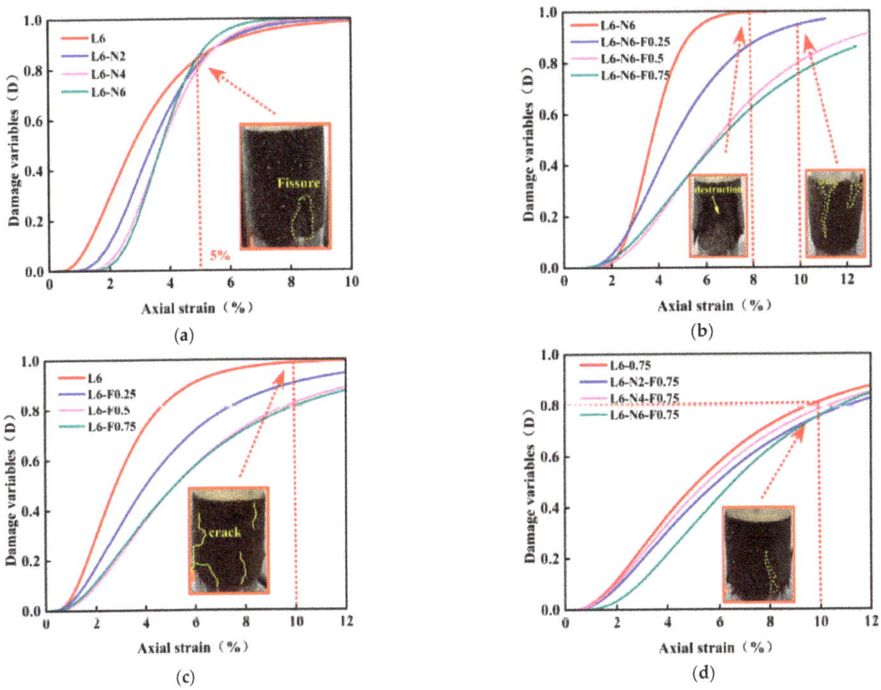

Figure 13. Mesoscopic damage evolution: (**a**) change of NLS with nano clay content; (**b**) changes of NFLS (N = 6%) with PPF content; (**c**) changes of FLS with PPF content; (**d**) change of NFLS (F = 0.75%) with nano clay content.

4. Conclusions

The elastic modulus and damage stress–strain model of NFLS are studied through UCS and dynamic triaxial test, and the following conclusions can be obtained.

(1) Polypropylene fiber and nano clay can significantly modify the strength of LS. The growth rate of UCS and residual strength of NFLS is the most significant when the content of polypropylene fiber is 0.75% and the content of nano clay is 6%. On the one hand, nano clay can promote the reaction between lime and soil, and on the other hand, it can react with LS to produce cementitious materials such as hydrated calcium silicate and hydrated calcium aluminate. Nano clay and the generated cementitious material can fill the pores between fiber and soil, enhance the interfacial friction between fiber and soil, and improve the strength of NFLS. UCS can be increased by up to 103%. Polypropylene fiber enhances the ductility of NFLS, slows down the development of cracks, and improves its residual strength. The residual strength can be increased by 827%.

(2) The static and dynamic elastic modulus of NLS, FLS, and NFLS are in functional relationship with the content of polypropylene fiber and nano clay. The addition of nano clay can improve the static and dynamic elastic modulus of NFLS and enhance the ability to resist deformation. In addition, the static and dynamic elastic modulus of NLS, FLS, and NFLS conform to linear, exponential, and logarithmic relationships respectively.

(3) The meso random damage model can characterize the stress–strain relationship of NFLS under axial load. At the same time, the damage variable *D* can further explain the relationship between strain and NFLS damage and failure, and reflect the damage and failure of soil. When the strain is 10%, the rising trend of damage variable *D* of NFLS slows down with the increase of polypropylene fiber and nano clay content, and the damage variable *D* decreases with the increase of nano clay content, indicating that nano clay can improve the strength of NFLS. When the damage variable *D* is 0.8, the axial strain of NFLS increases with the increase of the content of polypropylene fiber and nano clay, and the axial strain increases from 8% to 12% with the increase of the content of polypropylene fiber, indicating that polypropylene fiber can improve the ductility of NFLS. Adding polypropylene fiber and nano clay can effectively improve the damage resistance of NFLS.

The UCS and residual strength of NFLS can be significantly improved by modifying LS with nano clay and polypropylene fiber, and the brittle failure mode of lime soil can be improved. Subsequent microscopic tests can also be carried out to further explain the reasons for the enhancement of NFLS mechanical properties from the perspective of chemical mechanism, providing a theoretical basis for the practical application of NFLS in road engineering.

Author Contributions: Conceptualization, P.J.; investigation, W.Z.; writing—original draft preparation, Z.W.; writing—review and editing, C.L. All authors have read and agreed to the published version of the manuscript.

Funding: This research was funded by the National Natural Science Foundation of China (grant number 41772311), Zhejiang Provincial Natural Science Foundation of China (Grant number LQ20E080042), the Open Research Fund of State Key Laboratory of Geomechanics and Geotechnical Engineering, Institute of Rock and Soil Mechanics, Chinese Academy of Science (Grant number Z017013), and the Scientific Research Projects of Zhejiang Department of Housing and Urban and Rural Construction of China (Grant number 2017K179, 2019K171).

Institutional Review Board Statement: Not applicable.

Informed Consent Statement: Not applicable.

Data Availability Statement: Not applicable.

Conflicts of Interest: The authors declare no conflict of interest.

References

1. Liu, Y.; Wang, Q.; Liu, S.; ShangGuan, Y.; Fu, H.; Ma, B.; Chen, H.; Yuan, X. Experimental investigation of the geotechnical prop-erties and microstructure of lime-stabilized saline soils under freeze-thaw cycling. *Cold Reg. Sci. Technol.* **2019**, *161*, 32–42. [CrossRef]
2. Keybondori, S.; Abdi, E. Lime stabilization to improve clay-textured forest soil road subgrades. *Int. J. For. Eng.* **2021**, *32*, 112–118. [CrossRef]
3. Dash, S.; Hussain, M. Lime Stabilization of Soils: Reappraisal. *J. Mater. Civ. Eng.* **2012**, *24*, 707–714. [CrossRef]
4. Lemaire, K.; Deneele, D.; Bonnet, S.; Legret, M. Effects of lime and cement treatment on the physicochemical, microstructural and mechanical characteristics of a plastic silt. *Eng. Geol.* **2013**, *166*, 255–261. [CrossRef]
5. Pu, S.; Zhu, Z.; Song, W.; Wan, Y.; Wang, H.; Xu, X. Comparative Study of Compressibility and Deformation Properties of Silt Stabilized with Lime, Lime, and Cement, and SEU-2 Binder. *Arab. J. Sci. Eng.* **2020**, *45*, 4125–4139. [CrossRef]
6. Kutanaei, S.S.; Choobbasti, A.J. Triaxial behavior of fiber-reinforced cemented sand. *J. Adhes. Sci. Technol.* **2016**, *30*, 579–593. [CrossRef]

. Jiang, P.; Chen, Y.; Wang, W.; Yang, J.; Wang, H.; Li, N.; Wang, W. Flexural behavior evaluation and energy dissipation mechanisms of modified iron tailings powder incorporating cement and fibers subjected to freeze-thaw cycles. *J. Clean. Prod.* **2022**, *351*, 131527. [CrossRef]

.. Khan, S.Z.; Rehman, Z.u.; Khan, A.H.; Qamar, S.; Haider, F. Effect of Polypropylene Fibers and Cement on the Strength Improvement of Subgrade Lying on Expansive Soil. *IJST-T Civ. Eng.* **2022**, *46*, 343–352. [CrossRef]

.. Abdi, M.R.; Ghalandarzadeh, A.; Shafiei Chafi, L. An investigation into the effects of lime on compressive and shear strength characteristics of fiber-reinforced clays. *J. Rock Mech. Geotech. Eng.* **2021**, *13*, 885–898. [CrossRef]

10. Ghasabkolaei, N.; Janalizadeh Choobbasti, A.; Roshan, N.; Ghasemi, S.E. Geotechnical properties of the soils modified with nanomaterials: A comprehensive review. *Arch. Civ. Mech. Eng.* **2017**, *17*, 639–650. [CrossRef]

11. Niu, X.J.; Li, Q.B.; Hu, Y.; Tan, Y.S.; Liu, C.F. Properties of cement-based materials incorporating nano-clay and calcined nano-clay: A review. *Constr. Build. Mater.* **2021**, *284*, 122820. [CrossRef]

12. Gao, L.; Ren, K.Y.; Ren, Z.; Yu, X.J. Study on the shear property of nano-MgO-modified soil. *Mar. Georesour. Geotechnol.* **2018**, *36*, 465–470. [CrossRef]

13. Lin, D.F.; Luo, H.L.; Chen, C.T.; Cai, D. Study properties of soft subgrade soil stabilized by sewage sludge/lime and nano-SiO2. *Geomech. Eng.* **2016**, *10*, 793–806. [CrossRef]

14. Theodoridou, M.; Charalambous, E.; Maravelaki-Kalaitzaki, P.; Ioannou, I. Amelioration of crushed brick—Lime composites using nano-additives. *Cem. Concr. Comp.* **2016**, *68*, 77–87. [CrossRef]

15. Duran, A.; Navarro-Blasco, I.; Fernández, J.M.; Alvarez, J.I. Long-term mechanical resistance and durability of air lime mortars with large additions of nanosilica. *Constr. Build. Mater.* **2014**, *58*, 147–158. [CrossRef]

16. Zhao, G.; Yan, H.; Yan, Y.; Gao, S.; Fan, D.; He, Y.; Wang, Y. Experimental study on the dynamic properties of lime-solidified soft soil under cyclic loading. *Arab. J. Geosci.* **2020**, *13*, 671. [CrossRef]

17. Lu, Z.; Zhao, Y.; Xian, S.; Yao, H. Experimental Study on Dynamic Resilient Modulus of Lime-Treated Expansive Soil. *Adv. Mater. Sci. Eng.* **2020**, *2020*, 3272681. [CrossRef]

18. Zhou, B.; Kong, L.-W.; Guo, A.G. Stress-strain-strength behaviour and constitutive description of lime-treated expansive soil. *Rock Soil Mech.* **2012**, *33*, 999–1005.

19. Poinard, C.; Piotrowska, E.; Malecot, Y.; Daudeville, L.; Landis, E. Compression triaxial behavior of concrete: The role of the mesostructure by analysis of X-ray tomographic images. *Eur. J. Environ. Civ. Eng.* **2012**, *16*, 115–136. [CrossRef]

20. Jiang, P.; Zhou, L.; Zhang, W.; Wang, W.; Li, N. Unconfined Compressive Strength and Splitting Tensile Strength of Lime Soil Modified by Nano Clay and Polypropylene Fiber. *Crystals* **2022**, *12*, 285. [CrossRef]

21. Wang, M.; Xu, X.; Li, J.; Shen, F.; Yafeng, L. An experiment study on stress relaxation of unsaturated lime-treated expansive clay. *Environ. Earth Sci.* **2017**, *76*, 1–12. [CrossRef]

22. Research Institute of Highway Ministry of Transport. *Test Methods of Materials Stabilized with Inorganic Binders for Highway Engineering*; China Communications Press: Beijing, China, 2009.

23. Research Institute of Highway Ministry of Transport. *Test Methods of Soils for Highway Engineering*; China Communications Press: Beijing, China, 2020.

24. Li, J.; Yang, W. Elastoplastic stochastic damage constitutive law for concrete. *China Civ. Eng. J.* **2009**, *42*, 31–38.

25. Li, J.; Ren, X. Recent developments on stochastic damage mechanics for concrete. *J. Build. Struct.* **2014**, *35*, 20–29.

26. Pakravan, H.; Latifi, M.; Jamshidi, M. Ductility improvement of cementitious composites reinforced with polyvinyl alcohol-polypropylene hybrid fibers. *J. Ind. Text.* **2014**, *45*, 637–651. [CrossRef]

27. Jiang, P.; Zhou, L.; Mao, T.; Yuan, J.; Wang, W.; Li, N. Damage model and time effect of cement-modified waste slurry. *J. Jilin Univ.* **2021**, *51*, 1–8.

28. Jiang, P.; Zhou, L.; Wang, Y.; Qian, B.; Wang, W.; Li, N.; Zhang, F. Freeze–Thaw Damage Model of Polypropylene Fiber Reinforced Cement Stabilized Waste Construction Slurry under Uniaxial Action. *Minerals* **2021**, *11*, 743. [CrossRef]

29. Deb, S.; Mitra, N.; Majumder, S.B.; Maitra, S. Improvement in tensile and flexural ductility with the addition of different types of polypropylene fibers in cementitious composites. *Constr. Build. Mater.* **2018**, *180*, 405–411. [CrossRef]

30. Wang, W.; Zhu, X. Study on strength property of nanometer silica fume reinforced cemented soil and reinforcement mechanism. *Rock Soil Mech.* **2004**, *25*, 922–926.

31. Qian, B.; Yu, W.; Lv, B.; Kang, H.; Shu, L.; Li, N.; Wang, W. Mechanical Properties and Micro Mechanism of Nano-Clay-Modified Soil Cement Reinforced by Recycled Sand. *Sustainability* **2021**, *13*, 7758. [CrossRef]

32. Song, X.; Xu, H.; Zhou, D.; Yao, K.; Tao, F.; Jiang, P.; Wang, W. Mechanical Performance and Microscopic Mechanism of Coastal Cemented Soil Modified by Iron Tailings and Nano Silica. *Crystals* **2021**, *11*, 1331. [CrossRef]

33. Zheng, D.; Cui, H.; Tang, W.; Sang, G.; Lo, T.Y. Influence and mechanisms of active silica in solid waste on hydration of tricalcium aluminate in the resulting composite cement. *Mater. Today Commun.* **2021**, *27*, 102262. [CrossRef]

34. Wang, J.; Chang, Z.; Tang, Y. Dynamic triaxial test analysis of reinforced gravel soil under cyclic loading. *Rock Soil Mech.* **2020**, *41*, 2851–2860.

35. Wang, W.; Kang, H.B.; Li, N.; Guo, J.; Girma, D.Y.; Liu, Y. Experimental investigations on the mechanical and microscopic behavior of cement-treated clay modified by nano-MgO and fibers. *Int. J. Geomech. ASCE* **2022**, *22*, 04022059. [CrossRef]

36. Kang, S.; Lee, B.; Kim, B.; Kim, Y. The effect of fibre distribution characteristics on the flexural strength of steel fibre-reinforced ultra high strength concrete. *Constr. Build. Mater.* **2011**, *25*, 2450–2457. [CrossRef]
37. Jin, J.; Li, C.; Yuan, S.; Sun, Q.; Yang, H. Effect of fiber on early strength and interface stiffness of cemented tailings backfill. *Mater. Res. Express.* **2022**, *9*, 045202. [CrossRef]

polymers

MDPI

Article

Study on Compressive Properties and Dynamic Characteristics of Polypropylene-Fiber-and-Cement-Modified Iron-Ore Tailing under Traffic Load

Ping Jiang [1,2], Yewen Chen [1], Xinjiang Song [3], Na Li [1], Wei Wang [1] and Erlu Wu [2,*]

1 School of Civil Engineering, Shaoxing University, Shaoxing 312000, China; jiangping@usx.edu.cn (P.J.); ccw181898@163.com (Y.C.); lina@usx.edu.cn (N.L.); wellswang@usx.edu.cn (W.W.)
2 Key Laboratory of Ministry of Education for Geomechanics and Embankment Engineering, Hohai University, Nanjing 210098, China
3 Anhui and Huaihe River Water Resources Research Institute, Bengbu 233000, China; sxj06@163.com
* Correspondence: 170804010004@hhu.edu.cn

Abstract: Using polypropylene (PP) fiber and cement to modify iron-ore tailing and applying it to road engineering is an effective way to reuse iron-ore tailing. The compressive properties and deformation characteristics of PP-fiber-and-cement-modified iron-ore tailing (FCIT) under traffic load were studied by the unconfined-compressive-strength (UCS) test and the dynamical-triaxial (DT) test. The test results indicated that the UCS and residual strength both increased with increasing PP-fiber content, and tensile and toughness properties were positively correlated with PP-fiber content. Moreover, the dynamic elastic modulus and damping of FCIT both showed a negative linear relationship with cycle time. It can be found from the test results that 0.75% was the best PP-fiber content to modify iron tailing sand in this work. Lastly, a prediction model was developed to describe the relationship between the cumulative plastic strain, PP-fiber content and cycle time, which can effectively capture the evolution law of the cumulative plastic strain with cycle time of FCITs at different PP-fiber contents.

Keywords: iron-ore tailing; polypropylene fiber; unconfined compressive properties; dynamic properties; cumulative plastic strain

Citation: Jiang, P.; Chen, Y.; Song, X.; Li, N.; Wang, W.; Wu, E. Study on Compressive Properties and Dynamic Characteristics of Polypropylene-Fiber-and-Cement-Modified Iron-Ore Tailing under Traffic Load. *Polymers* **2022**, *14*, 1995. https://doi.org/10.3390/polym14101995

Academic Editor: Markus Gahleitner

Received: 21 April 2022
Accepted: 11 May 2022
Published: 13 May 2022

1. Introduction

The stacking of iron-ore tailing has a negative impact on the surrounding environment and people. The resource application of iron-ore tailing is an effective way to reduce the break risk of iron tailing dams. Therefore, some investigations have been focused on the resource application of iron-ore tailing. Yuan [1], Suthers [2] and Lv [3] et al. studied the application of iron-ore tailing in the recovery of valuable metal. Yao [4] and Shettim [5] studied the application of iron-ore tailing as aggregate in the preparation of concrete. Luo [6] and Kuranchi [7] studied the use of iron-ore tailing in brick making. In addition, there have also been some studies and applications of iron-ore tailing in cement composites [8] and filling materials [9,10]. As road-engineering construction needs a large amount of subgrade filler, the large-scale exploitation of river sand and natural sand as road-base materials has seriously affected the natural ecological environment. Zeng [11] and Bastosh [12] et al. investigated the mechanical properties of iron-ore tailing and found that iron-ore tailing can be used as subgrade road filler after chemical stabilization, which can replace river sand and natural sand.

At present, iron-ore tailing needs to be modified with materials so that it can be used as subgrade road filler. Quite a few researchers have mainly focused on modifying materials and investigating the mechanical properties of modified iron-ore tailing. Cement is an inorganic cementitious material commonly used to modify soil [13]. Therefore, some scholars modified iron-ore tailing with cement, such as Li et al. [14], who studied the physical and mechanical properties of cement-modified iron-ore tailings (CITP) and found that the cementation of cement can enhance the bonding between iron-ore tailings and improve its compressive strength and shear strength. Liu et al. [15] verified that CITP can meet the specification requirements of highway pavement base through the unconfined-compressive-strength (UCS) test and pointed out that the strength characteristics of CITP and cement soil are similar. Although the addition of cement can improve the UCS of iron-ore tailing, it can be seen from the studies of many scholars on CITP that the failure mode of CITP is mostly brittle failure, which can lead to serious consequences. Therefore, the fiber is added to CITP in order to change the failure mode. Yang et al. [16] used fiber and cement to modify loess and found that the failure mode of modified loess transitions from brittle failure to ductile failure with the increase in fiber content. Jin et al. [17] studied the UCS of iron-ore tailing modified with PP fiber, and found that the PP fiber can indeed improve the UCS, but the content of PP fiber has little effect on the UCS of modified iron-ore tailing. Jiang et al. [18] mixed polypropylene fiber with iron tailing sand for improvement and explained that PP fiber improves the shear performance of iron-ore tailing from the perspective of energy dissipation.

Random discrete fiber reinforcement is an efficient technology to improve the mechanical properties of soil, which can improve the strength and deformation characteristics of soil under static and dynamic loads to a certain extent. The existing studies mainly focus on the static properties of fiber-reinforced soil; only a few researchers have studied the dynamic response characteristics of fiber-reinforced soil, especially the cumulative strain of reinforced soil. Zhang et al. [19] mainly analyzed the development law of cumulative strain with dynamic stress amplitude and vibration time of fiber-reinforced soil and established an exponential model to describe the relationship between the cumulative strain, amplitude and cycle times. Zhao et al. [20] conducted cyclic loading tests on fiber-reinforced sand and derived an empirical formula between the maximum dynamic shear modulus and damping ratio. Wang et al. [21] put forward a functional relationship between the cumulative strain and cycle time of fiber-reinforced soil. Moreover, the fiber content has a significant impact on the dynamic response characteristics of modified soil. For example, Maher et al. [22] found that both the damping ratio and dynamic shear modulus of fiber-reinforced sand increased nearly linearly with the increase in fiber content. Li et al. [23] found that the dynamic elastic modulus of fiber-reinforced soil increases with the increase in fiber content in the dynamic-triaxial experiment. Orakoglu et al. [24] found that the addition of fiber increases the damping ratio and dynamic shear modulus and established a functional relationship between dynamic shear stress and dynamic shear modulus.

It can be found from the existing research that there are few studies on the mechanical properties of FCIT, especially the strength and deformation characteristics under traffic load. That is, a large number of experiments on the mechanical properties still need to be conducted for the application of FCIT in road engineering. To this end, the unconfined-compressive-strength test and dynamic-triaxial test were conducted to investigate the strength and deformation characteristics of FCIT under traffic load in this paper, which can provide the experimental and theoretical support for the application of modified iron-ore tailing in road engineering.

2. Materials and Methods

2.1. Materials

The iron-ore tailings used in the test were obtained from the Lizhu iron-tailing plant in Shaoxing, China, as shown in Figure 1. The liquid limit and plasticity index of the iron-ore tailing were 36.95% and 29.35, respectively, and the main compounds are shown in Table 1. The P.O. 42.5 cement used in the test was produced by Shaoxing Zhaoshan Building Materials Co., Ltd. in Shaoxing, China, as shown in Table 2. The PP fiber used in the test was produced by Shaoxing Fiber High Tech Co., Ltd. in Shaoxing, China, and had good dispersion with a length of 6 mm and a diameter of 48 μm, as shown in Table 3.

Figure 1. Iron-ore tailings for test.

Table 1. Main chemical composition and physical properties of iron-ore tailing.

Composition (%)			Liquid Limit (%)	Plasticity Index (%)
CaO	Fe_2O_3	SiO_2		
24.8	22.9	21.2	36.95	29.35

Table 2. Physical and mechanical properties of P.O. 42.5 cement.

Initial Setting Time (Min)	Final Setting Time (Min)	Compressive Strength		Break Off Strength	
		3 d	28 d	3 d	28 d
≥45	≤600	≥17	≥42.5	≥3.5	≥5.5

Table 3. Physical properties of polypropylene fiber.

Diameter (mm)	Specific Gravity (g/cm^3)	Tensile Strength (MPa)	Elastic Modulus (MPa)
0.048	0.91	340	4200

2.2. Test Scheme

The natural environment of iron-ore tailing is convenient for the actual comprehensive utilization of subgrade engineering. With reference to the liquid limit of iron tailing sand, the design moisture content was 30%. Considering economic cost and meeting the road-base requirements of mechanical properties, the cement content was 10%, and the contents of PP fiber were 0%, 0.25%, 0.5%, 0.75% and 1%. The water content, cement content, and PP-fiber content were defined as the ratio of the mass of water to the mass of dry iron-ore tailing, the ratio of the mass of cement to the mass of dry iron-ore tailing, and the ratio of the mass of PP fiber to the mass of dry iron-ore tailing, respectively.

UCS is the most commonly used mechanical index in road engineering. Therefore, the UCS test was used to study the influence of fiber content on the unconfined compressive performance of FCIT under static load. Furthermore, the dynamic-triaxial test was performed to investigate the dynamic characteristics of FCIT under cyclic traffic load. For general highway traffic, the low cyclic stress ratio (σ_d/σ_3) is in the range of 0.1 to 0.75 [25]. σ_d is taken as the amplitude of dynamic load, and σ_3 is the confining pressure. According to the amplitude of traffic load, σ_d during the test was set as 75 kPa. Moreover, the frequency, confining pressure and number of cycles were set as 1 Hz, 100 kPa and 1000, respectively by considering the normal driving density and speed under general road conditions, the lateral pressure on the general subgrade, and the number of cycles on the subgrade under general traffic flow. The test details are indicated in Table 4.

Table 4. FCIT test scheme.

Test	PP Content (%)	Loading Rate (mm/min)	Loading Frequency (Hz)	Cycle Time	Amplitude (kPa)	Confining Pressure (kPa)
UCS	0 0.25 0.50 0.75 1	1	—	—	—	0
DT	0 0.25 0.50 0.75 1	—	1	1000	75	100

2.3. Specimen Preparation

As shown in Figure 2, the required weight of iron-ore tailing, cement and fiber were poured into a mixing drum. First, the mixture was blended until the three ingredients were well combined. Then, corresponding quality of water was weighed and poured into the mixing bucket. After blending, the FCIT mixture was obtained. The FCIT mixture was poured into the specimen mold with a diameter of 39.1 mm and a height of 80 mm in two batches. After each addition, the FCIT mixture was vibrated up and down with the mold in order to vibrate the air bubbles out. After specimen preparation, the specimen needs to sit for an hour to develop the specimen shape. Next, the white bearing specimen was removed using filter papers to wrap both ends of the specimen, the number was marked, and it was placed in water. Finally, the specimens were placed in a standard curing chamber at 20 ± 2 °C with humidity of 95% for 7 d and 28 d.

Figure 2. Sample preparation flow.

2.4. Test Apparatus

The UCS test applies axial pressure to the specimen with a constant load P. Figure 3a shows the stress state of UCS test specimen. The instrument used for the UCS test was a TKA-WCY-1F full-automatic multifunctional unconfined-compressive-strength tester, as seen in Figure 3b. Three parallel specimens were used for the test, and the test loading rate was 1 mm/min.

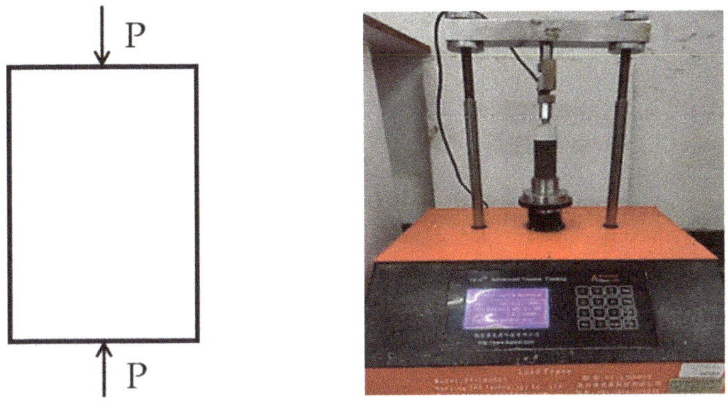

(a) (b)

Figure 3. UCS test. (**a**) Unconfined-stress state of specimen; (**b**) Test apparatus.

The DT test is a commonly used method to test the dynamic characteristics of soil. The stress state of the DT-test specimen is indicated in Figure 4a. The instrument used in the DT test was a GDS dynamic-triaxial instrument, as shown in Figure 4b. The dynamic load with the form of a sine wave was adopted in the DT test and 20 data points were collected in each cycle.

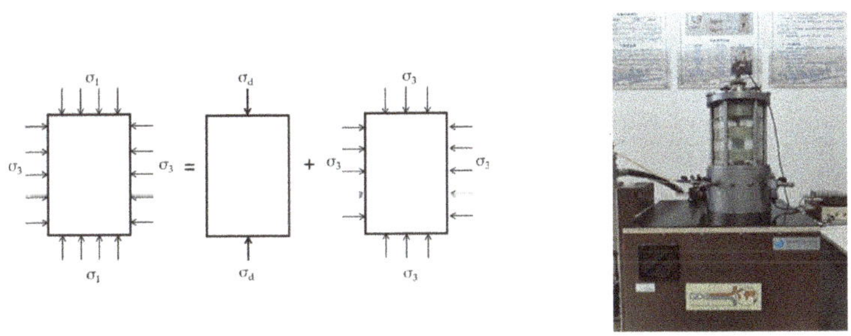

(a) (b)

Figure 4. DT test. (**a**) Triaxial-stress state of specimen; (**b**) Test apparatus.

3. Results and Discussion

3.1. Unconfined Compressive Properties

3.1.1. Characteristics of Stress–Strain Curve

Figure 5 gives the stress–strain curves of FCITs with different PP-fiber contents at the curing ages of 7 d and 28 d. It can be seen from Figure 5 that there are four stages in the stress–strain curve of FCIT: the linear-elastic stage, plastic stage, stress-attenuation stage and residual-strength stage. Moreover, with the increase in PP-fiber content, the brittleness of the specimen decreases and the plasticity increases, indicating that the PP fiber has an obvious strengthening effect on the plasticity of the cement-stabilized iron tailing.

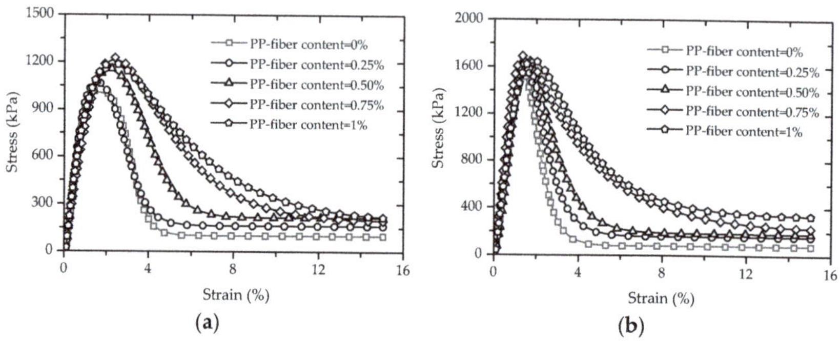

Figure 5. Stress–strain behaviors of specimen in unconfined-compressive test: (**a**) 7 d; (**b**) 28 d.

In the UCS-test process, the deformation of the specimen is effectively weakened due to the friction and tensile action of the PP fiber. The bonding action between the PP fiber and the soil particle inhibits the relative sliding of the soil particles and further delays the cracking of the specimen. The bonding force between the PP fiber and soil interface increases with increasing PP-fiber content, and a spatial-network structure is formed between staggered PP fibers, which can restrict the sliding and rolling of soil particles. That is, the stability of the specimen is improved at the spatial level and the plasticity is increased.

3.1.2. UCS and Residual Strength

From the stress–strain curves in Figure 5, the UCSs of FCITs with different PP-fiber contents and curing ages can be obtained, as indicated in Figure 6. The UCSs of FCITs with different PP-fiber contents at the age of 7 d are greater than 1 MPa which meets the strength requirement of relevant specification [26] for cement-stabilized material. As indicated in Figure 6, the UCSs of FCITs with different PP-fiber contents all increase by more than 35% with increasing curing age. The reason for the increasing strength is the increased hydration products of cement, such as C-S-H and C-A-S-H. On the one hand, these hydration products fill the pores between particles. On the other hand, the bonding between particles is strengthened through cementation, as well as the bonding between particles and PP fibers. Therefore, the addition of PP fiber can improve the UCS of FCIT. Taking the specimens with the curing age of 7 d as an example, compared with the specimen without fiber, the UCSs of the specimens with PP-fiber contents of 0.25%, 0.5%, 0.75% and 1% are increased by 5%, 13%, 20% and 17%, respectively. The UCS of the specimen with the curing age of 28 d has a similar change law as the specimen with the curing age of 7 d. The addition of PP fiber improves the ability of the specimen to resist external force, so as to improve the strength of the specimen. The UCS first increases and then decreases with the increasing content of PP fiber. The reason for this change is that the fiber will present a spatial-network structure in the specimen, and the agglomeration phenomenon will occur in the specimen with undue PP fiber, which will cause the cement particles and iron-ore-tailing particles to be unable to fully contact, thereby decreasing the UCS.

Figure 7 shows the relationship between the residual strength and the PP-fiber content of the FCIT specimen. It can be seen from Figure 7 that the PP-fiber contents of 0.25% and 0.5% have little reinforcement effects on the residual strength of FCIT. On the contrary, there is an obvious improvement effect on the residual strength of FCIT with the PP-fiber content of 0.75%. Compared to fiberless samples, the residual strength of the samples with 0.75% fiber content increases by 320% and 580% at 7 d and 28 d, respectively. Significantly, when the PP-fiber content increases from 0.75% to 1%, there is little improvement effect on the residual strength of FCIT, especially the residual strength at 28 d. In fact, the PP fibers in the specimen can form a staggered spatial-network structure, and this network structure restricts the slip of specimen particles so that the deformation resistance of the specimen is enhanced. Especially, it can be seen from Figures 6 and 7 that the reinforcement effect is best when the PP-fiber content is 0.75%.

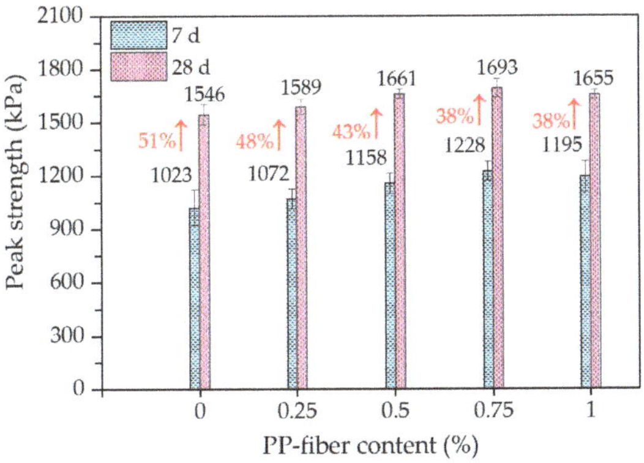

Figure 6. Unconfined compressive strengths of FCIT at different ages.

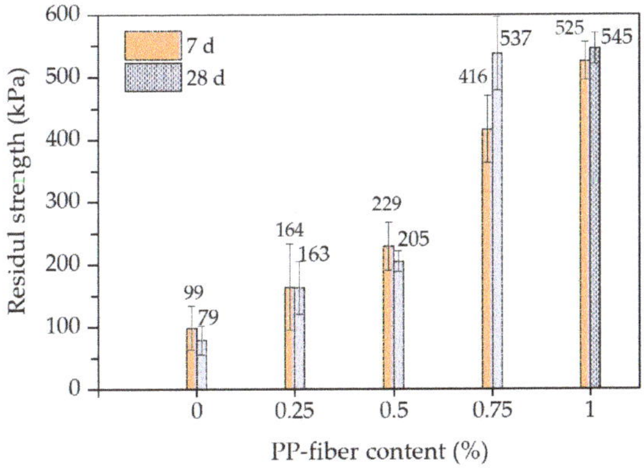

Figure 7. Residual strengths of FCIT at different curing ages.

3.1.3. Brittleness Index, Toughness Index, and Modulus Strength Ratio

Brittle materials often fail suddenly without warning when impacted by external forces. Although cement can better improve the strength of iron-ore tailing, its brittleness is also greatly increased [27]. In order to evaluate the effect of PP-fiber content on the brittle failure of FCIT, the brittleness index (I_B) is introduced to quantitatively describe the failure mode of FCIT. The value of I_B can be calculated according to Equation (1) [28]. The calculated values of I_B are indicated in Figure 8. As shown in Figure 8, added PP fiber can effectively reduce the brittleness index of FCIT and gradually change its failure mode from brittleness to plasticity, so as to improve the safety of the pavement base. Obviously, the value of the brittleness index linearly decreases with the increase in PP-fiber content, and the PP-fiber content and brittleness index of FCIT meet the linear relationship, as shown in Equation (2).

$$I_B = \frac{q_u - q_c}{q_u} \tag{1}$$

where q_u is the unconfined compressive strength, and q_c is the residual strength.

$$I_B = I_{B0} - \Delta I_B f \tag{2}$$

where f is PP-fiber content, I_{B0} is the brittleness index when f = 0, and ΔI_B is a material parameter and reflects the decreasing rate of I_B with increasing f.

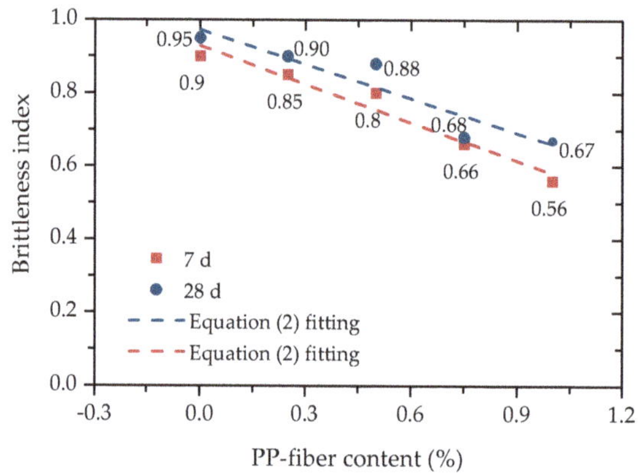

Figure 8. Relationship between brittleness index and PP–fiber content.

In order to describe the mechanical properties of FCIT after yield, the energy dissipa tion of FCIT after yield can be quantitatively characterized by the toughness index (T_I). Th higher the toughness index, the better the ability of the absorbing energy after yield [29 T_I can be calculated according to Equation (3), and then the calculated values of T_I ar shown in Figure 9. Apparently, there is an exponential relationship between the conter of PP fiber and T_I, as shown in Equation (4). It can be found that there is an increase i the difference between T_I of 7 d and 28 d with increasing PP-fiber content. The reason fo this is the more sufficient hydration reaction of cement and the more obvious interfacia bonding between the PP fiber and iron-ore-tailing particles so that FCIT can absorb mor strain energy after yield.

$$T_I = \frac{\int_{\varepsilon_q}^{\varepsilon_q+5\%} \sigma(\varepsilon)\mathrm{d}\varepsilon}{5\%} \quad (3)$$

where, T_I is the strain corresponding to the maximum value of stress.

$$T_I = T_{I0}e^{\Delta T_I f} \quad (4)$$

where T_{I0} is the toughness index when $f = 0$, and ΔT_I is a material parameter.

Due to the tensile action of PP fiber, the crack expands slowly in the specimen afte the FCIT specimen yields. The broken test block does not immediately fall but is bonded t the specimen (Figure 10), which can continue to bear the load. That is, the broken test bloc can absorb part of the external energy and improve the toughness index of FCIT.

The modulus strength ratio can be used to characterize the tensile performance o FCIT. The modulus strength ratio refers to the ratio of soil modulus to strength [30,31], a given by

$$\eta = \frac{E_{50}}{q_u} \quad (5)$$

where, E_{50} is the secant modulus.

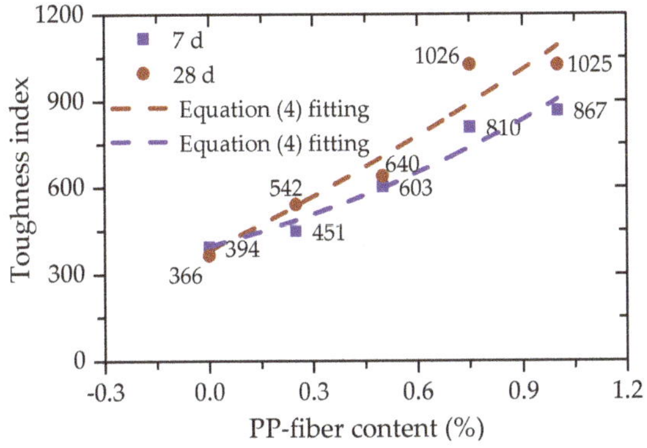

Figure 9. Relationship between toughness index and PP–fiber content.

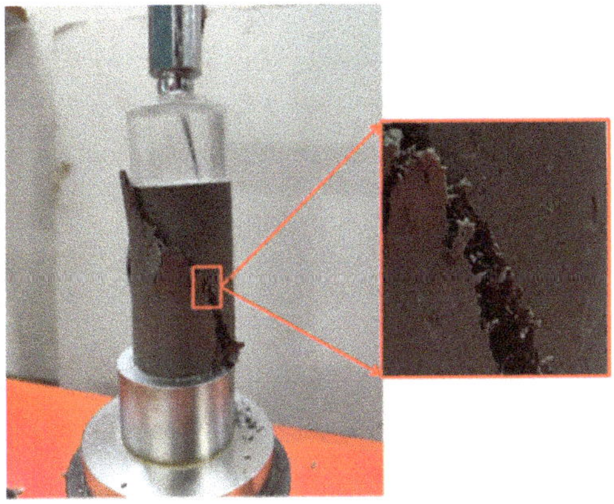

Figure 10. Damaged surface of specimen.

There is a negative correlation between the modulus strength ratio and the tensile performance of the material. The modulus strength ratio of FCIT can be calculated through Equation (5), as shown in Figure 11. It can be seen from Figure 11 that the modulus strength ratio of FCIT decreases with the increase in fiber content, indicating that the addition of fiber can improve the tensile properties of FCIT.

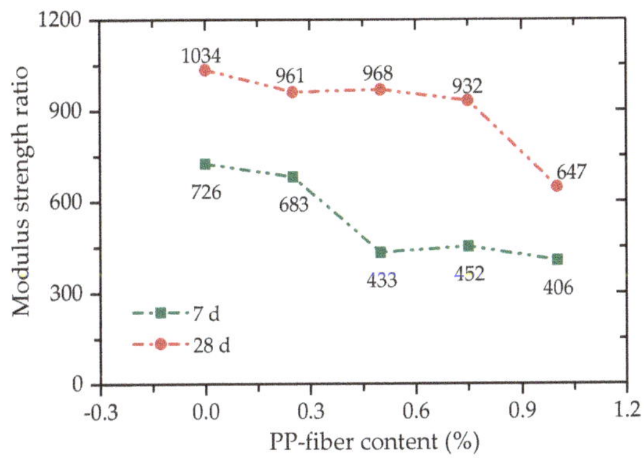

Figure 11. Relationship between modulus strength ratio and PP-fiber content.

4. Dynamic Characteristics

When the pavement base is subjected to cyclic traffic load, the base material will produce strain accumulation, and the cumulative strain linearly increases with the increasing number of cycles. Finally, it will evolve into two possible results. One is a strain-softening type in which the soil deformation reaches the limit value and then fails. The other is a strain-hardening type in which the material strength increases with increasing strain and the dynamic equilibrium is achieved when the final strain reaches a certain value. Dynamic stress–strain characteristics, elastic modulus, damping ratio and cumulative strain can be used to characterize the dynamic characteristics of FCIT under cyclic traffic load.

4.1. Dynamic Stress–Strain Characteristics

Figure 12 shows the stress–strain hysteretic curve of FCIT under cyclic loading. The red dotted arrow in Figure 12 indicates the test loading process. A complete cycle is formed by applying compressive stress to the specimen and releasing the stress. The hysteresis loop area can effectively reflect the damping and energy dissipation of soil. The fuller the hysteresis loop in Figure 12, the greater the damping ratio of the specimen, that is, the greater the energy dissipation. According to the hysteresis loop in the figure, the dynamic elastic modulus E and damping ratio λ can be calculated.

$$E = \frac{L_{\mathrm{AB}}}{L_{\mathrm{OB}}} \tag{6}$$

$$\lambda = \frac{S_{\mathrm{h}}}{4\pi S_{\mathrm{a}}} \tag{7}$$

where L_{AB} is the length of AB, and L_{OB} is the length of OB; S_{h} represents the area of hysteresis loop, that is, the energy loss of loading one cycle, and S_{a} represents the triangular AOB area.

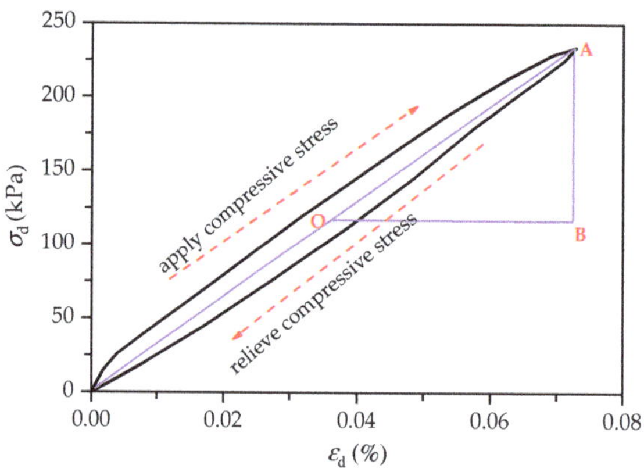

Figure 12. Stress–strain hysteretic curve of FCIT under cyclic loading.

The stress–strain hysteretic curves of FCIT are shown in Figure 13. It can be seen from the figure that the hysteretic circle moves in the direction of increasing strain with the increasing number of cycles N. Obviously, there is an accumulation of the strain of FCIT. At the same time, it can be determined from Figure 14 that the distance between hysteretic loops gradually decreases with the increasing number of cycles, which indicates that the cumulative strain rate gradually decreases and eventually tends to zero with the increasing number of cycles.

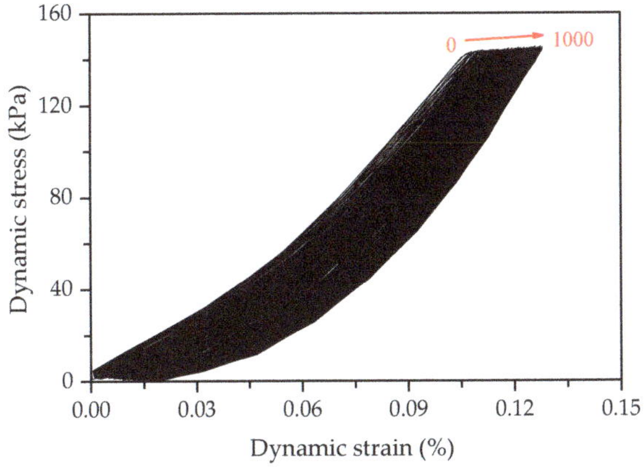

Figure 13. Hysteresis loop.

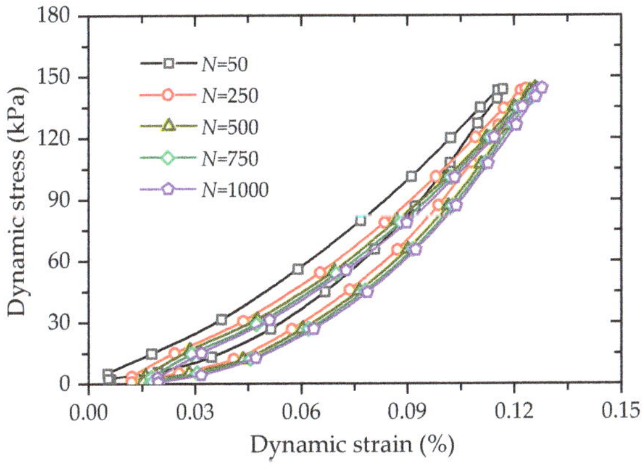

Figure 14. Variation of hysteresis loop with increasing cycle time.

Figure 15 shows the hysteresis loops ($N = 500$) of FCITs with different PP-fiber contents. It is found from Figure 15 that PP-fiber content has an obvious impact on the dynamic elastic modulus and damping ratio of FCIT.

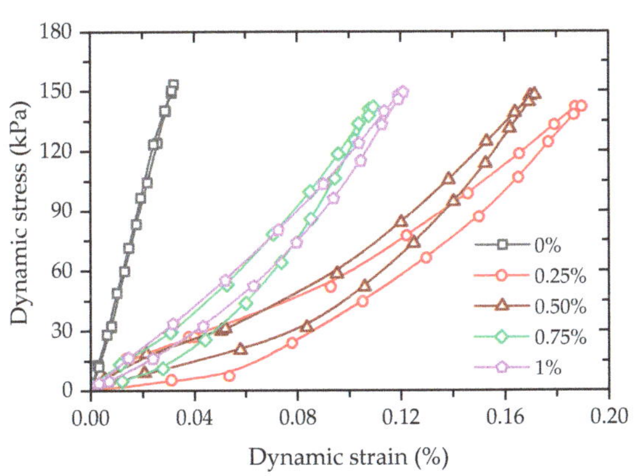

Figure 15. Hysteresis loops of FCIT at different PP-fiber contents ($N = 500$).

4.2. Evolution Law of Dynamic Characteristics with Cycle Times

4.2.1. Dynamic Elastic Modulus

The dynamic elastic modulus is an important mechanical index of pavement base. The computed values of dynamic elastic modulus are plotted against cycle times (see Figure 16). As illustrated in Figure 16, the dynamic elastic modulus decreases with the increase in the number of cycles, and there exists a linear relationship between the dynamic elastic modulus and the number of cycles. It can also be found that the dynamic elastic modulus of FCIT without PP fiber decreases faster. The reason for this is that the fiberless specimen is brittle material, and the internal damage of the specimen without the reinforcement effect of PP fiber develops faster when subjected to external force. That is, the addition of PP fiber can reduce the internal-damage rate of the specimen. Additionally, it can be seen that the dynamic elastic modulus of the specimen with PP fiber first increases and then decreases with increasing PP-fiber content. Obviously, the dynamic elastic modulus of the specimen with the PP-fiber content of 0.75% is the greatest among the specimens with PP fiber, which is similar to the change law of UCS.

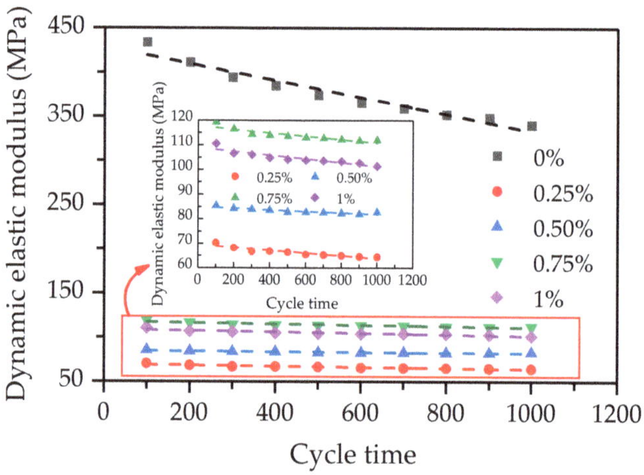

Figure 16. Relationship between dynamic elastic modulus and cycle time.

4.2.2. Damping Ratio

The damping ratio is used to describe the speed of energy dissipation of the pavement base in the process of traffic-load vibration. Figure 17 gives the test data between the damping ratios and cycle times of specimens with different PP-fiber contents. It can be seen that there is a linear relationship between the damping ratio and the number of cycles. Especially, added PP fiber can effectively improve the damping ratio of FCIT. This is because the addition of fiber allows the specimen to absorb more energy during vibration and promote vibration attenuation.

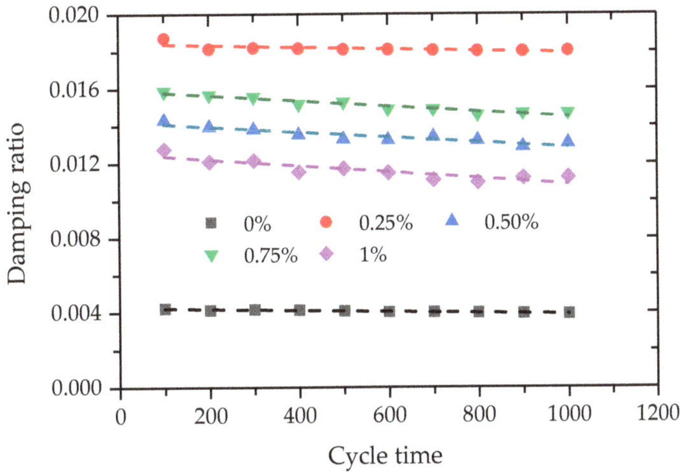

Figure 17. Relationship between damping ratio and cycle time.

4.3. Development Law of Cumulative Plastic Strain with the Number of Cycles

4.3.1. Porosity

The porosities of FCIT specimens with different PP-fiber contents are different, and the porosity has a significant impact on the dynamic characteristics of FCIT. The volume of the specimen is supposed as 1, and the porosity n can be calculated based on Equation (8).

$$\begin{cases} V_s = \frac{m_s}{G_s} = \frac{\rho_d}{(1+f+c)G_s} \\ V_f = \frac{m_f f}{G_f} \\ V_c = \frac{m_c c}{G_c} \\ n = 1 - V_s - V_f - V_c \end{cases} \quad (8)$$

where V_s, V_f and V_c are the volumes of iron-ore tailing sand, PP fiber and cement, respectively; m_s, m_f and m_c are the qualities of iron-ore tailing sand, PP fiber and cement, respectively; G_s, G_f and G_c are the proportions of iron-ore tailing sand, PP fiber and cement, respectively; c is the cement content, f is the PP-fiber content, and ρ_d is the dry density of specimen.

The porosity of FCIT before the test are then calculated according to Equation (8), as shown in Figure 18. It is intuitively found from Figure 18 that the porosity increases with the increase in PP-fiber content. The reason for this is that the PP fibers are randomly distributed in the specimen, and the PP fibers are in the state of intersection and interleaving, which increases the porosity of the specimen.

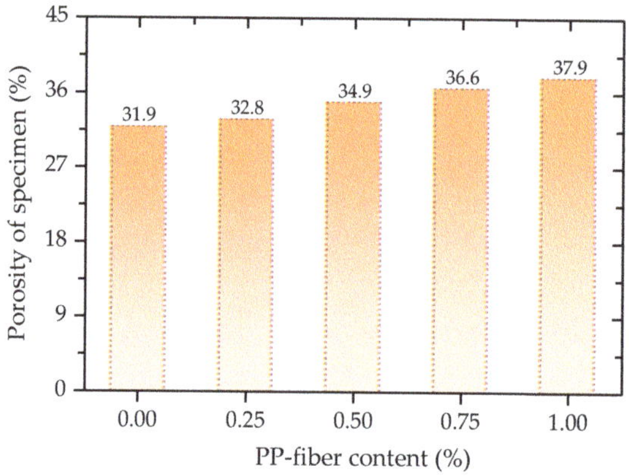

Figure 18. Porosity of specimen at different fiber contents.

4.3.2. Analysis of Cumulative Deformation Characteristics

Cumulative plastic strain refers to the plastic strain produced in the whole process of loading–unloading–loading, and it is suitable for evaluating the internal-damage evolution degree of FCIT under traffic load. Figure 19 shows the cumulative plastic strain ε_p, elastic strain ε_e and total strain ε_d. The total strain is composed of cumulative plastic strain and elastic strain [32]. The cumulative plastic strain is shown in the red curve in Figure 19, which is the valley value in each sine wave.

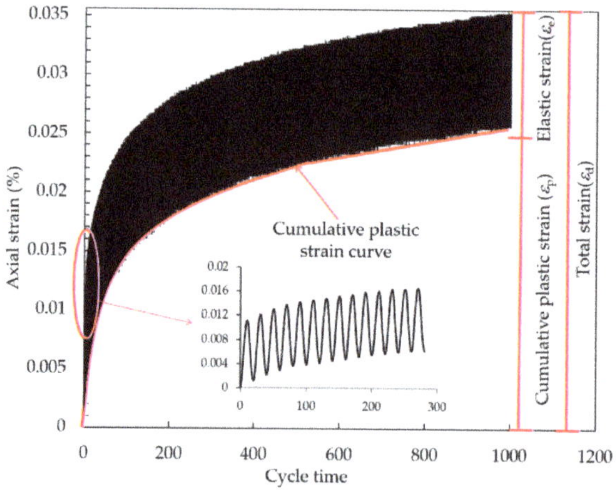

Figure 19. Strain curves with increasing cycle time.

The test results show that there is an accumulation of the deformation of FCIT with increasing cycle time. Figure 20 shows the test data between the cumulative plastic strain and cycle time of FCIT at different PP-fiber contents. From Figure 20, it can be seen that the cumulative plastic strain increases at a decreasing rate with increasing cycle time and increases with the increase in PP-fiber content. As previously discussed, more fibers can cause more pores in the specimen, and this is the reason for the cumulative plastic strain increasing with increasing PP-fiber content. That is, there is a positive relation between the cumulative plastic strain and PP-fiber content.

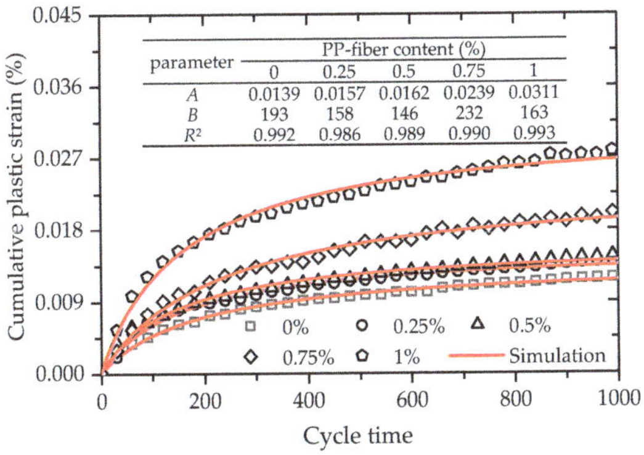

Figure 20. Relationship between the cumulative plastic strain and cycle time.

From Figure 20, it is apparent that the strain curve is the strain-hardening type. That is, the cumulative plastic strain increases at a decreasing rate with increasing cycle time, and eventually tends to be constant. However, the semi-logarithmic function is used to describe the relationship between the cumulative plastic strain and cycle time [33,34], in which the cumulative plastic strain will increase with increasing cycle time. For this reason, the hyperbolic function is adopted to describe the relationship between the cumulative plastic strain and cycle time, as given by

$$\varepsilon_p = \frac{AN}{B+N} \quad (9)$$

where A and B are the two material parameters, and A is regarded as the final cumulative plastic strain.

Equation (9) was used to fit the test data between the cumulative plastic strain and cycle time, as illustrated in Figure 20. As presented in Figure 20, the coefficients of determination R^2 are all greater than 0.980, which indicates that Equation (9) can effectively express the relationship between the cumulative plastic strain and cycle time.

Given that there is a positive relation between the cumulative plastic strain and PP-fiber content, the values of cumulative plastic strain were then plotted against the PP-fiber content at cycle times of 200, 600 and 1000, as shown in Figure 21. It is illustrated in Figure 21 that an exponential relationship exists between the cumulative plastic strain and PP-fiber content, as given by

$$\varepsilon_p = \exp(\alpha + \beta f^{1.5}) \quad (10)$$

where α and β are the two material parameters.

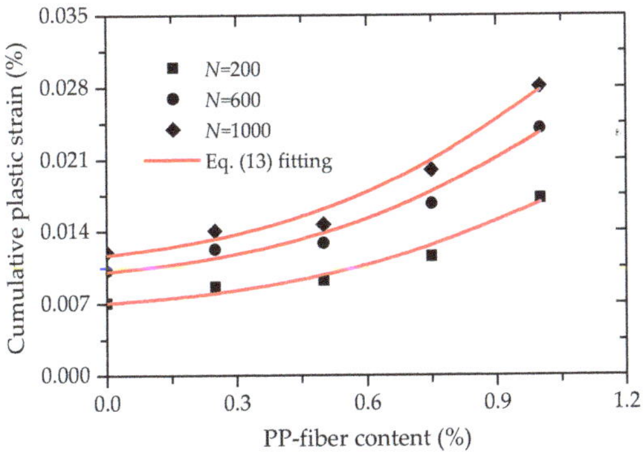

Figure 21. Relationship between the cumulative plastic strain and PP-fiber content.

Equation (10) was used to fit the test data between the cumulative plastic strain and PP-fiber content (Figure 21). Obviously, a good agreement can be found between the test data and curves fitted by Equation (10) from Figure 21. Based on Equations (9) and (10), a prediction model for the cumulative plastic strain considering cycle time and PP-fiber content can be expressed as

$$\varepsilon_p = \exp(\alpha + \beta f^{1.5}) \frac{N}{N+\omega} \tag{11}$$

where ω are the two material parameters

In order to verify the validity of the proposed prediction model, Equation (11) was used to simulate the deformation behaviors of FCIT at different cycle times and PP-fiber contents, as presented in Figure 22. The coefficient of determination R^2 fitted by Equation (11) is 0.981, which indicates that good effects in simulation have been achieved. At the same time, it can be found that there is a satisfactory agreement between the test data and model simulations. As observed, the proposed model can effectively capture the evolution law of the cumulative plastic strain with increasing cycle time at different PP-fiber contents of FCIT.

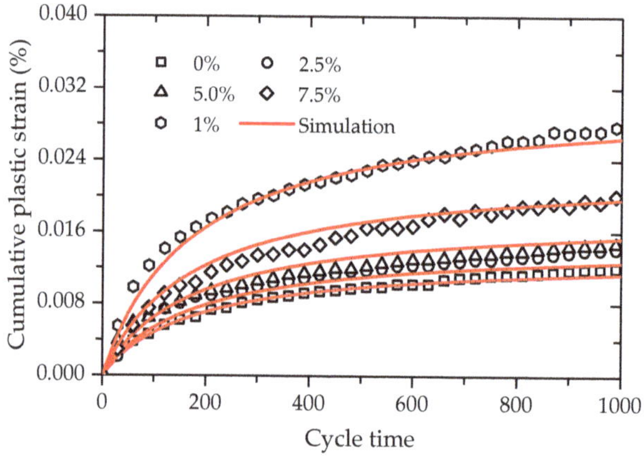

Figure 22. Simulations of the cumulative plastic strain.

5. Conclusions

Modifying iron-ore tailing with PP fiber and cement can improve its mechanical properties so that modified iron-ore tailing can be applied in road engineering, which realizes the reuse of iron-ore tailing. This paper investigated the strength and deformation behaviors of modified iron-ore tailing through the unconfined-compressive-strength test and the dynamic-triaxial test, which aims to provide a theoretical basis for the application of modified iron-ore tailing in road engineering. The main conclusions are as follows:

(1) The addition of PP fiber increases the UCS and residual strength of FCIT, and 0.75% has the best effect on improving the UCS and residual strength among all PP-fiber contents.

(2) The brittleness index and modulus strength ratio both decrease with increasing PP-fiber content, and the addition of PP fiber efficiently improves the tensile and toughness properties of FCIT.

(3) The dynamic elastic modulus and damping ratio of FCIT both meet the linear relationship with cycle time. The dynamic elastic modulus of FCIT first increases and then decreases, and the damping ratio first decreases, then increases, and finally decreases among specimens with increasing PP-fiber content from 0.25% to 1%.

(4) The deformation characteristics of FCIT under cycle load are related to the PP-fiber content, and the cumulative plastic strain increases with increasing PP-fiber content.

(5) A prediction model is developed for simulating the deformation behaviors of the dynamic-triaxial test, which can effectively capture the evolution law of the cumulative plastic strain with cycle time of FCIT at different PP-fiber contents.

Author Contributions: Conceptualization, W.W. and X.S.; methodology, N.L. and E.W.; investigation, Y.C.; writing—original draft preparation, P.J. and Y.C.; writing—review and editing, E.W. All authors have read and agreed to the published version of the manuscript.

Funding: This research was funded by the National Natural Science Foundation of China (grant number 41772311), Zhejiang Provincial Natural Science Foundation of China (Grant number LQ20E080042), the Open Research Fund of Key Laboratory of Ministry of Education for Geomechanics and Embankment Engineering of Hohai University (Grantnumber 2019020), the Open Research Fund of State Key Laboratory of Geomechanics and Geotechnical Engineering, Institute of Rock and Soil Mechanics, Chinese Academy of Science (Grant number Z017013) and the Scientific Research Projects of Zhejiang Department of Housing and Urban and Rural Construction of China (Grant number 2017K179, 2019K171).

Institutional Review Board Statement: Not applicable.

Informed Consent Statement: Not applicable.

Data Availability Statement: Not applicable.

Conflicts of Interest: The authors declare no conflict of interest.

References

1. Yuan, S.; Zhou, W.T.; Han, Y.X.; Li, Y.J. Efficient enrichment of iron concentrate from iron tailings via suspension magnetization roasting and magnetic separation. *J. Mater. Cycles Waste Manag.* **2020**, *22*, 1152–1162. [CrossRef]
2. Suther, S.; Pinto, P.; Nunna, V.; Nguyen, A.V. Experimental study of dry desliming iron ore tailings by air classification. *Min. Proc. Ext. Met. Rev.* **2019**, *40*, 344–355. [CrossRef]
3. Lv, X.S.; Cui, F.H.; Ning, Z.Q.; Free, M.L.; Zhai, Y.C. Mechanism and Kinetics of Ammonium Sulfate Roasting of Boron-Bearing Iron Tailings for Enhanced Metal Extraction. *Processes* **2019**, *7*, 812. [CrossRef]
4. Yao, G.; Wang, Q.; Wang, Z.; Wang, J.; Lv, X. Activation of hydration properties of iron ore tailings and their application as supplementary cementitious materials in cement. *Powder Technol.* **2020**, *360*, 863–871. [CrossRef]
5. Shettima, A.U.; Hussin, M.W.; Ahmad, Y.; Mirza, J. Evaluation of iron ore tailings as replacement for fine aggregate in concrete. *Constr. Build. Mater.* **2016**, *120*, 72–79. [CrossRef]
6. Luo, L.; Li, K.; Fu, W.; Liu, C.; Yang, S. Preparation, characteristics and mechanisms of the composite sintered bricks produced from shale, sewage sludge, coal gangue powder and iron ore tailings. *Constr. Build. Mater.* **2020**, *232*, 117250.1–117250.8. [CrossRef]
7. Kuranchie, F.A.; Shukla, S.K.; Habibi, D. Utilisation of iron ore mine tailings for the production of geopolymer bricks. *Int. J. Min. Reclam. Env.* **2016**, *30*, 90–114. [CrossRef]
8. Huang, X.; Ranade, R.; Ni, W.; Li, V.C. Development of green engineered cementitious composites using iron ore tailings as aggregates. *Constr. Build. Mater.* **2013**, *44*, 757–764. [CrossRef]
9. Wang, A.N.; Liu, G.C.; Xu, X.H.; Li, X.G.; Li, Y.L. Evaluation of soil quality in iron tailing ore wastelands of various reclamation periods. *J. Beijing For. Univ.* **2020**, *42*, 104–113.
10. Li, N.; Lv, S.W.; Wang, W.; Guo, J.; Liu, Y. Experimental Investigations on the Mechanical Behavior of Iron Tailings Powder with Compound Admixture of Cement and Nano-Clay. *Constr. Build. Mater.* **2020**, *254*, 119259. [CrossRef]
11. Zeng, Y.; Pan, J.P.; Zeng, B. Application of iron tailings in road base materials. *Appl. Chem. Ind.* **2018**, *47*, 358–364.
12. Bastos, D.C.; Silva, G.C.; Mendes, J.C.; Peixoto, R.A.F. Using Iron Ore Tailings from Tailing Dams as Road Material. *J. Mater. Civil. Eng.* **2016**, *28*, 4016102. [CrossRef]
13. Wang, W.; Kang, H.B.; Li, N.; Guo, J.; Girma, D.Y.; Liu, Y. Experimental investigations on the mechanical and microscopic behavior of cement-treated clay modified by nano-MgO and fibers. *Int. J. Geomech.* **2022**, *22*, 04022059. [CrossRef]
14. Li, X.G.; Jing, S.S.; Ma, Y.P. Mechanical properties and pore structure of cement mortar with iron ore tailings. *Concrete* **2014**, 121–128. [CrossRef]
15. Liu, J.L.; Wang, Y.F.; Liu, H.; Zhang, J.; Huang, Q.L. Characterization Parameter of Unconfined Compressive Strength of Cement Stabilized Iron Tailings. *Sci. Technolo. Eng.* **2018**, *18*, 275–281.
16. Yang, B.H.; Weng, X.Z.; Liu, J.Z.; Kou, Y.N.; Fu, J.; Jiang, L.; Li, H.L. Mechanical properties of modified polypropylene fiber reinforced cement stabilized loess. *J. Build. Mater.* **2016**, *19*, 694–701.
17. Jin, J.X.; Cui, H.Z.; Zhou, H.Y.; Zhou, Y.B.; Zhang, E.J. Experimental Study on the Polypropylene Fibers on Mechanical Properties of Reinforced Tailing Sand. *Non-Met. Mines* **2017**, *40*, 29–31.

18. Jiang, P.; Lv, S.W.; Wang, Y.; Li, N.; Wang, W. Investigation on Direct Shear and Energy Dissipation Characteristics of Iron Tailings Powder Reinforced by Polypropylene Fiber. *Appl. Sci.* **2019**, *9*, 5098. [CrossRef]
19. Zhang, Y.; Sun, L.; Sheng, L.; Qian, K.F. Deformation Characteristics of Fiber Reinforced Subgrade Soil Based on Cyclic triaxial Test. *J. Jixi Univ.* **2020**, *20*, 113–116.
20. Zhao, Y.Y.; Yang, Y.; Ling, X.Z.; Gong, W.M. Dynamic behavior of natural sand soils and fiber reinforced soils in heavy-haul railway embankment under multistage cyclic loading. *Transp. Geotech.* **2021**, *28*, 100507. [CrossRef]
21. Wang, J.Q.; Chang, Z.C.; Huang, S.B.; Wang, Q. Experimental Analysis of Dynamic Characteristics of Geogrid Reinforced Gravel Soil under Cyclic Dynamic Load. *Sci. Technolo. Eng.* **2019**, *19*, 350–357.
22. Maher, M.H.; Woods, R.D. Dynamic response of sand reinforced with randomly distributed fibers. *J. Geotech.* **1990**, *116*, 1116–1131 [CrossRef]
23. Li, J.; Ding, D.W. Nonlinear elastic behavior of fiber-reinforced soil under cyclic loading. *Soil Dyn. Earthq. Eng.* **2002**, *22*, 977–983. [CrossRef]
24. Orakoglu, M.E.; Liu, J.; Niu, F. Dynamic behavior of fiber-reinforced soil under freeze-thaw cycles. *Soil Dyn. Earthq. Eng.* **2017**, *101*, 269–284. [CrossRef]
25. Cai, Y.; Zhao, L.; Cao, Z.; Gu, C. Experimental study on dynamic characteristics of unbound granular materials under cyclic loading with different frequencies. *Chin. J. Rock Mech. Eng.* **2017**, *36*, 1238–1246.
26. Highway Research Institute of the Ministry of Transport. *Construction of Highway Pavement Base*; China Communications Press: Beijing, China, 2015.
27. Lu, Q.; Guo, S.L.; Wang, M.M.; Gao, M. Experimental study of mechanical properties of fiber cement soil. *Rock Soil Mech.* **2016**, *37*, 421–426.
28. Huang, B.X.; Liu, J.W. The effect of loading rate on the behavior of samples composed of coal and rock. *Int. J. Rock Mech. Min.* **2013**, *61*, 23–30. [CrossRef]
29. Khattak, M.J.; Alrashidi, M. Durability and mechanistic characteristics of fiber reinforced soil-cement mixtures. *Int. J. Pavement Eng.* **2006**, *7*, 53–62. [CrossRef]
30. Gu, C.; Wang, J.; Zhang, T.T.; Cai, Y.Q. Influence of stress path on secant modulus of soft saturated clay. *Rock Soil Mech.* **2013**, *34*, 3394–3402.
31. Jamsawang, P.; Suansomjeen, T.; Sukontasukkul, P.; Jongpraist, P.; Bergado, D.T. Comparative flexural performance of compacted cement-fiber-sand. *Geotext. Geomembr.* **2018**, *46*, 414–425. [CrossRef]
32. Li, Y.F.; Nie, R.S.; Leng, W.M.; Guo, Y.P.; Dong, J.L.; Sun, B.L. Cumulative permanent strain and critical dynamic stress of silty filler for subgrade subjected to intermittent cyclic loading of trains. *B. Eng. Geol. Environ.* **2021**, *80*, 3079–3096. [CrossRef]
33. Ahari, A.S.; Forough, S.A.; Khodaii, A. Modeling the primary and secondary regions of creep curves for sbs-modified asphalt mixtures under dry and wet conditions. *J. Mater. Civil. Eng.* **2014**, *26*, 904–911. [CrossRef]
34. Hao, F.; Liu, Q.T.; Mo, L.T.; Javilla, B.; Sun, B.N.; Wu, S.P. Characterization of three-stage rutting development of asphalt mixtures. *Constr. Build. Mater.* **2017**, *154*, 340–348.

polymers

MDPI

Article

Triaxial Mechanical Properties and Mechanism of Waterborne Polyurethane-Reinforced Road Demolition Waste as Road Bases

Beifeng Lv, Yinuo Zhao, Na Li *, Yanfei Yu, Yanting Wu and Miaojie Gu

School of Civil Engineering, Shaoxing University, Shaoxing 312000, China; 20020852047@usx.edu.cn (B.L.); cosm1cc@163.com (Y.Z.); yuyanfei@usx.edu.cn (Y.Y.); 19020852065@usx.edu.cn (Y.W.); gmjzoe@163.com (M.G.)
* Correspondence: lina@usx.edu.cn

Abstract: The recycling and reuse of construction waste have not only effectively protected natural resources but also promoted the sustainable development of the environment. Therefore, in this article, waterborne polyurethane (WPU) as a promising new polymer reinforcement material was proposed to reinforce the road demolition waste (RDW), and the mechanical performance of WPU-reinforced RDW (named PURD) was investigated using triaxial unconsolidated and undrained shear (UU) and Scanning Electron Microscope (SEM) tests. The results showed that under the same curing time and confining pressure, the shear strength of PURD increased with the increase in WPU content. When the WPU content was 6%, the WPU presented the best reinforcement effect on RA. The failure strain of PURD increased with the increase in confining pressure, but increased first and then reduced with the increase in WPU content. The specimens with 5% WPU content showed the best ductility. At the curing time of 7 and 28 days, the internal friction angle and cohesion of PURD increased with the increase in WPU content, and they reached a maximum when the WPU content was 6%. The internal friction angle barely budged, but the cohesion increased obviously. The enhancement effect of WPU was attributed to the spatial reticular membrane structure produced by wrapping and bonding particles with the WPU film. Microscopic analysis showed that with the increase in WPU content, the internal pore and crack size of PURD gradually decreased. As the WPU content increased, the WPU film became increasingly thicker, which increased the adhesion between WPU and RA particles and made the structure of PURD become increasingly denser.

Keywords: recycled aggregate; waterborne polyurethane; mechanical properties; reinforcement mechanism

Citation: Lv, B.; Zhao, Y.; Li, N.; Yu, Y.; Wu, Y.; Gu, M. Triaxial Mechanical Properties and Mechanism of Waterborne Polyurethane-Reinforced Road Demolition Waste as Road Bases. *Polymers* **2022**, *14*, 2725. https://doi.org/10.3390/polym14132725

Academic Editor: Sándor Kéki

Received: 8 June 2022
Accepted: 27 June 2022
Published: 3 July 2022

1. Introduction

Over the past few decades, infrastructure construction has flourished due to widespread urbanization, the world population soaring, and the changing economic landscape of the developing countries [1,2]. This has led to a sharp increase in the amount of demolition waste from old buildings, thereby posing a serious burden on the ecological environment and human health [3–5]. Additionally, due to the construction of a large number of new construction projects, the serious shortage of natural resources and the extensive mining methods have destroyed the ecological balance of the production area [6–10]. The world has been persistently striving to create a development environment with steady progress in material, energy, production, economy, environment, and efficiency. The construction industry can achieve this by effectively reusing demolition waste as building materials [11,12]. Therefore, the construction demolition waste is used to replace natural building materials in current society, which will make great contributions to the sustainable development [13–15].

In the process of pursuing sustainable development in the construction industry, cement production consumes a lot of energy and emits a large amount of greenhouse gases, and the total emission accounts for about 8% of the total global carbon dioxide emissions [16–18]. The high-molecular polymers present fast hardening, micro-expansion, stable strength, and good mechanical properties, and they can be used to replace cement materials and widely used in infrastructure construction fields such as road construction, slope stabilization, foundation, and roadbed treatment [19–21]. Polyurethane as a high-molecular polymer is formed by the reaction of diisocyanate and polyol (or equivalent) in the presence of a catalyst. Polyurethane has good biocompatibility, hydrolysis stability, and chemical resistance [22–24]. According to the preparation process, polyurethane can be divided into solvent and waterborne. Compared with traditional solvent polyurethane, waterborne polyurethane (WPU) is more environmentally friendly, and there is a great demand in the fields of coatings, thickeners, sealants, and adhesives [25,26].

Currently, the high-molecular polymers have been applied in the reinforcement of building materials, among which polyurethane is particularly prominent in improving strength and stability [27–30]. Wei et al. [31] conducted a series of unconfined compressive strength tests and scanning electron microscope tests with different WPU contents to explore the reinforcement effect of WPU on sand, showing that with the increase in WPU content, the unconfined compressive strength and residual strength of the sand increase, and the strength and ductility of sand can be effectively improved. Because WPU has a unique reticular membrane structure, the WPU can improve the cohesion between sand particles. In order to study the influence of polyurethane on the permeability of sand, Liu et al. [32] carried out a series of reinforcement layer form tests, single-hole permeability tests, and porous permeability tests on the sand reinforced by polyurethane. The test results showed that the impermeability of the sand reinforced by polyurethane is improved. With the increase in polyurethane concentration, the thickness and complete degree of the reinforcement layer increase, and the permeability coefficient decreases. Zhang et al. [33] studied the effect of polyurethane content on the mechanical properties and microstructure of cement mortar, with the results showing that with the increase in polyurethane content, the compressive strength of cement mortar gradually decreases but the flexural strength gradually increases. The polyurethane helps to optimize the microstructure of cement mortar and inhibit crack propagation. Samaila et al. [34] investigated the marine clay treated with polyurethane (PU); it was found that the compressive strength of marine clay specimens is effectively improved; and the failure strain, compression index, and expansion index are reduced by PU. The microstructure of marine clay specimens treated with PU becomes more compact. Liu et al. [35] used cyclic triaxial tests to investigate the effect of PU on the dynamic characteristics of crushed stone, and in this work, the effects of confining pressure, PU content, consolidation stress ratio, and loading frequency on the shear modulus and damping ratio of polyurethane-reinforced crushed stone were considered. The results showed that increasing the confining pressure, polyurethane content, and consolidation stress ratio can increase the maximum shear modulus and reduce the damping ratio of PU-reinforced crushed stone. However, the shear modulus of PU-reinforced crushed stone slightly increases with the increase in loading frequency, but the damping ratio is not sensitive to it. In order to explore the effect of waterborne polyurethane on the mechanical properties, durability, and microstructure of concrete, Fan et al. [36] carried out a series of experimental research and theoretical analysis. The test results showed that the compressive strength, splitting tensile strength, flexural strength, and elastic modulus of concrete can be improved by adding an appropriate amount of waterborne polyurethane. Meanwhile, under the action of waterborne polyurethane, the chloride penetration resistance, impermeability, freezing and thawing resistance, and microstructure of concrete have also been significantly improved.

In summary, it is found that due to the pollution problems caused by the use of cement and the accumulation of construction waste, as well as the shortcomings of high energy consumption in the mining of construction materials, the current society urgently needs to find new environmental protection reinforcement materials and resource utilization methods of construction waste to alleviate the pressure on the environment and resources.

At present, a large number of research works mainly focuses on natural building materials such as sand, marine clay, gravel, and concrete, while the research on recycled building materials is relatively rare. Therefore, using waterborne polyurethane as a new environmental protection reinforcing material to study its effect on the mechanical properties of recycled aggregate can realize the engineering recycling of building materials, to achieve the social benefits of protecting the environment and saving resources. Meanwhile, the abundant research work has only explored the compressive strength, flexural strength, shear modulus, and damping ratio of various building materials, and has not yet conducted in-depth research on shear strength.

The objective of this paper was to explore the effect of WPU on the triaxial shear strength of RA by unconsolidated undrained shear (UU) tests on the WPU-reinforced RA (PURD) specimens under different ages, confining pressures, and contents of waterborne polyurethane. Meanwhile, the microstructure of PURD was characterized by scanning electron microscope (SEM) tests to explore the WPU reinforcement mechanism on RA.

2. Experimental Materials, Schemes, and Methods

2.1. Materials

The RA was collected from the abandoned road section of Erhuan North Road, Shaoxing City, Zhejiang Province, which was the mixed gravel after the crushing of the roadbed base material. The main chemical components of RA are SiO_2 and $CaCO_3$. The acquisition process of RA required for the test is shown in Figure 1. According to the Test Methods of Materials Stabilized with Inorganic Binders for Highway Engineering (JTG E51-2009) [37], fine-grained gravel soil with a particle size of less than 4.75 mm was selected as the RA used in the test, and the particle grading curve is shown in Figure 2. According to the Test Methods of Soils for Highway Engineering (JTG 3430 2020) [38], the physical performance indexes of RA were tested, and the results are shown in Table 1. The WPU is produced by Shenzhen, China, Jitian Chemical Co., Ltd. The model is F0410 and its main performance indexes are shown in Table 2.

Figure 1. Acquisition process of RA required for test.

Table 1. Physical performance indexes of RA.

Physical Index	Natural Moisture Content (%)	Plasticity Index	Proportion (g/cm^3)	Mud Content (%)	Apparent Density (kg/m^3)
Index value	16	16.6	2.68	20.5	2.67

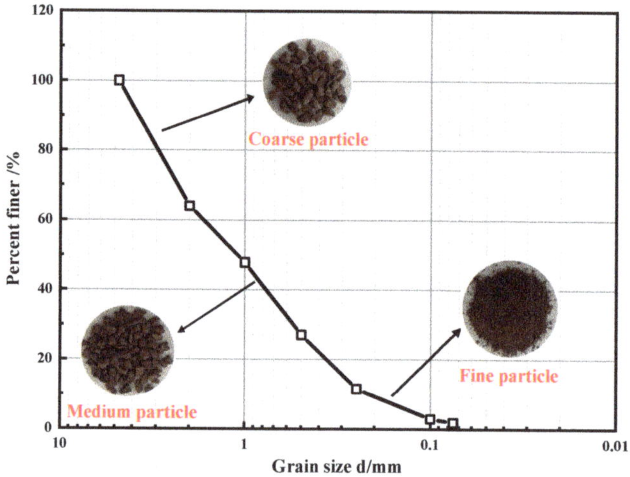

Figure 2. Particle grading curve.

Table 2. Performance indexes of WPU solution.

Performance Index	Appearance	Solid Content (%)	pH Value (25 °C)	Viscosity (mPa·S)	Film Forming Temperature (°C)
Index value	Milky white liquid	40	6~8	800~1500	20

2.2. Experimental Scheme

The RA was reinforced by WPU with four contents. The WPU solutions were diluted with water to a solution with a low content, and then the RA was mixed with WPU. The specific test scheme is shown in Table 3. For the convenience of expression, PURD-X is used to represent the specimen number, where X represents the content of WPU (%). Meanwhile, the SEM tests were performed to reveal the strengthening mechanism of mechanical properties of WPU-reinforced RA.

Table 3. Test scheme for RA reinforced with different contents of WPU.

Specimen Number	Dry Density (g/cm^3)	WPU Content (%)	Moisture Content (%)	Curing Time (Day)
PURD-3	1.6	3	16	7, 28
PURD-4		4		
PURD-5		5		
PURD-6		6		

The WPU content in the test scheme is the mass ratio of WPU and RA, and the solid content of the solution needs to be considered. The moisture content of the specimen was set to 16%, the main purpose was to keep consistent with the natural moisture content of the RA, and the moisture contained in the WPU solution should be considered. The calculation process of the test mix proportion is shown in Equations (1) and (2).

$$m_{\mathrm{WPU}} = \frac{m_{\mathrm{RA}} \times r_{\mathrm{WPU}}}{40\%} \quad (1)$$

$$m_{\mathrm{w}} = (m_{\mathrm{RA}} + m_{\mathrm{WPU}} \times 40\%) \times 16\% - m_{\mathrm{WPU}} \times (1 - 40\%) \quad (2)$$

where m_{WPU} is the mass of the WPU solution (g), r_{WPU} is the content of the WPU (%), m_{RA} is the mass of RA (g), and m_{W} is the mass of water (g).

2.3. Specimen Preparation

The preparation process of the UU test specimens in this study is divided into the following steps (see Figure 3) as per the Standard for geotechnical testing method (GB/T50123-2019) [39] and the designed experimental scheme.

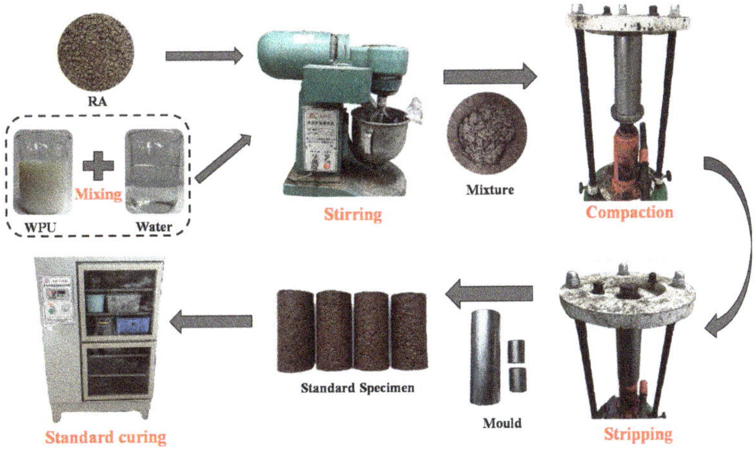

Figure 3. Specimen preparation process.

(1) The RA was placed in a constant-temperature oven at 105 °C for 24 h to ensure that the moisture in the RA was completely dried, and then cooled to room temperature.
(2) The corresponding qualities of the RA, WPU solution, and water were weighed according to the mix proportion set in Table 3. Then, the weighed water was slowly poured into the WPU solution while stirring. Afterward, the WPU solution was mixed with the RA.
(3) An amount of 154 g of the mixture stirred evenly was shaped in a cylindrical mold with a height of 80 mm and a diameter of 39.1 mm. It must be fully compacted with a jack and statically pressed for 10 min.
(4) The specimens were demolded after being fully compacted. Then, the specimens were cured in a standard curing box for 7 and 28 days, respectively. The curing temperature was controlled at 20 ± 2 °C, and the relative humidity was >95%.

The samples for the SEM test were from the dried sample blocks of broken sample in the UU test. A layer of conductive adhesive was pasted tightly on the sample with a thin rod, and then the specimen particle powder on the surface was blown away with a suction balloon to avoid polluting the instrument lens.

2.4. Experimental Method

2.4.1. UU Test

The instrument used in the UU test was a fully automatic triaxial shearing instrument (TKA.TTS.3S) produced by Nanjing, China, TKA Technology Co., Ltd. The four confining pressures (100, 200, 300, and 400 kPa) were applied, and the loading rate was set to 1 mm/min. According to the relevant provisions of the Standard for geotechnical testing method (GB/T 50123-2019) [39], after the deviatoric stress reached the maximum, the axial strain increased by 3–5% and the test was stopped. In this article, the test was terminated when the axial strain reached 14%.

2.4.2. SEM Test

The equipment used for the SEM test was a high- and low-vacuum scanning electro microscope (JSM-6360LV), which is produced by Tokyo, Japan, Electronics Co., Ltd. Th specimens damaged from the UU test were placed in an oven for 24 h, and the test wa performed after the specimen was dry. Finally, the SEM images were taken under differen magnifications, and the magnifications of this study were selected as 500× and 2000×.

3. Results and Discussions

3.1. Deviatoric Stress–Strain Behaviors

According to the deviatoric stress q and the axial strain ε obtained from the test, the d viatoric stress–strain relationship curves of PURD specimens are drawn in Figures 4 and at 7 and 28 days of curing time, respectively. From Figures 4 and 5, the deviator stress–strain curves of PURD at 7 and 28 days of curing time presented strain softe ing behaviors, namely, the deviatoric stress first increased and then decreased with th increase in strain. The softening characteristics of PURD-6 became more obvious tha those of PURD-3, PURD-4, and PURD-5. The deviatoric stress of PURD-6 had an obviou decreasing trend after reaching the peak stress, and the strain softening behavior becam more obvious especially under a lower confining pressure.

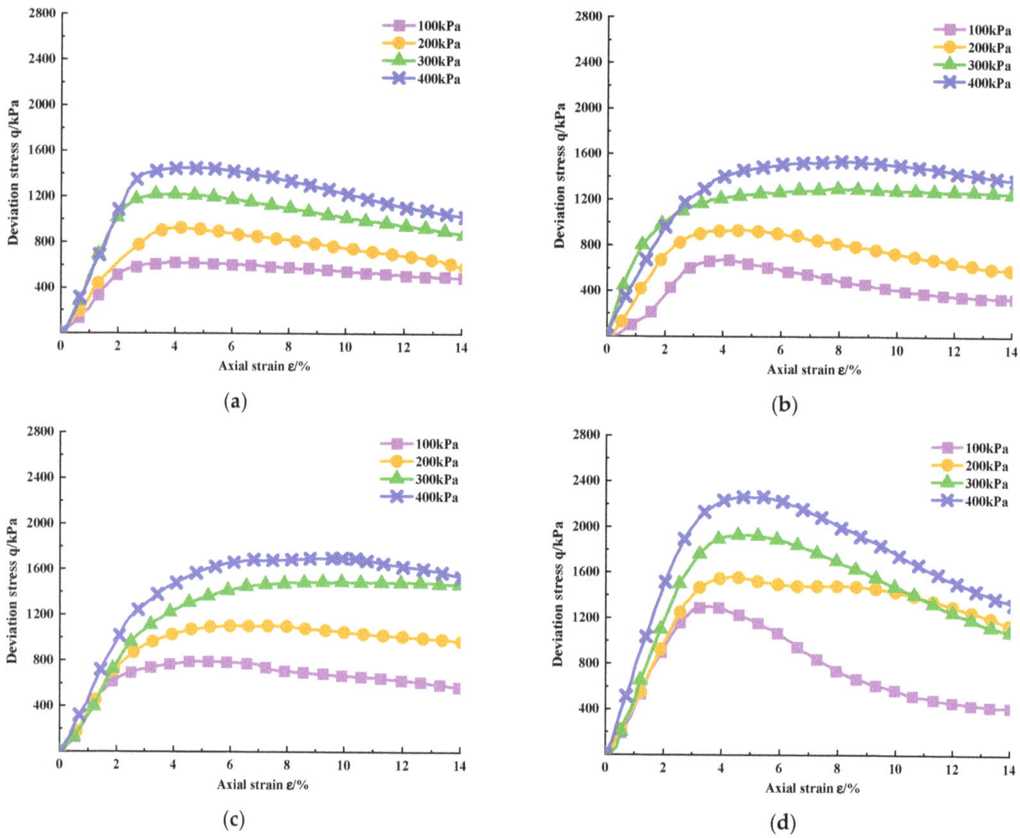

Figure 4. Deviatoric stress–strain curves of PURD at 7 days of curing time. (**a**) PURD-3; (**b**) PURD-4 (**c**) PURD-5; (**d**) PURD-6.

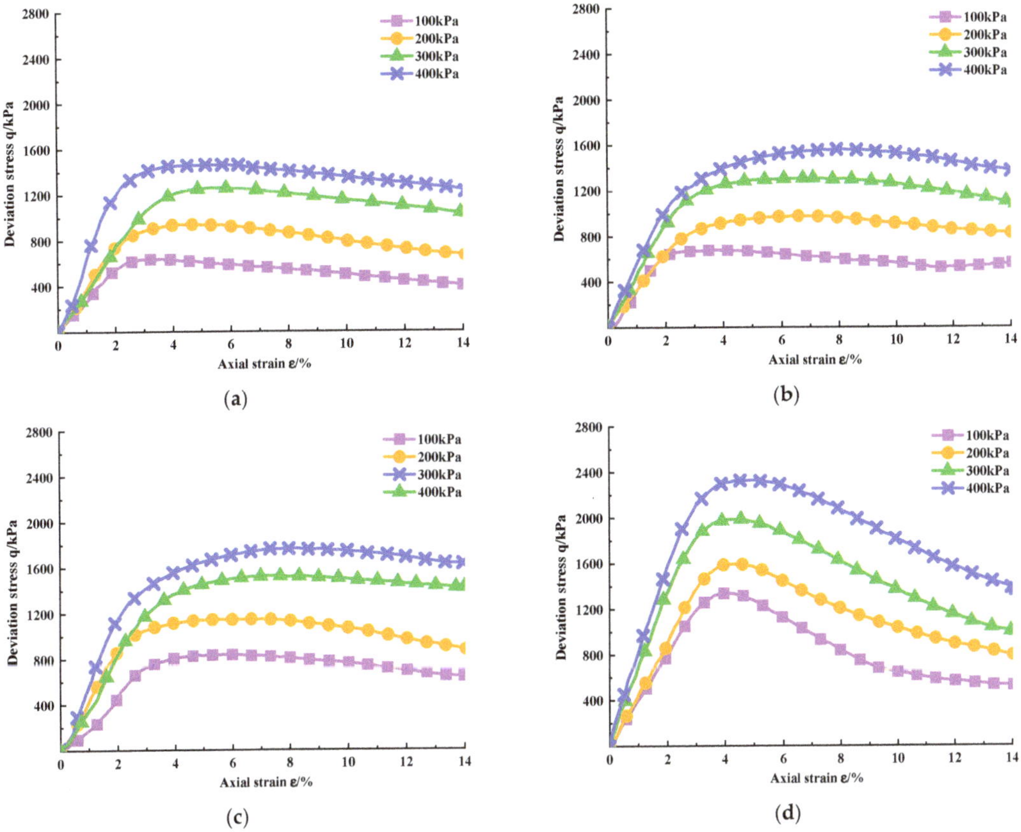

Figure 5. Deviatoric stress–strain curves of PURD at 28 days curing time. (**a**) PURD-3; (**b**) PURD-4; (**c**) PURD-5; (**d**) PURD-6.

Figure 6 shows the specimen failure modes of PURD specimens under different confining pressures. From Figure 6, the failure form of the specimens was bulging failure under different confining pressures. As the confining pressure increased, the bulging of the specimens became more obvious. When the confining pressure was 100 kPa, bulging cracks appeared on the surface of the specimen. First, this is mainly because the bonding and cementation effect of WPU can effectively inhibit the sliding of RA particles in the specimen to a certain extent [40]. However, under low confining pressure, the internal structure of the specimen has poor resistance to external loads. Under the action of external load, the friction and bite force between RA particles cannot inhibit its deformation, which will lead to the failure of the specimen [41]. Under high confining pressure, the specimen is constrained by large peripheral pressure, which makes the RA particles in the specimen subject to a certain binding effect, so it is not easy to slide and cause specimen damage [42,43].

Figure 6. Failure mode of specimens.

3.2. Shear Strength

According to the deviatoric stress–strain curves of PURD from Figure 5, the peak deviatoric stress under different confining pressures can be obtained, which is defined as the peak strength q_{max}, kPa, which can reflect the shear strength of PURD. Figure 7 shows the variation law of shear strength of PURD at different curing times. From Figure 7, it is found that the peak strength of PURD with different WPU contents increased gradually with the increase in confining pressure. Meanwhile, it can be found that the change trend of peak strength curves of PURD-3, PURD-4, PURD-5, and PURD-6 at 7 and 28 days was basically the same. Therefore, the PURD with different WPU contents had the same variation law of peak strength.

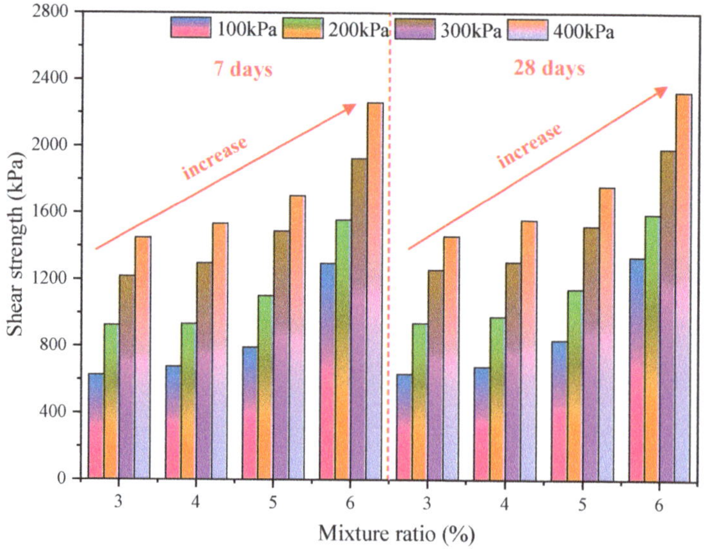

Figure 7. Shear strength.

By further analyzing Figure 7, the variation law of the peak strength of PURD with WPU content is obtained. Under the same curing time and the same confining pressure, the peak strength of PURD increases with the increase in WPU content. Meanwhile, the increase range of peak strength becomes increasingly more obvious with the increase in WPU content. At the curing time of 7 and 28 days, under the same confining pressure, the peak strength of PURD-4 increases by less than 10% compared with that of PURD-3. The peak strength of PURD-5 increases by 10% to 25% compared with that of PURD-4. The peak strength of PURD-6 increases by 30% compared with that of PURD-5, and when the confining pressure is 100 kPa, the rising range reaches 64%. Therefore, a 6% content of WPU can make PURD have the highest shear strength and the best reinforcement effect on RA. The important effect of WPU content on the shear strength of PURD is mainly because with the increase in WPU content, the polyurethane itself will produce a higher bonding effect and participate in more curing reactions, which will affect the RA. The bonding between particles becomes increasingly stronger, which makes the internal connection of PURD closer, so the shear strength of WPU will gradually increase with the increase in WPU content.

The curing time of cement-based materials has an important influence on the strength of the specimen. Cement usually completes almost all hydration reactions when curing for 28 days, and the strength of the specimen will increase significantly compared with that under the curing time of 7 days. Therefore, it is necessary to explore the effect of curing time on WPU-strengthening RA [44,45]. Compared with Figure 7, the peak strength of PURD with each content at the curing time of 28 days increases slightly compared with 7 days, and the increase range is less than 6%, which is almost negligible. It can be concluded that the curing time has little effect on WPU strengthening RA, and also has little effect on the shear strength. The above analysis shows that WPU can directly exert all its strength effects in a short period of time, and does not require a long curing period. Therefore, WPU has relatively low requirements on construction time and is relatively flexible in engineering applications, which is a very effective and promising new type of high-molecular polymer reinforcement material.

3.3. Failure Strain

The strain corresponding to the peak strength of PURD in the deviatoric stress–strain curve in Figures 5 and 6 is called the failure strain [46], and Figure 8 shows the change law of failure strain of the specimens. The failure strain can describe the ductility of the specimen to a certain extent. The larger the failure strain value is, the later the shear failure of the specimen occurs [47,48].

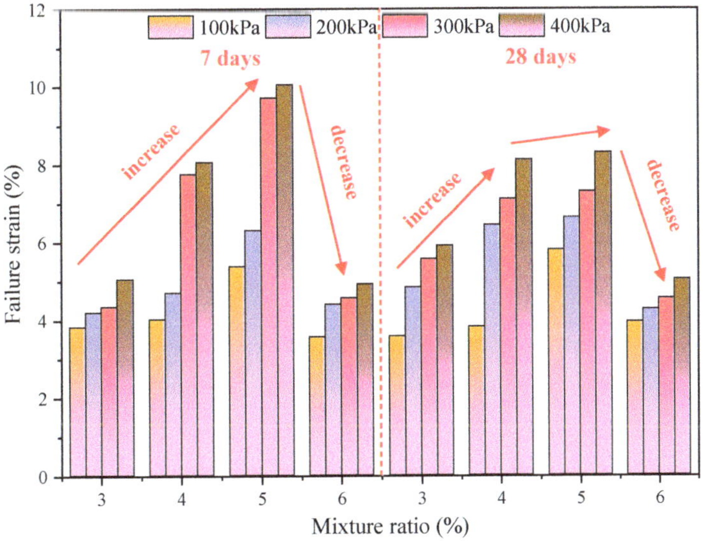

Figure 8. Failure strain.

According to Figure 8, the failure strain of PURD with different WPU contents increase with the increase in confining pressure at 7 and 28 days of curing time. On the whole, the change range of the failure strain of PURD relative to the confining pressure is relatively gentle. Only the failure strains of PURD-4 and PURD-5 show a significant increase between 100 kPa to 200 kPa and 200 kPa to 300 kPa confining pressures. However, the effect of curing time on the failure strain of PURD does not show a particularly obvious rule, and the overall difference is not significant.

Comparing the failure strain of PURD with different WPU contents, it can be found that the failure strain of PURD increases first and then decreases with the increase in WPU content under the curing time of 7 and 28 days. Under two curing times and four confining pressures, the failure strain of PURD-5 reaches the maximum value. Compared with PURD-3, the failure strain increases by 41%, 50%, 124%, and 99% at the curing time of 7 days, and increases by 61%, 37%, 31%, and 41% at the curing time of 28 days, respectively. The growth rate is relatively large. The above analysis shows that PURD-5 exhibits the best ductility, mainly because the increase in WPU content will make the cementation between the RA particles more obvious, and there will be a certain "pulling" effect between particles to prevent brittle failure [49]. However, when the content of WPU is too much, a large number of agglomerates will appear in the PURD, and the dispersion is not uniform, which inhibits the ductility of the specimen.

3.4. Shear Strength Parameters

Taking the normal stress σ as the abscissa, the shear stress τ as the ordinate, $(\sigma_{1f} + \sigma_{3f})/2$ as the center, and $(\sigma_{1f} - \sigma_{3f})/2$ as the radius, the limit Mohr stress circle is drawn on the τ-σ stress plane diagram. Then, the common tangents of the Mohr circle under four different confining pressures are drawn, which are the shear strength envelopes of the limit Mohr stress circle, where σ_1 represents the large principal stress, σ_3 represents the small principal stress, and the subscript f represents the limit equilibrium state, that is, the molar stress circle is tangent to the shear strength envelope, which means that the shear stress on the τ-σ stress plane is equal to the shear strength. Based on this, the limit Mohr stress circles and shear strength envelopes of PURD with different WPU contents under 7 and 28 days of curing time were drawn, respectively (see Figures 9 and 10).

Figure 11 shows the variation law of the internal friction angle φ and cohesion c of PURD specimens obtained from the shear strength envelopes. As can be seen from Figure 11, at the curing time of 7 and 28 days, the internal friction angle and cohesion of PURD show an increasing trend with the increase in WPU content and reaches the maximum when the WPU content is 6%. The change range of internal friction angle is relatively small, with an increase of only about 5%, while the increase in cohesion is more obvious. Under 7 days of curing time, the cohesion of PURD-6 increases by 144%, 142%, and 91% compared with those of PURD-3, PURD-4, and PURD-5, respectively. Under 28 days of curing time, the cohesion of PURD-6 increases by 141%, 134%, and 77% compared with those of PURD-3, PURD-4, and PURD-5, respectively. Meanwhile, compared with the test results in Section 2.2, it can be found that the variation rules of the internal friction angle and cohesion of PURD are highly consistent with the variation rules of the shear strength. However, the content of WPU has a limited effect on the internal friction angle of PURD. As a polymer curing time, WPU has a high degree of fluidity, and its consolidation effect cannot significantly change the internal friction angle of RA. Therefore, it can be known that WPU mainly improves the shear strength by increasing the cohesion of the specimen.

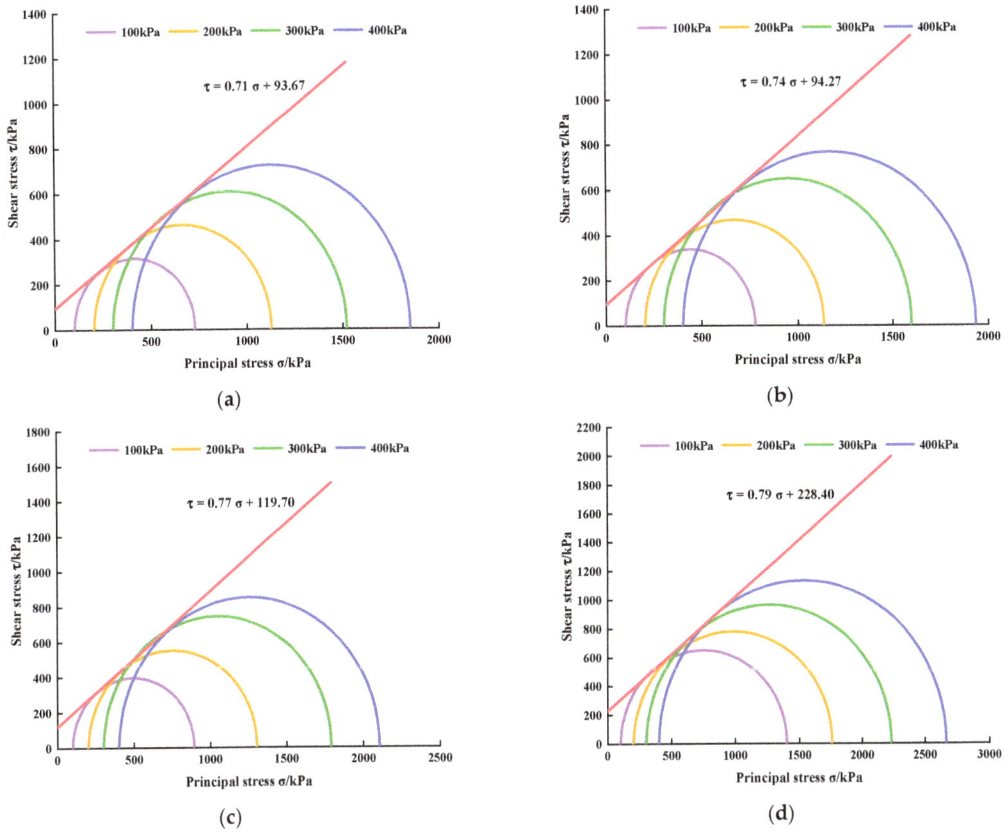

Figure 9. Limit Mohr stress circles and shear strength envelopes of PURD at 7 days of curing time. (**a**) PURD-3; (**b**) PURD-4; (**c**) PURD-5; (**d**) PURD-6.

Under the same WPU content, the internal friction angle and cohesion of PURD both increase with curing time, but not obviously. The strength of PURD is mainly provided by the bite force and friction generated by the mutual extrusion of RA particles and the bonding force generated by the curing reaction of WPU. The increase in internal friction is mainly due to the curing reaction between WPU and water in the specimen, and the curing reaction will gradually end with the increase in curing time, resulting in the decrease in water in the specimen. The internal friction angle will increase with the decrease in water content, and the decrease in water content will lead to the increase in bite force and friction between RA particles [50]. Additionally, WPU itself is also prone to oxidative consolidation reaction, which will produce high-strength adhesion and can better bond loose RA particles together. Moreover, a large number of long-chain macromolecules and isocyanate groups in WPU make it have a special polymer reticular structure, which can significantly improve the cohesion of the specimen [31,32]. With the increase in curing time, the consolidation reaction between WPU and water and the oxidative consolidation reaction of WPU will proceed more completely, thereby further increasing the cohesion of PURD.

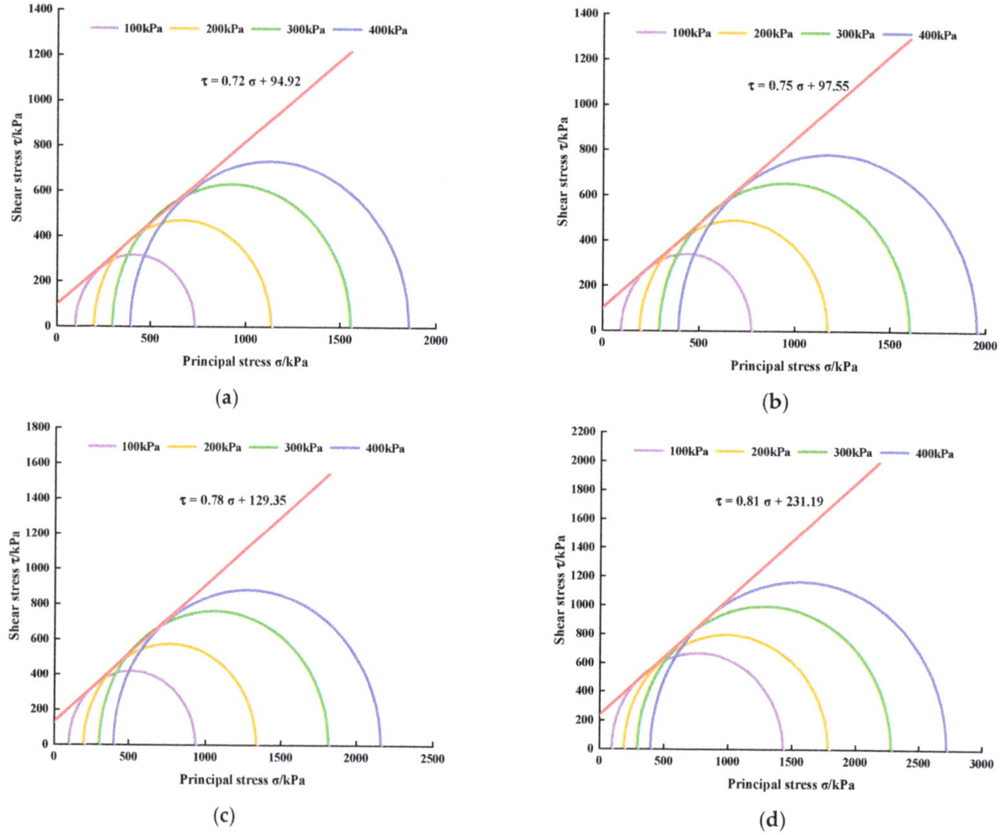

Figure 10. Limit Mohr stress circles and shear strength envelopes of PURD at 28 days of curing time. (**a**) PURD-3; (**b**) PURD-4; (**c**) PURD-5; (**d**) PURD-6.

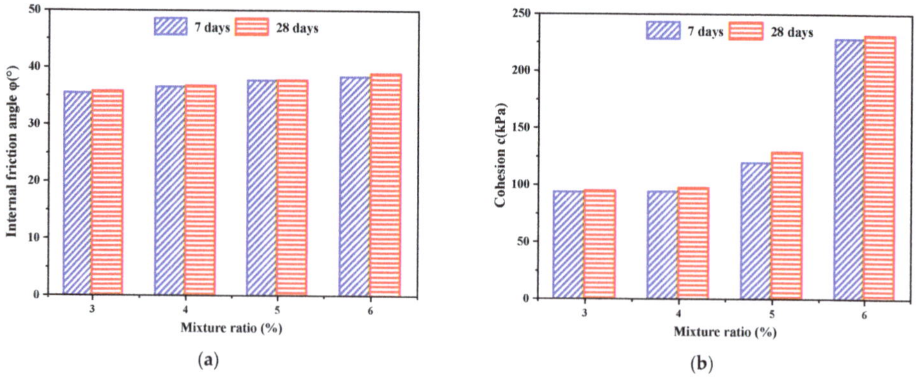

Figure 11. Shear strength parameters. (**a**) Internal friction angle φ; (**b**) cohesion c.

4. Mechanism Analysis

Through the above UU test results and analysis, it has been found that WPU has an important effect on the mechanical properties of RA, and the variation law of the shear strength of PURD with different WPU contents has been obtained. However, the above results are analyzed from the macro-mechanical level, which has the disadvantage of relative superficiality. Therefore, we can try to make an in-depth analysis of the mechanism of WPU enhancing RA from the micro-level. Therefore, the SEM test was performed on PURD with different WPU contents, and the SEM images at 500 times and 2000 times magnification were taken, respectively (see Figure 12).

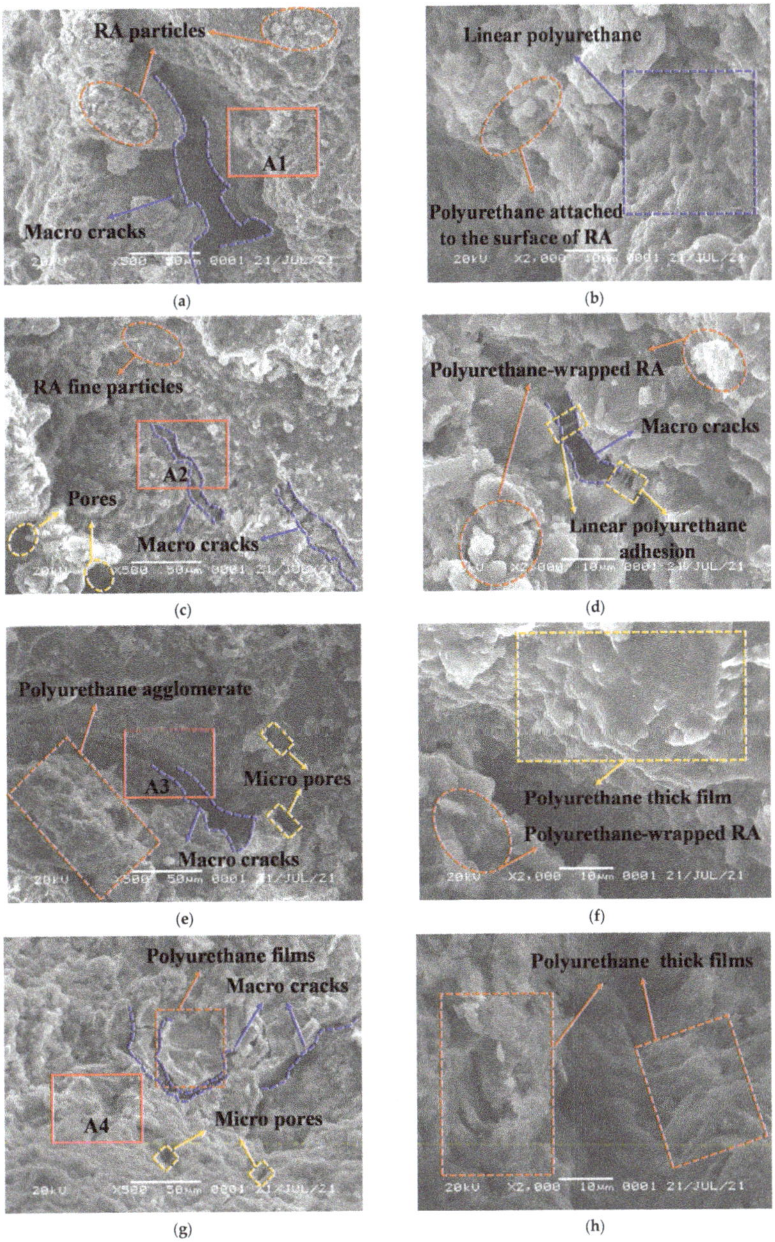

Figure 12. SEM images of PURD with different WPU contents. (**a**) PURD-3, 500 times; (**b**) PURD-3, 2000 times; (**c**) PURD-4, 500 times; (**d**) PURD-4, 2000 times; (**e**) PURD-5, 500 times; (**f**) PURD-5, 2000 times; (**g**) PURD-6, 500 times; (**h**) PURD-6, 2000 times.

Figure 12a shows the SEM image of PURD-3 under 500 times. It can be seen that PURD-3 has many macropores, the internal structure of the specimen is loose and not tight enough, and it can be clearly seen that there are many recycled aggregate particles on the surface of the specimen, which are bonded together through polyurethane. Part A1 in Figure 12a is enlarged to obtain Figure 12b. Figure 12b shows that a large amount of polyurethane adheres to the surface of RA, but the particle shape of RA can still be seen. The bonding between particles is not particularly tight. Meanwhile, the polyurethane exists in a linear form and plays a bridging role between RA particles, which is similar to the reinforcing effect of fiber in cement-based materials [51]. Figure 12c shows the SEM image of PURD-4 under 500 times. The image shows that there are some large pores and tiny cracks inside the specimen, but the pore size is reduced compared with the PURD-3, so the internal structure of the specimen is improved. Meanwhile, according to Figure 12d obtained by enlarging part A2 in Figure 12c, only some fine particles of RA can be seen on the surface of PURD-4, which is mainly because the RA particles are wrapped by a large amount of polyurethane, but due to the limited content of polyurethane, some fine particles of RA are exposed and adhered to the outside. It is more clearly seen that linear polyurethane has a strong adhesion effect, which can "pull" the various parts of the specimen together, providing a basis for the further connection of the specimen as a whole. Figure 12e is the SEM image time of the PURD-5 at 500 times. It is observed that the pores inside PURD-5 are smaller in size than those inside PURD-4, the internal structure is more compact, and the particle shape of the RA is also less visible. Due to the increase in the content of polyurethane, a large number of polyurethane-wrapped RA form polyurethane agglomerates, which make the integrity of the specimen more prominent [35,52]. Figure 12f is obtained by enlarging part A3 in Figure 12e. It can be seen from Figure 12f that a large amount of polyurethane wraps the RA to form an agglomerate structure, and a large number of polyurethane thick films appear, that is, the polyurethane covers and wraps the surface of the RA particles in the form of a film. Figure 12g is the SEM image of the PURD-6 at 500 times. It can be found that PURD-6 has only a few tiny pores, the overall structure of the specimen is quite complete, and the thin film structure of polyurethane can be clearly seen. The A4 part in Figure 12g is enlarged to obtain Figure 12h, from which it can be found that PURD-6 is covered with more thick polyurethane film than PURD-5, and almost no pores and cracks in the specimen can be seen. Moreover, the thickness of the polyurethane film also increases to a certain extent, which is mainly caused by the increase in the content of polyurethane.

In summary, the reinforcement effect of polymer can be attributed to the polymer film wrapping and binding the particles to create a spatially reticular membrane structure [27]. The above SEM microscopic image analysis shows that with the increase in WPU content, the internal pore and crack size of the PURD gradually decreases, and the number of both also decreases. Meanwhile, the polyurethane-wrapped RA particles gradually form agglomerates and finally produce a large number of polyurethane films. With the gradual increase in WPU content, the polyurethane film becomes increasingly thicker, which makes the bonding effect between the polyurethane and the RA particles become increasingly stronger, so the overall structure of the PURD becomes increasingly denser. This phenomenon can more fully explain the mechanical results that the shear strength of PURD increases with the increase in WPU content in the UU test. Due to the limitation of the SEM test magnification, the specific size of pores cannot be determined. In order to accurately determine the nanometer size of pores, a further transmission electron microscopy (TEM) test and specific surface area (BET) test are required.

5. Conclusions

To explore new environmentally friendly reinforcement materials and realize the recycling and utilization of construction waste, a new reinforcement method using PU to enhance RDW was proposed in this study, thus reducing the environmental pressure and resource shortage. The UU test and SEM test were conducted to investigate the reinforcement performance and mechanism of PU. The main conclusions are presented as follows:

(1) WPU can obviously improve the shear strength of RA. The shear strength of PURD increases with the increase in WPU content. When the WPU content is 6%, the shear strength reaches the maximum value, which is nearly 30% higher than that of PURD-5. However, the curing time has little effect on the reinforcement effect of WPU, and the increase in shear strength can be ignored. Therefore, WPU can directly exert all strength effects in a short curing time, and is relatively flexible in engineering application, which is a very effective and promising new type of high-molecular polymer reinforcement material.

(2) WPU has great influence on the ductility of RA. The failure strain of PURD increases with the increase in confining pressure, and first increases and then decreases with the increase in WPU content. The failure strain of PURD-5 reaches the maximum value. Compared with PURD-3, the failure strain of PURD-5 increases by 41%, 50%, 124%, and 99% at 7 days and 61%, 37%, 31%, and 41% at 28 days, respectively. The growth range is large, so PURD-5 shows the best ductility.

(3) WPU can obviously increase the cohesion of RA. The internal friction angle and cohesion of PURD increase with the increase in WPU content, which is highly consistent with the change law of shear strength. However, the increase in internal friction angle is small, only about 5%, and the increase in cohesion is obvious. At the curing time of 7 days, the cohesion of PURD-6 increases by 144%, 142%, and 91%, respectively, compared with those of PURD-3, PURD-4, and PURD-5. At the curing time of 28 days, the cohesion of PURD-6 increases by 141%, 134%, and 77%, respectively, compared with those of PURD-3, PURD-4, and PURD-5.

(4) The enhancement effect of the polymer can be attributed to the spatial reticular membrane structure produced by wrapping and bonding particles with the polymer film. With the increase in WPU content, the internal pore and crack size of PURD gradually decreases, and the number of them also decreases. Meanwhile, the polyurethane-wrapped RA particles gradually form agglomerates and finally produce a large number of polyurethane films. With the gradual increase in WPU content, the polyurethane film becomes increasingly thicker, which makes the bonding effect of the polyurethane on the RA particles become increasingly stronger, and the overall structure of the PURD becomes increasingly denser.

(5) In this test, the mechanical strength of RA is obtained on the basis that the mud content is 20.5%. If different practical projects need to obtain higher mechanical parameters, it is necessary to reduce the mud content of RA. Similarly, if the engineering has low requirements for mechanical parameters, the requirements for mud content can be appropriately relaxed. Thus, the processing steps for RA can be appropriately reduced.

Author Contributions: Conceptualization, B.L. and N.L.; formal analysis, Y.W.; investigation, Y.Z. and M.G.; writing—original draft preparation, B.L.; writing—review and editing, Y.Y. and N.L. All authors have read and agreed to the published version of the manuscript.

Funding: This research was funded by the National Natural Science Foundation of China (Grant No. 52179107), China Scholarship Council (201607910002), and International scientific and technological cooperation projects of Shaoxing University (Project No. 2019LGGH1007).

Institutional Review Board Statement: Not applicable.

Informed Consent Statement: Not applicable.

Data Availability Statement: Not applicable.

Conflicts of Interest: The authors declare no conflict of interest.

References

1. Liu, C.; Liu, H.; Zhu, C.; Bai, G. On the mechanism of internal temperature and humidity response of recycled aggregate concrete based on the recycled aggregate porous interface. *Cement Concrete Comp.* **2019**, *103*, 22–35. [CrossRef]
2. Golafshani, E.M.; Behnood, A.; Hosseinikebria, S.S.; Arashpour, M. Novel metaheuristic-based type-2 fuzzy inference system for predicting the compressive strength of recycled aggregate concrete. *J. Clean. Prod.* **2021**, *320*, 128771. [CrossRef]
3. Raza, A.; Rafique, U. Efficiency of GFRP bars and hoops in recycled aggregate concrete columns: Experimental and numerical study. *Compos. Struct.* **2021**, *255*, 112986. [CrossRef]
4. Tang, Z.; Li, W.; Tam, V.W.; Luo, Z. Investigation on dynamic mechanical properties of fly ash/slag-based geopolymeric recycled aggregate concrete. *Compos. Part B-Eng.* **2020**, *185*, 107776. [CrossRef]
5. Yue, Y.; Zhou, Y.; Xing, F.; Gong, G.; Hu, B.; Guo, M. An industrial applicable method to improve the properties of recycled aggregate concrete by incorporating nano-silica and micro-$CaCO_3$. *J. Clean. Prod.* **2020**, *259*, 120920. [CrossRef]
6. Shaban, W.M.; Elbaz, K.; Yang, J.; Thomas, B.S.; Shen, X.; Li, L.; Du, Y.; Xie, J.; Li, L. Effect of pozzolan slurries on recycled aggregate concrete: Mechanical and durability performance. *Constr. Build. Mater.* **2021**, *276*, 121940. [CrossRef]
7. Wang, Y.; Zhang, S.; Niu, D.; Su, L.; Luo, D. Strength and chloride ion distribution brought by aggregate of basalt fiber reinforced coral aggregate concrete. *Constr. Build. Mater.* **2020**, *234*, 117390. [CrossRef]
8. Wang, Y.; Huang, J.; Wang, D.; Liu, Y.; Zhao, Z.; Liu, J. Experimental study on hygrothermal characteristics of coral sand aggregate concrete and aerated concrete under different humidity and temperature conditions. *Constr. Build. Mater.* **2020**, *230*, 117034. [CrossRef]
9. Bai, G.; Zhu, C.; Liu, C.; Liu, B. An evaluation of the recycled aggregate characteristics and the recycled aggregate concrete mechanical properties. *Constr. Build. Mater.* **2020**, *240*, 117978. [CrossRef]
10. Zhang, Y.; Luo, W.; Wang, J.; Wang, Y.; Xu, Y.; Xiao, J. A review of life cycle assessment of recycled aggregate concrete. *Constr. Build. Mater.* **2019**, *209*, 115–125. [CrossRef]
11. Zhang, L.; Sojobi, A.; Liew, K. Sustainable CFRP-reinforced recycled concrete for cleaner eco-friendly construction. *J. Clean. Prod.* **2019**, *233*, 56–75. [CrossRef]
12. Zhang, L.; Sojobi, A.; Kodur, V.; Liew, K. Effective utilization and recycling of mixed recycled aggregates for a greener environment. *J. Clean. Prod.* **2019**, *236*, 117600. [CrossRef]
13. Mi, R.; Pan, G.; Liew, K.; Kuang, T. Utilizing recycled aggregate concrete in sustainable construction for a required compressive strength ratio. *J. Clean. Prod.* **2020**, *276*, 124249. [CrossRef]
14. Tang, Y.; Feng, W.; Feng, W.; Chen, J.; Bao, D.; Li, L. Compressive properties of rubber-modified recycled aggregate concrete subjected to elevated temperatures. *Constr. Build. Mater.* **2021**, *268*, 121181. [CrossRef]
15. Zhu, P.; Hao, Y.; Liu, H.; Wang, X.; Gu, L. Durability evaluation of recycled aggregate concrete in a complex environment. *J. Clean. Prod.* **2020**, *273*, 122569. [CrossRef]
16. Tang, Z.; Li, W.; Tam, V.W.; Yan, L. Mechanical behaviors of CFRP-confined sustainable geopolymeric recycled aggregate concrete under both static and cyclic compressions. *Compos. Struct.* **2020**, *252*, 112750. [CrossRef]
17. Nanayakkara, O.; Gunasekara, C.; Sandanayake, M.; Law, D.W.; Nguyen, K.; Xia, J.; Setunge, S. Alkali activated slag concrete incorporating recycled aggregate concrete: Long term performance and sustainability aspect. *Constr. Build. Mater.* **2021**, *271*, 121512. [CrossRef]
18. Azúa, G.; González, M.; Arroyo, P.; Kurama, Y. Recycled coarse aggregates from precast plant and building demolitions: Environmental and economic modeling through stochastic simulations. *J. Clean. Prod.* **2019**, *210*, 1425–1434. [CrossRef]
19. Shi, C.; Zou, X.; Yang, L.; Wang, P.; Niu, M. Influence of humidity on the mechanical properties of polymer-modified cement-based repair materials. *Constr. Build. Mater.* **2020**, *261*, 119928. [CrossRef]
20. Li, M.; Fang, H.; Du, M.; Zhang, C.; Su, Z.; Wang, F. The behavior of polymer-bentonite interface under shear stress. *Constr. Build. Mater.* **2020**, *248*, 118680. [CrossRef]
21. Mohammed, A.; Mahmood, W.; Ghafor, K. TGA, rheological properties with maximum shear stress and compressive strength of cement-based grout modified with polycarboxylate polymers. *Constr. Build. Mater.* **2020**, *235*, 117534. [CrossRef]
22. Joseph, J.; Patel, R.; Wenham, A.; Smith, J. Biomedical applications of polyurethane materials and coatings. *Trans. IMF* **2018**, *96*, 121–129. [CrossRef]
23. Adak, B.; Butola, B.S.; Joshi, M. Effect of organoclay-type and clay-polyurethane interaction chemistry for tuning the morphology, gas barrier and mechanical properties of clay/polyurethane nanocomposites. *Appl. Clay Sci.* **2018**, *161*, 343–353. [CrossRef]
24. Kemona, A.; Piotrowska, M. Polyurethane recycling and disposal: Methods and prospects. *Polymers* **2020**, *12*, 1752. [CrossRef] [PubMed]
25. Chen, R.-D.; Huang, C.-F.; Hsu, S.-H. Composites of waterborne polyurethane and cellulose nanofibers for 3D printing and bioapplications. *Carbohyd. Polym.* **2019**, *212*, 75–88. [CrossRef]
26. Wan, T.; Chen, D. Mechanical enhancement of self-healing waterborne polyurethane by graphene oxide. *Prog. Org. Coat.* **2018**, *121*, 73–79. [CrossRef]
27. Bai, Y.; Liu, J.; Song, Z.; Chen, Z.; Jiang, C.; Lan, X.; Shi, X.; Bu, F.; Kanungo, D.P. Unconfined compressive properties of composite sand stabilized with organic polymers and natural fibers. *Polymers* **2019**, *11*, 1576. [CrossRef]
28. Qi, C.; Bai, Y.; Liu, J.; Bu, F.; Kanungo, D.P.; Song, Z.; He, X. Desiccation cracking behavior of polyurethane and polyacrylamide admixed clayey soils. *Polymers* **2020**, *12*, 2398. [CrossRef] [PubMed]

29. Li, M.; Du, M.; Wang, F.; Xue, B.; Zhang, C.; Fang, H. Study on the mechanical properties of polyurethane (PU) grouting material of different geometric sizes under uniaxial compression. *Constr. Build. Mater.* **2020**, *259*, 119797. [CrossRef]
30. Liu, J.; Bu, F.; Bai, Y.; Chen, Z.; Kanungo, D.P.; Song, Z.; Wang, Y.; Qi, C.; Chen, J. Study on engineering properties of sand strengthened by mixed fibers and polyurethane organic polymer. *B. Eng. Geol. Environ.* **2020**, *79*, 3049–3062. [CrossRef]
31. Wei, J.; Kong, F.; Liu, J.; Chen, Z.; Kanungo, D.P.; Lan, X.; Jiang, C.; Shi, X. Effect of sisal fiber and polyurethane admixture on the strength and mechanical behavior of sand. *Polymers* **2018**, *10*, 1121. [CrossRef] [PubMed]
32. Liu, J.; Qi, X.; Zhang, D.; Feng, Q.; Wang, Y.; Kanungo, D.P. Study on the permeability characteristics of polyurethane soil stabilizer reinforced sand. *Adv. Mater. Sci. Eng.* **2017**, *2017*, 5240186. [CrossRef]
33. Zhang, X.; Fang, H.; Du, M.; Shi, M.; Zhang, C. Experimental Study on the Mechanical Properties of the Fiber Cement Mortar Containing Polyurethane. *Adv. Mater. Sci. Eng.* **2021**, *2021*, 9956897. [CrossRef]
34. Saleh, S.; Yunus, N.Z.M.; Ahmad, K.; Ali, N.; Marto, A. Micro-level analysis of marine clay stabilised with polyurethane. *KSCE J. Civ. Eng.* **2020**, *24*, 807–815. [CrossRef]
35. Liu, P.; Meng, M.; Xiao, Y.; Liu, H.; Yang, G. Dynamic properties of polyurethane foam adhesive-reinforced gravels. *Sci. China Technol. Sc.* **2021**, *64*, 535–547. [CrossRef]
36. Fan, G.; Sha, F.; Yang, J.; Ji, X.; Lin, F.; Feng, C. Research on working performance of waterborne aliphatic polyurethane modified concrete. *J. Build. Eng.* **2022**, *51*, 104262. [CrossRef]
37. *JTG E51-2009*; Test Methods of Materials Stabilized with Inorganic Bingers for Highway Engineering. Industrial Standards of the People's Republic of China: Beijing, China, 2009.
38. *JTG 3430-2020*; Test Methods of Soils for Highway Engineering. Industrial Standards of the People's Republic of China: Beijing, China, 2020.
39. *GBT 50123-2019*; Standard for Geotechnical Testing Method. National Standards of the People's Republic of China: Beijing, China, 2019.
40. Wu, L.; Qian, W.; Liu, J.; Song, Z.; Kanungo, D.P.; Bai, Y.; Bu, F. Sisal Fiber-Polymer–Treated Sand Mechanical Properties in Triaxial Test. *Environ. Eng. Geosci.* **2020**, *26*, 227–242. [CrossRef]
41. Xiao, Y.; Stuedlein, A.W.; Chen, Q.; Liu, H.; Liu, P. Stress-strain-strength response and ductility of gravels improved by polyurethane foam adhesive. *J. Geotech. Geoenviron.* **2018**, *144*, 04017108. [CrossRef]
42. Liu, J.; Bai, Y.; Feng, Q.; Song, Z.; Wei, J.; Sun, S.; Kanungo, D.P. Strength properties of sand reinforced with a mixture of organic polymer stabilizer and polypropylene fiber. *J. Mater. Civil. Eng.* **2018**, *30*, 4018330. [CrossRef]
43. Yuan, B.; Li, Z.; Chen, Y.; Ni, H.; Zhao, Z.; Chen, W.; Zhao, J. Mechanical and microstructural properties of recycling granite residual soil reinforced with glass fiber and liquid-modified polyvinyl alcohol polymer. *Chemosphere* **2022**, *286*, 131652. [CrossRef]
44. Li, C.; Jiang, L. Utilization of limestone powder as an activator for early-age strength improvement of slag concrete. *Constr. Build. Mater.* **2020**, *253*, 119257. [CrossRef]
45. Zhang, L.; Jiang, Z.; Zhang, W.; Peng, S.; Chen, P. Flexural properties and microstructure mechanisms of renewable coir-fiber-reinforced magnesium phosphate cement-based composite considering curing ages. *Polymers* **2020**, *12*, 2556. [CrossRef] [PubMed]
46. Wang, W.; Kang, H.; Li, N.; Guo, J.; Girma, D.Y.; Liu, Y. Experimental investigations on the mechanical and microscopic behavior of cement-treated clay modified by nano-MgO and fibers. *Int. J. Geomech.* **2022**, *22*, 04022059. [CrossRef]
47. Kumarappa, D.B.; Peethamparan, S. Stress-strain characteristics and brittleness index of alkali-activated slag and class C fly ash mortars. *J. Build. Eng.* **2020**, *32*, 101595. [CrossRef]
48. Dong, Q.; Wei, H.; Ma, G. Failure mechanism of S-shaped fissure in brittle materials under uniaxial tension: Experimental and numerical analyses. *Int. J. Solids Struct.* **2020**, *191*, 486–496. [CrossRef]
49. Liu, J.; Wang, Y.; Kanungo, D.P.; Wei, J.; Bai, Y.; Li, D.; Song, Z.; Lu, Y. Study on the brittleness characteristics of sand reinforced with polypropylene fiber and polyurethane organic polymer. *Fiber Polym.* **2019**, *20*, 620–632. [CrossRef]
50. Dijkstra, T.; Rogers, C.; Smalley, I.; Derbyshire, E.; Li, Y.J.; Meng, X.M. The loess of north-central China: Geotechnical properties and their relation to slope stability. *Eng. Geol.* **1994**, *36*, 153–171. [CrossRef]
51. Liu, J.; Bai, Y.; Song, Z.; Kanungo, D.P.; Wang, Y.; Bu, F.; Chen, Z.; Shi, X. Stabilization of sand using different types of short fibers and organic polymer. *Constr. Build. Mater.* **2020**, *253*, 119164. [CrossRef]
52. Masoumi, E.; Abtahi Forooshani, S.M.; Abdi Nian, F. Problematic soft soil improvement with both polypropylene fiber and polyvinyl acetate resin. *Geotech. Geol. Eng.* **2013**, *31*, 143–149. [CrossRef]

polymers MDPI

ystematic Review

Knowledge Mapping of the Literature on Fiber-Reinforced Geopolymers: A Scientometric Review

Hassan Ali Alkadhim [1], Muhammad Nasir Amin [1,*], Waqas Ahmad [2], Kaffayatullah Khan [1], Mohammed Najeeb Al-Hashem [1], Sara Houda [3], Marc Azab [3] and Zaher Abdel Baki [3]

1 Department of Civil and Environmental Engineering, College of Engineering, King Faisal University, Al-Ahsa 31982, Saudi Arabia
2 Department of Civil Engineering, COMSATS University Islamabad, Abbottabad 22060, Pakistan
3 College of Engineering and Technology, American University of the Middle East, Egaila 54200, Kuwait
* Correspondence: mgadir@kfu.edu.sa; Tel.: +966-13-589-5431; Fax: +966-13-581-7068

Abstract: This study examined the bibliographic data on fiber-reinforced geopolymers (FRGPs) using scientometrics to determine their important features. Manual review articles are inadequate in their capability to connect various segments of literature in an ordered and systematic manner. Scientific mapping, co-citation, and co-occurrence are the difficult aspects of current research. The Scopus database was utilized to find and obtain the data needed to achieve the study's aims. The VOSviewer application was employed to assess the literature records from 751 publications, including citation, bibliographic, keyword, and abstract details. Significant publishing outlets, keywords, prolific researchers in terms of citations and articles published, top-cited documents, and locations actively participating in FRGP investigations were identified during the data review. The possible uses of FRGP were also highlighted. The scientometric analysis revealed that the most frequently used keywords in FRGP research are inorganic polymers, geopolymers, reinforcement, geopolymer, and compressive strength. Additionally, 27 authors have published more than 10 articles on FRGP, and 29 articles have received more than 100 citations up to June 2022. Due to the graphical illustration and quantitative contribution of scholars and countries, this study can support scholars in building joint ventures and communicating innovative ideas and practices.

Keywords: fibers; geopolymers; fiber-reinforced geopolymers; bibliographic analysis

Citation: Alkadhim, H.A.; Amin, M.N.; Ahmad, W.; Khan, K.; Al-Hashem, M.N.; Houda, S.; Azab, M.; Baki, Z.A. Knowledge Mapping of the Literature on Fiber-Reinforced Geopolymers: A Scientometric Review. *Polymers* **2022**, *14*, 5008. https://doi.org/10.3390/polym14225008

Academic Editors: Wensheng Wang, Yongchun Cheng, Heping (Fred) Chen and Guojin Tan

Received: 3 October 2022
Accepted: 10 November 2022
Published: 18 November 2022

1. Introduction

Geopolymer (GP) is a type of inorganic silico-aluminum composite with a 3D network made of a SiO_4 and AlO_4 tetrahedral unit structure [1–3]. It is manufactured by the interaction of active low-calcium silico-alumina ingredients with alkaline activators [4,5]. Active solid aluminosilicates and activators comprising alkali silicates and metals are required for the production of GPs [6–8]. The alkaline solution functions as an activator of alkali, binder, and dispersant [9]. In comparison to cementitious materials, GPs offer the benefits of high initial-age strength, rapid hardening, and a large variety of raw ingredients [10,11]. GPs use less energy and emit fewer pollutants during manufacture, and they are regarded as the material having the greatest potential to substitute cement [12–14]. The concept of GPs was proposed in 1978 to explain inorganic aluminosilicate polymers made using natural ingredients [15–17]. The intention was to utilize alkali metal silicate solutions to encourage the formation of polymeric aluminum silicate materials from geological minerals under severe alkaline environments [18,19]. Consequently, various solid silicate raw materials such as fly ash, slag, silica fume, and other wastes were utilized to effectively produce GPs [20,21].

Conventional cementitious materials have reduced durability such as resistance to elevated temperatures and deterioration [22,23]. GP composites effectively solve this deficiency [24–26]. However, GPs are comparable to ceramics in that their tensile and flexural strengths are inadequate, and they are very susceptible to microcracks [27,28]. By integrating fibers, the brittleness of GPs may be addressed by increasing the toughness of composites [29–31]. The addition of fibers to the GP can inhibit the emergence of fractures while simultaneously enhancing its ductility, toughness, and tensile strength [32–34]. In recent years, several researchers have investigated the durability of GPs, which focused mostly on their resistance to abrasion, weathering, freeze–thaw, sulfate, water absorption, chloride ions, and various dry and wet impacts [35,36]. The mechanical performance and durability of composites are enhanced by altering the concentration of the alkaline solution, silicon to aluminum ratio, curing conditions, and the addition of fibers [37–39] The inclusion of fibers increases the material's fracture performance and flexural strength and enhances its toughening process [29,40]. GPs with fiber reinforcement are more durable than cementitious composites of the same grade [41].

Currently, natural fibers, inorganic fibers, synthetic fibers, and steel fibers are the most frequent types of fibers utilized in GP composites [42–44]. Numerous research studies have been conducted on synthetic-fiber-reinforced GPs, such as polypropylene (PP), polyvinyl alcohol (PVA), polyethylene (PE), etc., but their manufacturing method contaminates the atmosphere and has difficulty satisfying the needs of sustainable development [45]. The majority of natural fibers is cellulose or plant fibers. Natural fibers have low cost, are lightweight, have strong adhesion, have easy production methods, and are biodegradable and are attracting the attention of academics [46].

As scientists continue to study fiber-reinforced geopolymers (FRGPs), because of the growing worries regarding the initiation and development of cracks, there is an issue regarding knowledge limitations that might prevent the establishment of new research and academic relationships. Therefore, it is essential to develop and use a method that enables scholars to acquire important information from the highly trustworthy sources available. This issue could be resolved with the aid of a scientometric process. Therefore, the purpose of this study is to conduct a scientometric evaluation of the literature data on FRGP investigations that has been made accessible up until June 2022. A scientometric analysis is used to carry out a quantitative examination of vast bibliographic records using advanced tools. This is because the many aspects of the literature cannot be appropriately and completely linked in traditional review studies. Complex elements of advanced studies include scientific visualization, co-citations, and co-occurrence [47,48]. Scientometric evaluations highlight the regions actively engaged in a research issue as well as the outlets with the greatest publications, frequently used keywords, and highly cited researchers and papers. To achieve the objectives of the current study, data from 751 pertinent papers were found using the Scopus search engine. This data included abstracts, keywords, citations, and bibliographic details. Additionally, the restrictions related to the applications of FRGP in the building industry were explored, along with possible remedies to these limits. Due to the graphical interpretation and quantitative records of countries and scientists, this study will assist academics in developing collaborative developments and exchanging fresh concepts and techniques.

2. Review Strategy

This work identified the different facets of the literature using a scientometric assessment of bibliometric data. In scientometric studies, systematic visualization, a technique developed by experts for analyzing bibliographic records, is applied [49,50]. Data retrieval was done using the Preferred Reporting Items for Systematic Reviews and Meta-Analyses (PRISMA) approach. The PRISMA checklist is attached as Table S1 in supplementary materials. Since there are extensive articles published on the topic under investigation, it was vital to use a trustworthy database. For this reason, the very dependable databases Scopus and Web of Science were appropriate [51,52]. The Scopus database, which academics strongly suggest [53,54], was used to compile bibliometric data on FRGP research. As of June 2022, 949 results for the term "fiber-reinforced geopolymers" were returned by a Scopus search. There were several filter settings used to reduce unnecessary papers.

The whole PRISMA technique for data extraction, assessment, and the various restrictions and filters is shown in Figure 1. A similar strategy was used in some earlier investigations across other subject areas [55–57]. Finally, 751 records were used for further analysis using the appropriate tools. The literature data were compiled in Comma Separated Values (CSV) format, and VOSviewer software, version 1.6.18, was used to create the scientific representation and quantitative valuation of the information acquired. VOSviewer is an open-source visualization tool and freely accessible [58–60]. Therefore, by using VOSviewer, the intentions of the current study were achieved. The generated data (CSV files) were imported to the VOSviewer to allow for analysis. The systematic research looked at the publication outlets, the highly popular keywords, the highly referenced authors and publications, and countries' participation. Tables were created to provide quantitative data, while graphs were given to show the different traits, their interactions, and co-occurrence.

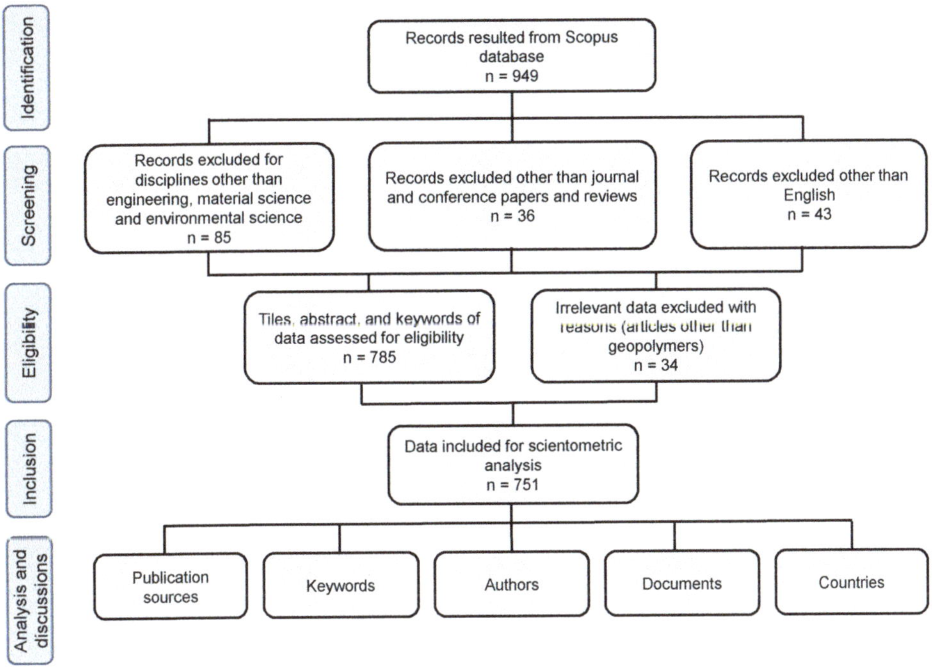

Figure 1. Flowchart of PRISMA technique for data retrieval, filters applied, and analysis.

3. Results and Discussion

3.1. Progress on the Research of FRGP

For research development and subject area evaluation, the Scopus analyzer was utilized. As illustrated in Figure 2, engineering and materials science were revealed to be the two top disciplines that produced the most articles, each field producing around 39% of the total papers in the FRGP research. Furthermore, documents related to the study field were searched for on Scopus (Figure 3). According to this study, almost 73%, 18%, 6%, and 3% of all the data comprised journal articles, conference papers, conference reviews, and journal reviews, respectively. Figure 4 shows the annual progress of papers published on FRGP studies from 1991 to June 2022. The initial publication of the FRGP research was discovered to be from 1991. With an average of nearly three articles published yearly up to 2010, there has been a very slight increase in the progress of publications in the area of FRGP investigations. Thereafter, the rate of publications grew gradually, averaging nearly 21 articles per year between 2011 and 2016, with 37 publications in 2015. A significant increase in publications occurred between 2017 and 2021, with a yearly average of about

94 papers and 136 papers in 2021. With 111 articles published so far this year (June 2022 the publications in the field of the subject under study are increasing every year.

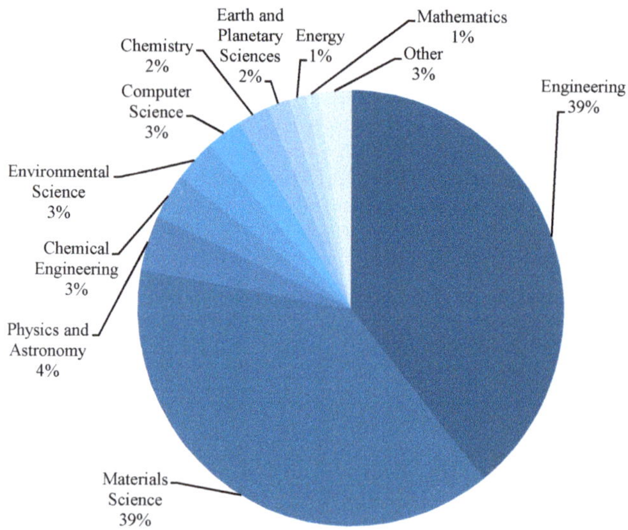

Figure 2. Relevant subject areas containing publications on FRGP studies.

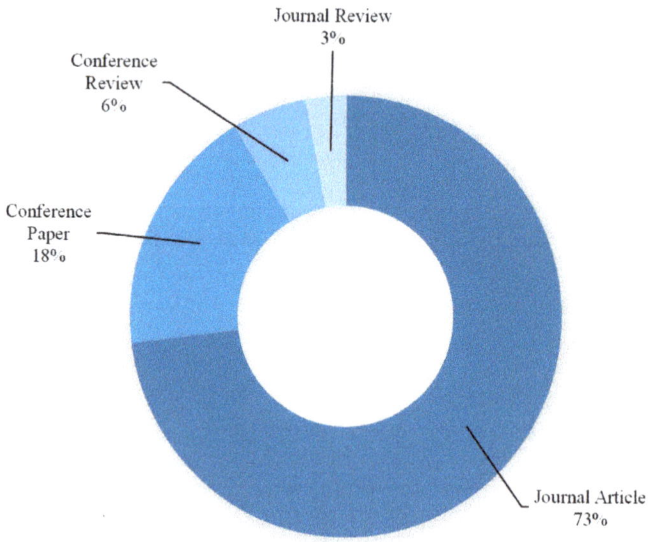

Figure 3. Kinds of documents available on FRGP studies.

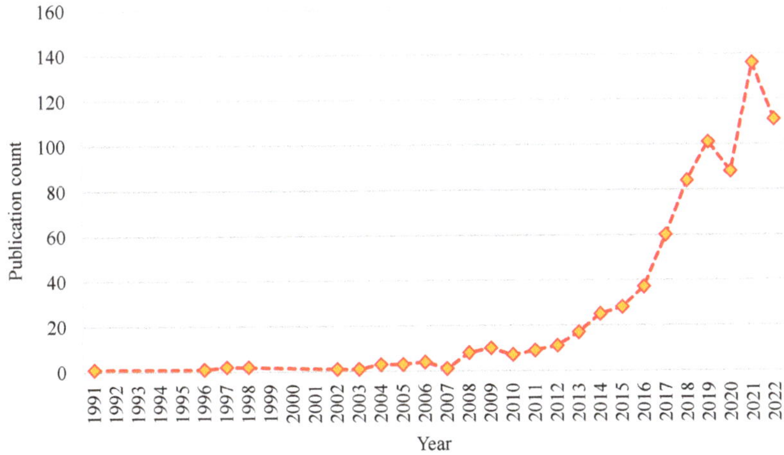

Figure 4. Yearly progress of publications on FRGP studies up to June 2022.

3.2. Knowledge Visualization of Publishing Outlets

Based on bibliographic records, VOSviewer was employed to assess publishing outlets (conferences/journals). A minimum limit of 10 articles was established for an outlet, and 17 of the 206 outlets complied with this requirement. Sources with at least 10 articles on FRGP till June 2022 are included in Table 1, along with the citations obtained during that period. The average citations for each outlet were calculated by dividing the citations with documents. With 95, 57, and 31 documents, respectively, *Construction and Building Materials (CBM)*, *IOP Conference Series: Materials Science and Engineering*, and *Materials* were found to be the leading three publication outlets in terms of total publications. Moreover, *CBM*, which obtained 3188 citations, *Composites Part B: Engineering*, which obtained 1615 citations, and *Ceramics International*, which received 690 citations, were the top three journals based on citations gained till June 2022. When the comparison of each outlet was made using average citations, the leading outlets were noted to be *Composites Part B: Engineering*, with nearly 77, *Cement and Concrete Composites*, with about 37, and *Journal of Materials in Civil Engineering*, with nearly 35 average citations. In particular, this study would serve as the foundation for the next scientometric analyses of FRGP research. Additionally, past standard review studies were unable to offer a comprehensive summary of the data published. An illustration of sources with at least 10 articles published is shown in Figure 5. The frame size in Figure 5a is connected to the source's contribution to the topic under investigation based on document count; a greater frame size denotes a stronger influence. For instance, *CBM* has a broader frame than the others, suggesting that it is a key outlet in the subject area. On the map, three clusters were created, each with its unique color (green, red, and blue). Clusters were created using the research outlet's breadth or how frequently they are mentioned together in relevant articles [61]. The co-citation rates of the sources were categorized by VOSviewer in publications. Ten articles, for instance, were included in the red cluster and were co-cited several times in similar articles. Moreover, in a cluster, linkages between close outlets were stronger than those between widely scattered frames. Compared to *Materials Today: Proceedings* or *MATEC Web of Conferences*, *CBM* showed a greater association with *Composites Part B: Engineering*. Varying colors corresponded to different density concentrations for an outlet, as noticed in Figure 5b. Red, yellow, green, and blue were in order of declining density concentration, with red having the maximum intensity. The red/yellow shades in *CBM*, *Materials*, and other well-known outlets signified a stronger commitment to FRGP investigations. Additionally, the text of some outlets was found to be smaller, faded, and unclear because of the low-density concentration, implying their lower contribution to the research of FRGP.

Table 1. Most contributing publication outlets in FRGP studies.

S/N	Publication Outlet	Documents	Citations	Average Citations
1	*Construction and Building Materials*	95	3188	34
2	*IOP Conference Series: Materials Science and Engineering*	57	198	3
3	*Materials*	31	498	16
4	*Ceramics International*	27	690	26
5	*Composites Part B: Engineering*	21	1615	77
6	*Ceramic Engineering and Science Proceedings*	21	195	9
7	*Structures*	20	263	13
8	*Materials Today: Proceedings*	20	55	3
9	*Cement and Concrete Composites*	16	598	37
10	*Composite Structures*	15	325	22
11	*Journal of Materials in Civil Engineering*	12	415	35
12	*Engineering Structures*	12	374	31
13	*Polymers*	12	85	7
14	*Journal of Building Engineering*	11	105	10
15	*Case Studies in Construction Materials*	11	41	4
16	*International Journal of Civil Engineering and Technology*	10	32	3
17	*MATEC Web of Conferences*	10	31	3

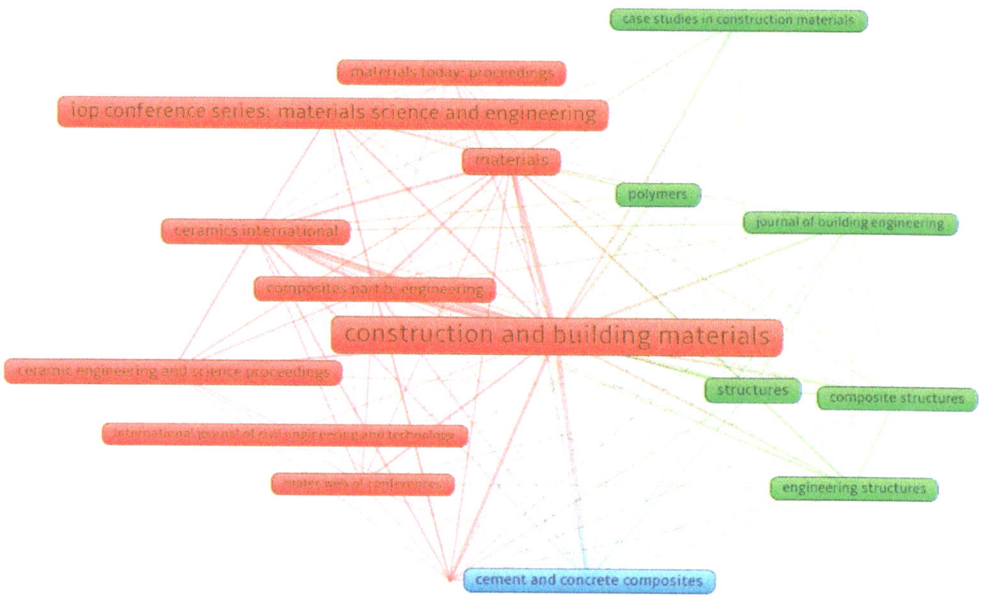

(**a**)

Figure 5. *Cont.*

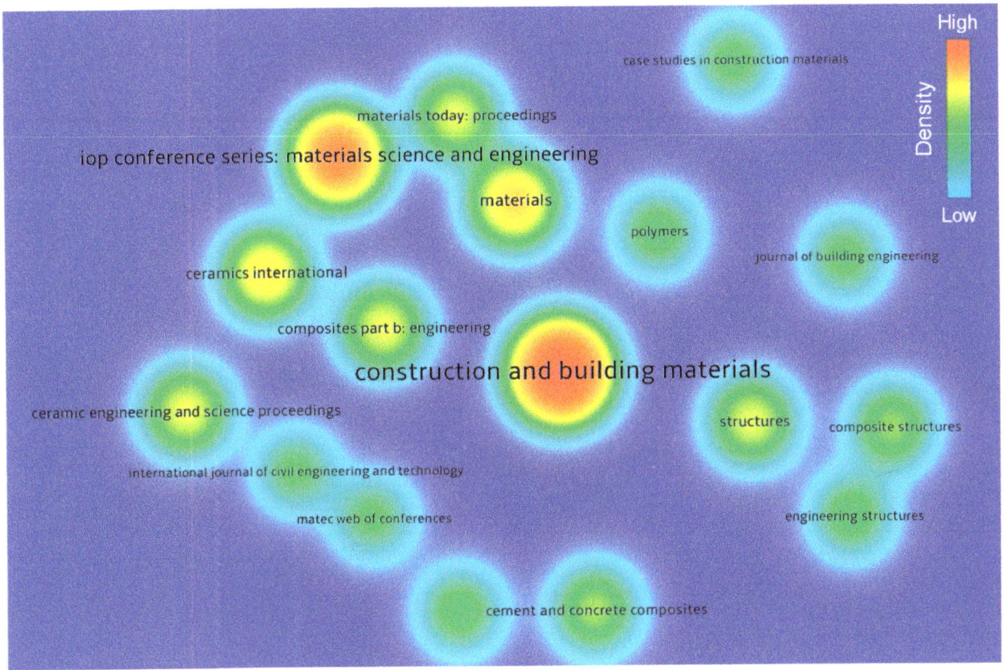

(**b**)

Figure 5. Systematic map of publication outlets: (**a**) network map; (**b**) density map.

3.3. Knowledge Visualization of Keywords

In research, keywords are vital because they differentiate and draw attention to the main domain of the study [62]. The least number of repeats for a term was set at 20, and 88 of the 4261 keywords met this requirement. The top 30 keywords that were used the most frequently in the literature are shown in Table 2. The five very frequent terms in the FRGP research included inorganic polymers, geopolymers, reinforcement, geopolymer, and compressive strength. FRGP has mainly been researched to increase mechanical performance and durability, mainly to decrease brittle behavior by bridging fractures, according to the keyword evaluation. Figure 6 displays a systematic map of terms with relationships, co-occurrences, and densities according to their frequency of occurrence. In Figure 6a, the size of a keyword node reveals its rate of recurrence, and its location reveals where it co-occurs in articles. The leading keywords also have wider nodes on the map than the rest, suggesting that they are important keywords for careful evaluation in the research of FRGP. The graph highlights clusters in a way that shows how frequently they appear together across different publications. The color-coded classification of keywords is based on their co-occurrence in articles. Six clusters of varying colors are shown in Figure 6a. Figure 6b illustrates how different hues correspond to various levels of keyword density. Indicating a higher number of occurrences, inorganic polymers, geopolymers, reinforcement, and other noteworthy keywords are shaded in red/yellow. This discovery will help ambitious scholars choose keywords that will make it simpler to find publications on a specific topic.

Table 2. List of 30 highly employed keywords in publications of FRGP studies.

S/N	Keyword	Occurrences
1	Inorganic polymers	514
2	Geopolymers	496
3	Reinforcement	326
4	Geopolymer	285
5	Compressive strength	216
6	Fly ash	214
7	Geopolymer concrete	205
8	Fibers	204
9	Geopolymer composites	180
10	Reinforced plastics	167
11	Steel fibers	158
12	Reinforced concrete	152
13	Mechanical properties	127
14	Tensile strength	126
15	Fiber reinforced plastics	120
16	Portland cement	97
17	Bending strength	90
18	Scanning electron microscopy	79
19	Concretes	72
20	Slags	71
21	Polypropylenes	63
22	Fiber reinforced materials	62
23	Concrete beams and girders	55
24	Curing	54
25	Basalt	53
26	Carbon fibers	53
27	Glass fibers	52
28	Fiber-reinforced	51
29	Ordinary Portland cement	51
30	Silicates	50

3.4. Knowledge Visualization of Researchers

Citations serve as evidence of a scholar's importance in a specific area of research [63]. The minimal publication requirement for a scientist was set at 10, and 27 of the 1583 writers reached this constraint. According to bibliometric data, Table 3 lists the authors of FRGP research with the most publications and citations. By dividing the overall citations by the publications, the average number of citations for a writer was calculated. When all factors, including the quantity of papers, the average citations, and the overall citations, were taken into account, it was difficult to assess a researcher's success. Instead, each element's evaluation for the scientist was evaluated separately. Shaikh F.U.A. was found to be the most productive scholar, based on the data analysis, with 26 papers, followed by Korniejenko K., with 25, and Ganesan N., with 18 publications. In the research of FRGP, Shaikh F.U.A. was first in the research area based on total citations with 1486, Alomayri T. was second with 714, and Jia D. was third with 638 total citations. However, Nematollahi B. may be placed at the top with about 62 average citations, Sanjayan J. may be in second place with about 58, and Shaikh F.U.A. may be in third place with approximately 57 average citations. The association between the most well-known writers and authors with at least 10 publications is seen in Figure 7. Figure 7a displays the scientific visualization of the researcher's co-authorship with at least 10 published articles in the investigation of FRGP. Additionally, Figure 7b shows the greatest number of citation-based related authors. A few FRGP scholars were connected by citations, as it was discovered that 8 of the 27 authors made up the largest group of connected authors based on citations.

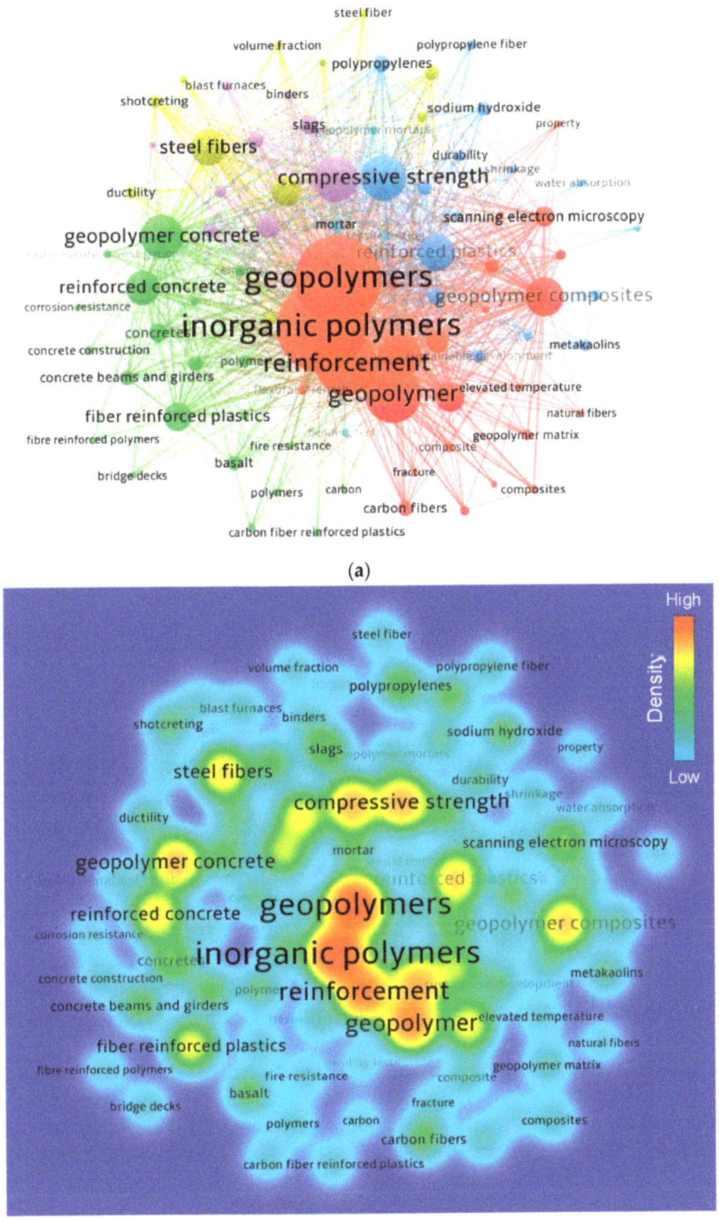

(a)

(b)

Figure 6. Systematic map of keywords in the research of FRGP: (**a**) visualization map; (**b**) density.

Table 3. Authors together with their publications and citations in FRGP research.

S/N	Researcher	Publication Count	Total Citations	Average Citations
1	Shaikh F.U.A.	26	1486	57
2	Korniejenko K.	25	274	11
3	Ganesan N.	18	220	12
4	Jia D.	17	638	38
5	Hao H.	17	323	19
6	Alomayri T.	16	714	45
7	He P.	16	638	40
8	Kriven W.M.	16	341	21
9	Elchalakani M.	16	186	12
10	Łach M.	15	87	6
11	Indira P.V.	13	69	5
12	Zhou Y.	12	382	32
13	Sanjayan J.	11	634	58
14	Benmokrane B.	11	409	37
15	Manalo A.C.	11	377	34
16	Yang Z.	11	195	18
17	Wang Y.	11	138	13
18	Louda P.	11	126	11
19	Nematollahi B.	10	621	62
20	Maranan G.B.	10	360	36
21	Zhang M.	10	273	27
22	Duan X.	10	193	19
23	Dong M.	10	149	15
24	Karrech A.	10	140	14
25	Hadi M.N.S.	10	97	10
26	Dai J.-G.	10	94	9
27	Mikuła J.	10	90	9

3.5. Knowledge Visualization of Documents

An article's importance in a particular academic field is shown by the number of citations it has obtained. In their respective academic domains, articles having the highest citations are interpreted as revolutionary. A minimum of a 50 citation limit was set for an article, and 84 of 751 articles met this requirement. The top five documents in the field of FRGP, together with their citation details, are shown in Table 4. The article "Geopolymers—Inorganic Polymeric New Materials", by Davidovits J. [64], had 2553 citations. Additionally, Yan L. [65] and Lyon R.E. [66] were in the top three, with 310 and 291 citations, respectively, for their publications. However, as of June 2022, only 29 papers had received more than 100 citations. The systematic map of papers and their ties to the subject area based on citations is also demonstrated in Figure 8. A map of connected articles with a minimum of 50 citations up to June 2022 is shown in Figure 8a. According to the data analysis, 82 out of 84 articles were linked by citations. As a result, the majority of important papers in the current study field is connected together by citations. The map of density for the linked articles based on citations is shown in Figure 8b. Clearly, papers with more citations showed larger density concentrations.

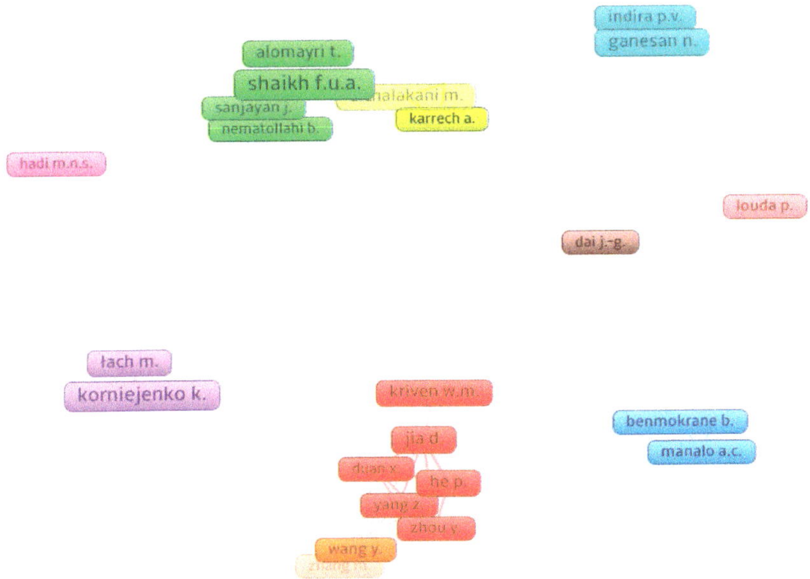

(a)

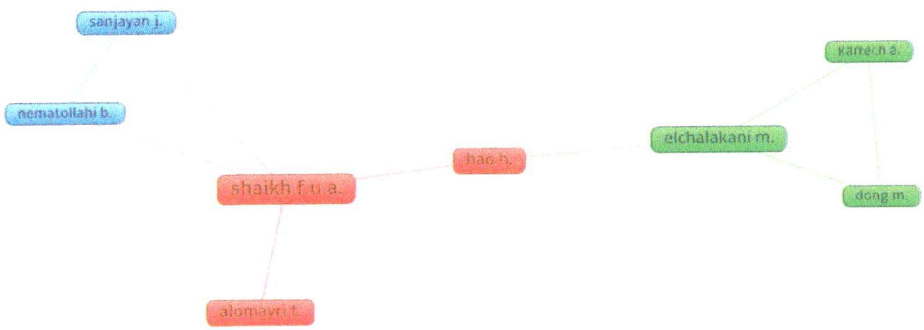

(b)

Figure 7. Systematic map indicating author's collaborations: (**a**) authors with at least 10 articles; (**b**) connected authors based on citations.

Table 4. List of documents having most citations received up to June 2022.

S/N	Article	Title	Citations
1	Davidovits J. [64]	Geopolymers—Inorganic Polymeric New Materials	2553
2	Yan L. [65]	A Review of Recent Research on the use of Cellulosic Fibres, their Fibre Fabric Reinforced Cementitious, Geo-polymer and Polymer Composites in Civil Engineering	310
3	Lyon R.E. [66]	Fire-Resistant Aluminosilicate Composites	291
4	Panda B. [67]	Anisotropic Mechanical Performance of 3D Printed Fiber Reinforced Sustainable Construction Material	265
5	Li W. [68]	Mechanical Properties of Basalt Fiber Reinforced Geopolymeric Concrete under Impact Loading	258

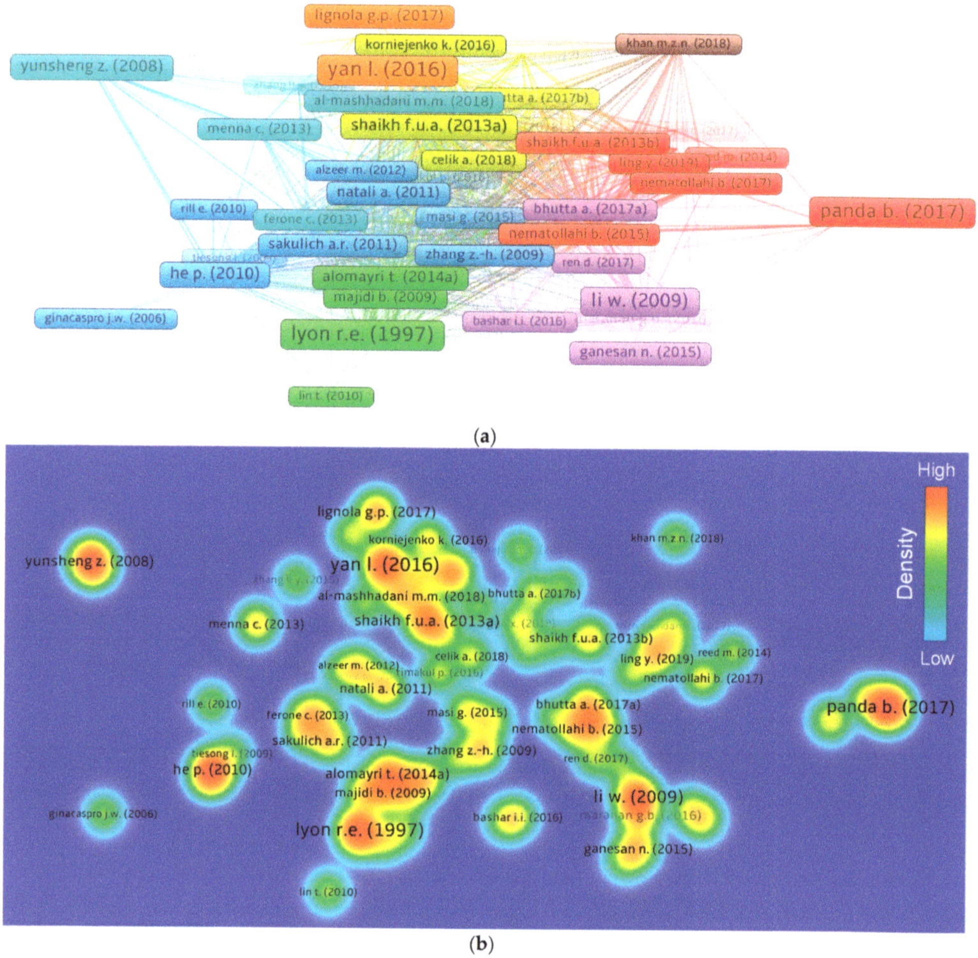

Figure 8. Systematic mapping of papers: (**a**) connected articles based on citations; (**b**) density of connected articles.

3.6. Knowledge Mapping of Countries

In comparison to other states, some have contributed more documents to the subject topic and plan to keep on doing so. Readers can view sections of the scientific graph that are specifically for the FRGP study. The least articles limit for a country was set at 10, and 20 countries complied with this condition. According to Table 5, the represented countries have published at least 10 articles on the FRGP study. With 139, 129, and 116 publications each, Australia, China, and India published the most research. The top three countries in terms of citations were also discovered to be Australia (4153 citations), China (3439 citations), and the United States (1864 citations). Figure 9 indicates the scientific framework and the density concentration of the countries connected by citations. According to the quantity of papers released, a country's influence on a subject is reflected in the frame size shown in Figure 9a. The regions with the top levels of engagement had a larger density concentration, as seen in Figure 9b. Young researchers will be able to develop scientific alliances, start joint businesses, and discuss novel ideas and techniques with the aid of the graphical interpretation and statistical data of the participating states. Researchers from nations with an interest in FRGP investigations can collaborate with experts in the research area and gain from their experience.

Table 5. Detail of regions actively devoted to FRGP investigations.

S/N	Country	Publications	Citations
1	Australia	139	4153
2	China	129	3493
3	India	116	853
4	United States	79	1864
5	Poland	41	530
6	Saudi Arabia	34	699
7	United Kingdom	33	802
8	Turkey	33	557
9	Czech Republic	30	313
10	Iraq	26	331
11	Brazil	25	545
12	Pakistan	24	320
13	Italy	23	820
14	Canada	23	689
15	Malaysia	23	491
16	Iran	18	438
17	Germany	17	599
18	Hong Kong	17	521
19	Portugal	13	396
20	Thailand	12	332

(a)

Figure 9. *Cont.*

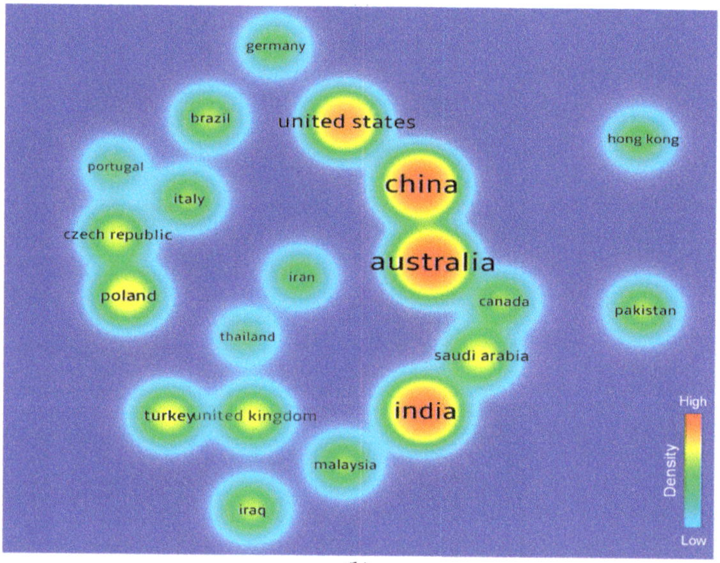

(**b**)

Figure 9. Systematic map of participating countries: (**a**) scientific visualization; (**b**) density.

4. Discussion and Potential Applications

Using literature data, this study carried out a systematic mapping and quantitative assessment of the FRGP research. Past conventional reviews lacked the ability to correctly and completely link various areas of the literature. This research identified the journals and conferences that presented the most articles, the commonly used keywords, the authors and publications that received the most citations, and the countries actively involved in FRGP investigations. The FRGP has mainly been researched to increase mechanical performance and durability, mostly to control brittle behavior by bridging cracks, according to keyword analysis. Additionally, bibliographic data were examined to sort out highly dedicated and prolific writers and countries based on publications and citations. New scientists will benefit from the graphical representation and statistical analysis of active nations and researchers as they establish joint ventures, form scientific alliances, and exchange new approaches and notions. Researchers from different countries motivated to further expand their studies on the application of FRGP can collaborate with specialists in the research field and benefit from their experience. This study explored and discussed the potential applications of FRGP based on an assessment of the literature data and a review of the highly pertinent documents.

As shown in Figure 10, the application of the material ranged from low-tech/low-cost to high-tech/high-cost, mostly dependent on the type of fiber and binder utilized to form the composites. It is essential to realize that correct conclusions on the effectiveness of each composite should be described separately for various uses. Below are some instances of the uses of FRGP composites.

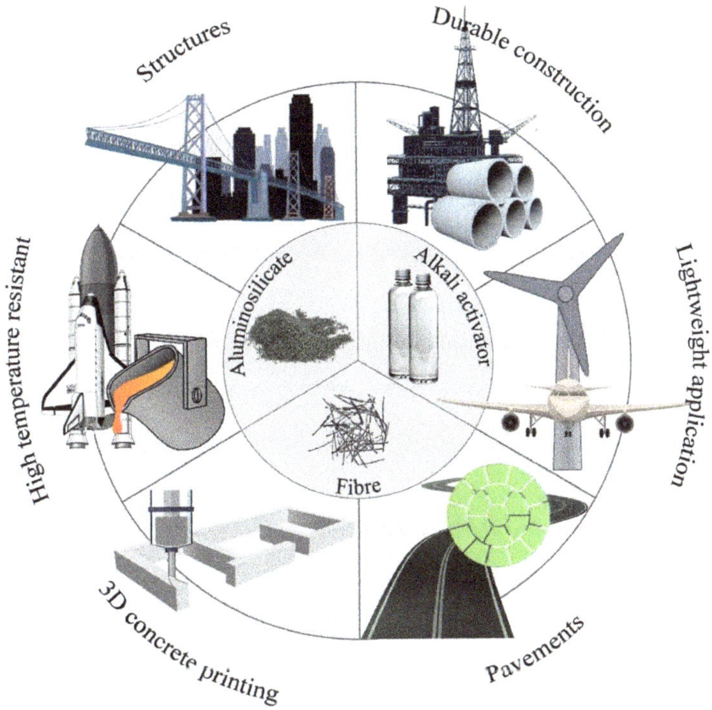

Figure 10. Fiber-reinforced geopolymer's possible applications [28].

Due to their accessibility and affordability, steel fibers have been frequently employed in cementitious materials for structural purposes. Steel fibers are incorporated to minimize shrinkage and enhance flexural performance, post-cracking performance, and energy absorption capability of GPs [69]. The highly alkaline nature of GPs maintains the passive condition of the steel reinforcement, hence making it a robust composite for several infrastructure applications [70,71]. Due to the high mechanical strength, flexibility, and hydrophilicity of PE and PVA fibers, niche applications have been examined for PE and PVA-FRGP, such as the progress of strain-hardening GP composites, which require a material with ultra-high impact resistance and ductility [28,72]. A similar use for PP-FRGP has been examined to make the composite more cost-efficient and environmentally friendly [73,74]. In addition, PP and PVA-FRGP have been explored for extruding-based 3D printing processes to construct formwork-less structures with complicated geometries and minimal water curing requirements [67,75,76]. The inflammability of GPs, coupled with the exceptional elevated-temperature resilience of inorganic and carbon fibers, can be utilized in the fabrication of materials where thermal resistance is necessary [66,77]. In addition to their great strength and low weight, carbon fibers have the possibility to be used in the production of lightweight, durable, and robust huge constructions. In addition, carbon-nanotubes were utilized not only to improve the fracture energy of the GPs but also to offer electrical conductivity and piezoresistive responses to examine micro-crack development [78,79]. Inorganic fibers such as silicon carbide and basalt are cost-efficient substitutes for carbon fibers in the production of elevated-temperature GP composites [80–82]. Natural fibers are often inexpensive and flexible, and they may be employed as reinforcement in GPs at high concentrations. Several production procedures have been developed and used to address the poor compaction of GPs reinforced with natural fibers. In the presence of 8.3% short cotton fiber, for instance, roller compaction was utilized to drive the GP binder into the fiber system, resulting in a material with a tensile strength of around

32 MPa [83]. However, there is a need for in-depth investigations, methods, and guidelines for large-scale practical applications of FRGP.

5. Conclusions

This study conducted a scientometric review of the available literature data on fiber-reinforced geopolymers (FRGP) research to assess several criteria. The Scopus search engine was explored for 751 relevant records and evaluated by employing VOSviewer software. The main findings of this study are as follows:

- The assessment of publication outlets presenting articles on FRGP studies showed that *CBM*, *IOP Conference Series: Materials Science and Engineering*, and *Materials*, are the top publishing outlets based on the number of publications with 95, 57, and 31 documents, respectively. In terms of total citations, the leading three publication outlets are *CBM*, with 3188; *Composites Part B: Engineering*, with 1615; and *Ceramics International*, with 690 citations.
- The assessment of keywords used in the research of FRGP discovered that inorganic polymers, geopolymers, reinforcement, geopolymer, and compressive strength are the five most commonly occurring terms. Additionally, based on the evaluation, it was disclosed that the FRGP has mainly been explored to increase mechanical performance and durability, mainly to reduce brittle behavior by bridging cracks.
- The evaluation of researchers showed that 27 writers had published at least 10 papers up to June 2022. In terms of the number of published documents, overall citations, and average citations, the top authors were considered. Shaikh F.U.A. was determined to be the most prolific author with the most publications (26) and total citations (1486). However, Nematollahi B. was placed at the top based on average citations (almost 62).
- The top countries were analyzed based on their participation in FRGP investigations, and it was found that only 20 countries published at least 10 articles. Australia, China, and India published 139, 129, and 116 articles, respectively. Additionally, Australia obtained 4153 citations, China obtained 3439 citations, and the United States obtained 1864 citations and were positioned as the leading three in terms of citations.
- The potential applications of FRGP composites include elevated temperature resistance, 3D printing, lightweight structures, bridges, and pavements. However, in-depth research, techniques, and guidelines are required for large-scale practical uses of FRGP.

Supplementary Materials: The following supporting information can be downloaded at: https://www.mdpi.com/article/10.3390/polym14225008/s1, Table S1. PRISMA checklist.

Author Contributions: H.A.A.: funding acquisition, visualization, validation, writing—reviewing and editing. M.N.A.: conceptualization, validation, investigation, project administration, funding acquisition, writing—reviewing and editing. W.A.: conceptualization, methodology, resources, supervision, software, validation, investigation, writing—original draft, reviewing, and editing. K.K.: data curation, visualization, writing—reviewing and editing. M.N.A.-H.: investigation, writing—reviewing and editing. S.H.: data curation, funding acquisition, formal analysis, writing—reviewing and editing. M.A.: resources, validation, investigation, writing—reviewing and editing. Z.A.B.: formal analysis, visualization, writing—reviewing and editing. All authors have read and agreed to the published version of the manuscript.

Funding: This work was supported by the Deanship of Scientific Research, Vice Presidency for Graduate Studies and Scientific Research, King Faisal University, Saudi Arabia (Project No. GRANT1697), through its KFU Research Summer Initiative.

Institutional Review Board Statement: Not applicable.

Informed Consent Statement: Not applicable.

Data Availability Statement: The data used in this research have been properly cited and reported in the main text.

Acknowledgments: The authors acknowledge the Deanship of Scientific Research, Vice Presidency for Graduate Studies and Scientific Research, King Faisal University, Saudi Arabia (Project No. GRANT1697), through its KFU Research Summer Initiative. The authors extend their appreciation for the financial support that made this study possible.

Conflicts of Interest: The authors declare no conflict of interest.

References

1. Mahmood, A.; Noman, M.T.; Pechočiaková, M.; Amor, N.; Petrů, M.; Abdelkader, M.; Militký, J.; Sozcu, S.; Hassan, S.Z. Geopolymers and Fiber-Reinforced Concrete Composites in Civil Engineering. *Polymers* **2021**, *13*, 2099. [CrossRef] [PubMed]
2. Wang, Q.; Ahmad, W.; Ahmad, A.; Aslam, F.; Mohamed, A.; Vatin, N.I. Application of Soft Computing Techniques to Predict the Strength of Geopolymer Composites. *Polymers* **2022**, *14*, 1074. [CrossRef] [PubMed]
3. Raza, A.; Azab, M.; Baki, Z.A.; El Hachem, C.; El Ouni, M.H.; Kahla, N.B. Experimental study on mechanical, toughness and microstructural characteristics of micro-carbon fibre-reinforced geopolymer having nano TiO_2. *Alex. Eng. J.* 2022; *in press*. [CrossRef]
4. Yang, X.; Zhang, Y.; Lin, C. Microstructure Analysis and Effects of Single and Mixed Activators on Setting Time and Strength of Coal Gangue-Based Geopolymers. *Gels* **2022**, *8*, 195. [CrossRef] [PubMed]
5. Yang, X.; Zhang, Y.; Lin, C. Compressive and Flexural Properties of Ultra-Fine Coal Gangue-Based Geopolymer Gels and Microscopic Mechanism Analysis. *Gels* **2022**, *8*, 145. [CrossRef]
6. Zou, Y.; Zheng, C.; Alzahrani, A.M.; Ahmad, W.; Ahmad, A.; Mohamed, A.M.; Khallaf, R.; Elattar, S. Evaluation of Artificial Intelligence Methods to Estimate the Compressive Strength of Geopolymers. *Gels* **2022**, *8*, 271. [CrossRef]
7. Alhawat, M.; Ashour, A.; Yildirim, G.; Aldemir, A.; Sahmaran, M. Properties of geopolymers sourced from construction and demolition waste: A review. *J. Build. Eng.* **2022**, *50*, 104104. [CrossRef]
8. Xu, Y.; Guo, P.; Akono, A.-T. Novel Wet Electrospinning Inside a Reactive Pre-Ceramic Gel to Yield Advanced Nanofiber-Reinforced Geopolymer Composites. *Polymers* **2022**, *14*, 3943. [CrossRef]
9. Han, R.; Guo, X.; Guan, J.; Yao, X.; Hao, Y. Activation Mechanism of Coal Gangue and Its Impact on the Properties of Geopolymers: A Review. *Polymers* **2022**, *14*, 3861. [CrossRef]
10. Valente, M.; Sambucci, M.; Sibai, A. Geopolymers vs. Cement Matrix Materials: How Nanofiller Can Help a Sustainability Approach for Smart Construction Applications—A Review. *Nanomaterials* **2021**, *11*, 2007. [CrossRef]
11. Lan, T.; Meng, Y.; Ju, T.; Chen, Z.; Du, Y.; Deng, Y.; Song, M.; Han, S.; Jiang, J. Synthesis and application of geopolymers from municipal waste incineration fly ash (MSWI FA) as raw ingredient—A review. *Resour. Conserv. Recycl.* **2022**, *182*, 106308. [CrossRef]
12. Ahmed, H.U.; Abdalla, A.A.; Mohammed, A.S.; Mohammed, A.A.; Mosavi, A. Statistical Methods for Modeling the Compressive Strength of Geopolymer Mortar. *Materials* **2022**, *15*, 1868. [CrossRef]
13. Amin, M.N.; Khan, K.; Javed, M.F.; Aslam, F.; Qadir, M.G.; Faraz, M.I. Prediction of Mechanical Properties of Fly-Ash/Slag-Based Geopolymer Concrete Using Ensemble and Non-Ensemble Machine-Learning Techniques. *Materials* **2022**, *15*, 3478. [CrossRef]
14. Chen, Y.-C.; Lee, W.-H.; Cheng, T.-W.; Chen, W.; Li, Y.-F. The Length Change Ratio of Ground Granulated Blast Furnace Slag-Based Geopolymer Blended with Magnesium Oxide Cured in Various Environments. *Polymers* **2022**, *14*, 3386. [CrossRef]
15. Davidovits, J. Geopolymers and geopolymeric materials. *J. Therm. Anal.* **1989**, *35*, 429–441. [CrossRef]
16. Yang, H.; Liu, L.; Yang, W.; Liu, H.; Ahmad, W.; Ahmad, A.; Aslam, F.; Joyklad, P. A comprehensive overview of geopolymer composites: A bibliometric analysis and literature review. *Case Stud. Constr. Mater.* **2022**, *16*, e00830. [CrossRef]
17. Davidovits, J.; Cordi, S.A. Synthesis of new high temperature geo-polymers for reinforced plastics/composites. *Spe. Pactec.* **1979**, *79*, 151–154.
18. Meesala, C.R.; Verma, N.K.; Kumar, S. Critical review on fly-ash based geopolymer concrete. *Struct. Concr.* **2020**, *21*, 1013–1028. [CrossRef]
19. Lo, K.-W.; Lin, W.-T.; Lin, Y.-W.; Cheng, T.-W.; Lin, K.-L. Synthesis Metakaolin-Based Geopolymer Incorporated with SiC Sludge Using Design of Experiment Method. *Polymers* **2022**, *14*, 3395. [CrossRef]
20. Farooq, F.; Jin, X.; Javed, M.F.; Akbar, A.; Shah, M.I.; Aslam, F.; Alyousef, R. Geopolymer concrete as sustainable material: A state of the art review. *Constr. Build. Mater.* **2021**, *306*, 124762. [CrossRef]
21. Ahmad, A.; Ahmad, W.; Chaiyasarn, K.; Ostrowski, K.A.; Aslam, F.; Zajdel, P.; Joyklad, P. Prediction of Geopolymer Concrete Compressive Strength Using Novel Machine Learning Algorithms. *Polymers* **2021**, *13*, 3389. [CrossRef] [PubMed]
22. Hay, R.; Celik, K. Hydration, carbonation, strength development and corrosion resistance of reactive MgO cement-based composites. *Cem. Concr. Res.* **2020**, *128*, 105941. [CrossRef]
23. Han, S.; Zhong, J.; Ding, W.; Ou, J. Strength, hydration, and microstructure of seawater sea-sand concrete using high-ferrite Portland cement. *Constr. Build. Mater.* **2021**, *295*, 123703. [CrossRef]
24. Amran, M.; Al-Fakih, A.; Chu, S.H.; Fediuk, R.; Haruna, S.; Azevedo, A.; Vatin, N. Long-term durability properties of geopolymer concrete: An in-depth review. *Case Stud. Constr. Mater.* **2021**, *15*, e00661. [CrossRef]
25. Aygörmez, Y.; Canpolat, O.; Al-mashhadani, M.M. Assessment of geopolymer composites durability at one year age. *J. Build. Eng.* **2020**, *32*, 101453. [CrossRef]

26. Khan, K.; Ahmad, W.; Amin, M.N.; Nazar, S. A Scientometric-Analysis-Based Review of the Research Development on Geopolymers. *Polymers* **2022**, *14*, 3676. [CrossRef]
27. Lyu, B.-C.; Ding, C.; Guo, L.-P.; Chen, B.; Wang, A.-G. Basic performances and potential research problems of strain hardening geopolymer composites: A critical review. *Constr. Build. Mater.* **2021**, *287*, 123030. [CrossRef]
28. Ranjbar, N.; Zhang, M. Fiber-reinforced geopolymer composites: A review. *Cem. Concr. Compos.* **2020**, *107*, 103498. [CrossRef]
29. Li, W.; Shumuye, E.D.; Shiying, T.; Wang, Z.; Zerfu, K. Eco-friendly fibre reinforced geopolymer concrete: A critical review on the microstructure and long-term durability properties. *Case Stud. Constr. Mater.* **2022**, *16*, e00894. [CrossRef]
30. Zhang, M.; Na, M.; Yang, Z.; Shi, Y.; Guerrieri, M.; Pan, Z. Study on mechanical properties and solidification mechanism of stabilized dredged materials with recycled GFRP fibre reinforced geopolymer. *Case Stud. Constr. Mater.* **2022**, *17*, e01187. [CrossRef]
31. Samal, S.; Blanco, I. An Application Review of Fiber-Reinforced Geopolymer Composite. *Fibers* **2021**, *9*, 23. [CrossRef]
32. Abbas, A.-G.N.; Aziz, F.N.A.A.; Abdan, K.; Nasir, N.A.M.; Huseien, G.F. A state-of-the-art review on fibre-reinforced geopolymer composites. *Constr. Build. Mater.* **2022**, *330*, 127187. [CrossRef]
33. Zhang, P.; Gao, Z.; Wang, J.; Wang, K. Numerical modeling of rebar-matrix bond behaviors of nano-SiO_2 and PVA fiber reinforced geopolymer composites. *Ceram. Int.* **2021**, *47*, 11727–11737. [CrossRef]
34. Raza, A.; Alashker, Y.; Azab, M.; Khan, Q.u.Z.; Abdallah, M.; Barakat, O.; Elhadi, K.M. Development of eco-friendly alkali-activated nanocomposites comprising micro-fibers at ambient curing conditions. *Case Stud. Constr. Mater.* **2022**, *17*, e01540. [CrossRef]
35. Lingyu, T.; Dongpo, H.; Jianing, Z.; Hongguang, W. Durability of geopolymers and geopolymer concretes: A review. *Rev. Adv. Mater. Sci.* **2021**, *60*, 1–14. [CrossRef]
36. de Oliveira, L.B.; de Azevedo, A.R.G.; Marvila, M.T.; Pereira, E.C.; Fediuk, R.; Vieira, C.M.F. Durability of geopolymers with industrial waste. *Case Stud. Constr. Mater.* **2022**, *16*, e00839. [CrossRef]
37. Wang, K.; Zhang, P.; Guo, J.; Gao, Z. Single and synergistic enhancement on durability of geopolymer mortar by polyvinyl alcohol fiber and nano-SiO_2. *J. Mater. Res. Technol.* **2021**, *15*, 1801–1814. [CrossRef]
38. Chen, K.; Wu, D.; Xia, L.; Cai, Q.; Zhang, Z. Geopolymer concrete durability subjected to aggressive environments—A review of influence factors and comparison with ordinary Portland cement. *Constr. Build. Mater.* **2021**, *279*, 122496. [CrossRef]
39. Łach, M.; Kluska, B.; Janus, D.; Kabat, D.; Pławecka, K.; Korniejenko, K.; Guigou, M.D.; Choińska, M. Effect of Fiber Reinforcement on the Compression and Flexural Strength of Fiber-Reinforced Geopolymers. *Appl. Sci.* **2021**, *11*, 10443. [CrossRef]
40. Ganesh, A.C.; Muthukannan, M. Development of high performance sustainable optimized fiber reinforced geopolymer concrete and prediction of compressive strength. *J. Clean. Prod.* **2021**, *282*, 124543. [CrossRef]
41. Natali Murri, A.; Medri, V.; Landi, E. Production and thermomechanical characterization of wool–geopolymer composites. *J. Am. Ceram. Soc.* **2017**, *100*, 2822–2831. [CrossRef]
42. Silva, G.; Kim, S.; Aguilar, R.; Nakamatsu, J. Natural fibers as reinforcement additives for geopolymers—A review of potential eco-friendly applications to the construction industry. *Sustain. Mater. Technol.* **2020**, *23*, e00132. [CrossRef]
43. Kozub, B.; Bazan, P.; Mierzwiński, D.; Korniejenko, K. Fly-Ash-Based Geopolymers Reinforced by Melamine Fibers. *Materials* **2021**, *14*, 400. [CrossRef]
44. Wanasinghe, D.; Aslani, F.; Ma, G. Effect of Carbon Fibres on Electromagnetic-Interference-Shielding Properties of Geopolymer Composites. *Polymers* **2022**, *14*, 3750. [CrossRef]
45. Farooq, M.; Bhutta, A.; Banthia, N. Tensile performance of eco-friendly ductile geopolymer composites (EDGC) incorporating different micro-fibers. *Cem. Concr. Compos.* **2019**, *103*, 183–192. [CrossRef]
46. Ramamoorthy, S.K.; Skrifvars, M.; Persson, A. A review of natural fibers used in biocomposites: Plant, animal and regenerated cellulose fibers. *Polym. Rev.* **2015**, *55*, 107–162. [CrossRef]
47. Zakka, W.P.; Lim, N.H.A.S.; Khun, M.C. A scientometric review of geopolymer concrete. *J. Clean. Prod.* **2021**, *280*, 124353. [CrossRef]
48. Udomsap, A.D.; Hallinger, P. A bibliometric review of research on sustainable construction, 1994–2018. *J. Clean. Prod.* **2020**, *254*, 120073. [CrossRef]
49. Markoulli, M.P.; Lee, C.I.; Byington, E.; Felps, W.A. Mapping Human Resource Management: Reviewing the field and charting future directions. *Hum. Resour. Manag. Rev.* **2017**, *27*, 367–396. [CrossRef]
50. Amin, M.N.; Ahmad, W.; Khan, K.; Sayed, M.M. Mapping Research Knowledge on Rice Husk Ash Application in Concrete: A Scientometric Review. *Materials* **2022**, *15*, 3431. [CrossRef] [PubMed]
51. Aghaei Chadegani, A.; Salehi, H.; Yunus, M.; Farhadi, H.; Fooladi, M.; Farhadi, M.; Ale Ebrahim, N. A comparison between two main academic literature collections: Web of Science and Scopus databases. *Asian Soc. Sci.* **2013**, *9*, 18–26. [CrossRef]
52. Afgan, S.; Bing, C. Scientometric review of international research trends on thermal energy storage cement based composites via integration of phase change materials from 1993 to 2020. *Constr. Build. Mater.* **2021**, *278*, 122344. [CrossRef]
53. Bergman, E.M.L. Finding citations to social work literature: The relative benefits of using Web of Science, Scopus, or Google Scholar. *J. Acad. Librariansh.* **2012**, *38*, 370–379. [CrossRef]
54. Meho, L.I. Using Scopus's CiteScore for assessing the quality of computer science conferences. *J. Informetr.* **2019**, *13*, 419–433. [CrossRef]

55. Mei, J.; Xu, G.; Ahmad, W.; Khan, K.; Amin, M.N.; Aslam, F.; Alaskar, A. Promoting sustainable materials using recycled rubber in concrete: A review. *J. Clean. Prod.* **2022**, *373*, 133927. [CrossRef]
56. Huang, S.; Wang, H.; Ahmad, W.; Ahmad, A.; Ivanovich Vatin, N.; Mohamed, A.M.; Deifalla, A.F.; Mehmood, I. Plastic Waste Management Strategies and Their Environmental Aspects: A Scientometric Analysis and Comprehensive Review. *Int. J. Environ. Res. Public Health* **2022**, *19*, 4556. [CrossRef]
57. Zhang, B.; Ahmad, W.; Ahmad, A.; Aslam, F.; Joyklad, P. A scientometric analysis approach to analyze the present research on recycled aggregate concrete. *J. Build. Eng.* **2022**, *46*, 103679. [CrossRef]
58. Zuo, J.; Zhao, Z.-Y. Green building research–current status and future agenda: A review. *Renew. Sustain. Energy Rev.* **2014**, *30*, 271–281. [CrossRef]
59. Darko, A.; Zhang, C.; Chan, A.P. Drivers for green building: A review of empirical studies. *Habitat Int.* **2017**, *60*, 34–49. [CrossRef]
60. Ahmad, W.; Khan, M.; Smarzewski, P. Effect of Short Fiber Reinforcements on Fracture Performance of Cement-Based Materials: A Systematic Review Approach. *Materials* **2021**, *14*, 1745. [CrossRef]
61. Wuni, I.Y.; Shen, G.Q.; Osei-Kyei, R. Scientometric review of global research trends on green buildings in construction journals from 1992 to 2018. *Energy Build.* **2019**, *190*, 69–85. [CrossRef]
62. Su, H.-N.; Lee, P.-C. Mapping knowledge structure by keyword co-occurrence: A first look at journal papers in Technology Foresight. *Scientometrics* **2010**, *85*, 65–79. [CrossRef]
63. Yu, F.; Hayes, B.E. Applying data analytics and visualization to assessing the research impact of the Cancer Cell Biology (CCB) Program at the University of North Carolina at Chapel Hill. *J. E-Sci. Librariansh.* **2018**, *7*, 4. [CrossRef]
64. Davidovits, J. Geopolymers: Inorganic polymeric new materials. *J. Therm. Anal. Calorim.* **1991**, *37*, 1633–1656. [CrossRef]
65. Yan, L.; Kasal, B.; Huang, L. A review of recent research on the use of cellulosic fibres, their fibre fabric reinforced cementitious, geo-polymer and polymer composites in civil engineering. *Compos. Part B Eng.* **2016**, *92*, 94–132. [CrossRef]
66. Lyon, R.E.; Balaguru, P.N.; Foden, A.; Sorathia, U.; Davidovits, J.; Davidovics, M. Fire-resistant aluminosilicate composites. *Fire Mater.* **1997**, *21*, 67–73. [CrossRef]
67. Panda, B.; Chandra Paul, S.; Jen Tan, M. Anisotropic mechanical performance of 3D printed fiber reinforced sustainable construction material. *Mater. Lett.* **2017**, *209*, 146–149. [CrossRef]
68. Li, W.; Xu, J. Mechanical properties of basalt fiber reinforced geopolymeric concrete under impact loading. *Mater. Sci. Eng. A* **2009**, *505*, 178–186. [CrossRef]
69. Ranjbar, N.; Mehrali, M.; Mehrali, M.; Alengaram, U.J.; Jumaat, M.Z. High tensile strength fly ash based geopolymer composite using copper coated micro steel fiber. *Constr. Build. Mater.* **2016**, *112*, 629–638. [CrossRef]
70. Mundra, S.; Bernal Lopez, S.; Criado, M.; Hlaváček, P.; Ebell, G.; Reinemann, S.; Gluth, G.; Provis, J. Steel corrosion in reinforced alkali-activated materials. *RILEM Tech. Lett.* **2017**, *2*, 33–39. [CrossRef]
71. Monticelli, C.; Natali, M.E.; Balbo, A.; Chiavari, C.; Zanotto, F.; Manzi, S.; Bignozzi, M.C. Corrosion behavior of steel in alkali-activated fly ash mortars in the light of their microstructural, mechanical and chemical characterization. *Cem. Concr. Res.* **2016**, *80*, 60–68. [CrossRef]
72. Nematollahi, B.; Qiu, J.; Yang, E.-H.; Sanjayan, J. Micromechanics constitutive modelling and optimization of strain hardening geopolymer composite. *Ceram. Int.* **2017**, *43*, 5999–6007. [CrossRef]
73. Ranjbar, N.; Talebian, S.; Mehrali, M.; Kuenzel, C.; Cornelis Metselaar, H.S.; Jumaat, M.Z. Mechanisms of interfacial bond in steel and polypropylene fiber reinforced geopolymer composites. *Compos. Sci. Technol.* **2016**, *122*, 73–81. [CrossRef]
74. Ranjbar, N.; Mehrali, M.; Behnia, A.; Javadi Pordsari, A.; Mehrali, M.; Alengaram, U.J.; Jumaat, M.Z. A comprehensive study of the polypropylene fiber reinforced fly ash based geopolymer. *PLoS ONE* **2016**, *11*, e0147546. [CrossRef]
75. Nematollahi, B.; Vijay, P.; Sanjayan, J.; Nazari, A.; Xia, M.; Naidu Nerella, V.; Mechtcherine, V. Effect of Polypropylene Fibre Addition on Properties of Geopolymers Made by 3D Printing for Digital Construction. *Materials* **2018**, *11*, 2352. [CrossRef]
76. Panda, B.; Tan, M.J. Experimental study on mix proportion and fresh properties of fly ash based geopolymer for 3D concrete printing. *Ceram. Int.* **2018**, *44*, 10258–10265. [CrossRef]
77. Hammell, J.A.; Balaguru, P.N.; Lyon, R.E. Strength retention of fire resistant aluminosilicate–carbon composites under wet–dry conditions. *Compos. Part B Eng.* **2000**, *31*, 107–111. [CrossRef]
78. Saafi, M.; Andrew, K.; Tang, P.L.; McGhon, D.; Taylor, S.; Rahman, M.; Yang, S.; Zhou, X. Multifunctional properties of carbon nanotube/fly ash geopolymeric nanocomposites. *Constr. Build. Mater.* **2013**, *49*, 46–55. [CrossRef]
79. Bi, S.; Liu, M.; Shen, J.; Hu, X.M.; Zhang, L. Ultrahigh self-sensing performance of geopolymer nanocomposites via unique interface engineering. *ACS Appl. Mater. Interfaces* **2017**, *9*, 12851–12858. [CrossRef]
80. Davidovits, J. Reinforced geopolymer composites: A critical review. *Mater. Today* **2016**, *31*.
81. Masi, G.; Rickard, W.D.A.; Bignozzi, M.C.; van Riessen, A. The effect of organic and inorganic fibres on the mechanical and thermal properties of aluminate activated geopolymers. *Compos. Part B Eng.* **2015**, *76*, 218–228. [CrossRef]
82. Mills-Brown, J.; Potter, K.; Foster, S.; Batho, T. The development of a high temperature tensile testing rig for composite laminates. *Compos. Part A Appl. Sci. Manuf.* **2013**, *52*, 99–105. [CrossRef]
83. Alomayri, T.; Shaikh, F.U.A.; Low, I.M. Synthesis and mechanical properties of cotton fabric reinforced geopolymer composites. *Compos. Part B Eng.* **2014**, *60*, 36–42. [CrossRef]

MDPI
St. Alban-Anlage 66
4052 Basel
Switzerland
Tel. +41 61 683 77 34
Fax +41 61 302 89 18
www.mdpi.com

Polymers Editorial Office
E-mail: polymers@mdpi.com
www.mdpi.com/journal/polymers

www.ingramcontent.com/pod-product-compliance
Lightning Source LLC
Chambersburg PA
CBHW061814210926
51850CB00006B/698